The Missile Defense Equation: Factors for Decision Making

The Missile Defense Equation: Factors for Decision Making

Peter J. Mantle
Mantle & Associates, LLC Vashon Island, Washington

American Institute of Aeronautics and Astronautics, Inc.
1801 Alexander Bell Drive
Reston, Virginia 20191-4344

Publishers since 1930

American Institute of Aeronautics and Astronautics, Inc., Reston, Virginia

Library of Congress Cataloging-in-Publication Data

Mantle, Peter J.
  The missile defense equation : factors for decision making / Peter J.
Mantle.
     p. cm.
Includes bibliographical references and index.
  ISBN 1-56347-609-6
1. Ballistic missile defense–Evaluation. 2. Cruise missile
defenses–Evaluation. 3. Air defenses–Evaluation. 4. Armed
Forces–Procurement–Decision making. I. Title.

  UG740.M35 2004
  358.1′74–dc22   2003020787

Cover design by Sara Bluestone

For
Kathleen, Tracy, Chris, Dawn, Jennifer, Amy, Lauren, and Evan
who make it all worthwhile

ACKNOWLEDGMENTS

Much of the material in this book was developed by the author during the last two decades of the last century while the world was changing from dealing with a threat of strategic missiles between two super powers to one where the tactical use of missiles by many nations was and is a distinct possibility in future armed conflicts. Early in this period, the author was the Secretary of the Navy's liaison to the newly created Strategic Defense Initiative (SDI) in 1983 and later was chairman of a NATO Industrial Advisory Group (NIAG SG-37) for Missile Defense from 1990–2000. This NATO group varied in membership over those 10 years, but during that period more than 100 technical experts served at different times from 13 nations of the Alliance, specifically Belgium, Canada, Czech Republic, France, Germany, Greece, Italy, Netherlands, Norway, Turkey, Spain, United Kingdom, and the United States.

As the group struggled with the new identity of NATO in this new world after the end of the Cold War, their work has influenced in one form or another the views of the author and the outcome in this book, and their effort is gratefully acknowledged. There are too many experts to mention individually, but this does not diminish their contribution to molding the author's views, and their help is gratefully acknowledged. The many military officers and staff who served in the NATO oversight committees of the Conference of National Armaments Directors (CNAD); the Army, Navy, and Air Force Main Armament Groups; the NATO Air Defence Committee in Brussels; the members of SHAPE in Mons; AGARD in Paris and the agencies of NACMA (Brussels) and NC3A (The Hague) have all contributed to the shaping of the author's views over this important period, and their help is gratefully acknowledged. Key people in the various defense departments in each of the 13 nations have all contributed advice at one time or another that has influenced the author's views in his preparation of this book.

The help of the many friends and associates at Lockheed Martin are gratefully acknowledged. Bijan Abassi, Nick Pekelsma, George Uchida, Dick Brewer deserve special mention. Similarly, within Raytheon, the many discussions with Jim Jernigan, Mike Shannon, Mike Potter, Ed Hadro, Pete McNeany, and Ernie Sylagi have all helped crystallize the author's views. In the U.S. Office of the Secretary of Defense many helpful discussions, while serving on the NATO studies, with J. David Martin, Robert Soofer, Tim Yopek, Phil Jamison, and others have been especially helpful.

Of course, although the work, advice and guidance of these many friends, associates and institutions has definitely helped mold some of the author's findings in this book, it is to be emphasized that the views expressed in this book are those of the author alone, and no official views from any government office or corporation should be assumed from the contents or conclusions drawn from this book.

ABOUT THE AUTHOR

Peter J. Mantle is President of Mantle & Associates, LLC, a small consulting firm specializing in aerospace engineering and management. He has held research, engineering and executive responsibilities in both industry and government in the United States, Canada, and the United Kingdom. Born and raised in England, he began his career working on several exciting projects from the *Princess* Flying Boats, *Black Knight* missile, and the mixed propulsion unit aircraft *SR.53*, to introduction of Sir Christopher Cockerell's revolutionary hovercraft. During his work in Canada at the Canadian Armament Research & Development Establishment (CARDE), he worked on the Canadian embryonic space program and participated heavily in the hypersonic research and testing of upper-atmosphere vehicles. In the United States he was the chief engineer of the pioneering research into high-speed wing-in-ground-effect machines and was test pilot of the prototype *VRC-1* for the Maritime Administration in 1964. His 1975 and 1980 books published by the GPO on air cushion craft development are considered a milestone compendium on the technology. He has served as the Director of Technology Assessment for the U.S. Navy in the Office of Secretary of the Navy through two administrations and received the Superior Civilian Service award in 1984 for his work in the President's Strategic Defense Initiative (SDI) Program (on missile defense); for streamlining Advanced Technology Demonstration management procedures; and for his work on directed energy weapons, and other initiatives. He has served on various Navy Study Board activities on key projects for the U.S. Navy. Mantle was the technical director and program manager of the U.S. Navy's *SES-100B*, the world's fastest military surface effect ship (*Guinness Book of Records*), and patentee of its novel propulsion system. In the Pentagon he was instrumental in development of the U.S. Navy's stealth ship program among other high-level initiatives. Before retiring from Lockheed Martin in 1997, he served in several senior roles from Director of Surveillance Programs for the stealth ship *Sea Shadow* program and other key classified activities; then finally as Director of European New Business Initiatives.

For 10 years (1990–2000) Mantle served as Chairman of NIAG SG-37, an industrial group conducting a major set of studies into missile defense and other matters for NATO conducted by experts from 13 nations in the Alliance. He has received numerous awards and presented papers on a wide range of subjects to many associations around the world. He is currently chairman of the U.S. Delegation to NATO Industrial Advisory Group.

He now lives, with his wife Kathleen, on Vashon Island, Washington.

TABLE OF CONTENTS

PREFACE

For almost the last quarter of a century since 1983 until the present day, the subject of missile defense has challenged the best minds in all corners of the political community, the defense establishments, and the general populace. It is highly probable that armed conflict with adversaries to the free world in the future will include either or both the ballistic missile (BM) and the cruise missile (CM) or some variants of each type. Debate will continue as to which will be more prevalent, and the rate at which technology will be introduced to give them greater reach and capability is also uncertain. But the uncertainty of their nature and characteristics is not sufficient reason *not* to provide a defense against them. The consequences of having either no defense or an inadequate defense are too awesome to consider delay in the decision process. The additional complication of the possible incorporation of nuclear, biological, or chemical warheads into such weapons truly raises the stakes in the frightening concept of weapons of mass destruction (WMD).

The concepts for missile defense have run the gamut from global energy shields (the original concept, which was reminiscent of the science fiction solutions against *any* threat) to more practical combinations of various defensive weapons. These defensive weapons have covered all possibilities from space-based weapons to the more Earth-bound solutions of land-, sea-, and air-based defensive weapons. The concepts have ranged from directed energy (the closest concept yet to an energy shield) to the more conventional concept of the familiar "missile," where the concept is to throw "mass" at the threat (instead of "energy"). Even in the more familiar missile solution, because of the high energy imparted at missile speeds, concepts vary between a missile with an explosive charge to destroy the threat (a solution that has its origins deep in history) to the concept of using the explosive energy available of high-speed direct impact: the so-called "*hitting a bullet with a bullet*" or "*hit-to-kill*." Some of these concepts are still the subject of scientific exploration, and some have been battle tested in recent conflicts around the world. Sometimes, the concepts are solutions looking for a problem, and some are tackling real-world problems of a variety of known or anticipated threats. The missile defense equation does not have a unique solution.

Decisions facing defenders of the realm in such circumstances demand that the correct decisions are made. Because of the long development times of defensive weapons, it is paramount that not only the *right* decisions be made but also that they be made *early* in the development process. Frequently, often to the dismay of analysts, this must be done with only imprecise information at hand.

This book builds upon the many existing treatments of the potential missile threat to either North America or to the nations of Europe or in the defense of deployed forces of the NATO Alliance in potential trouble spots around the world. The purpose of this book is to develop a method to *help* the decision maker make *early* decisions so that development and acquisition can proceed properly.

This book is designed to help the decision makers at all levels in the defense business. It is designed to help those elected officials in the political arms of the governments in the nations of the Alliance who must deal with broad issues associated with the protection of the civilian populace. It is designed to help the key procurement officials in the defense departments and ministries who must decide between major programs of defense weapon systems. It is designed to help the senior managers in the defense industrial firms who must seek proper choices between performance, cost, schedule, and risk. It is designed to guide military officers who must make intelligent choices involving the defense of their forces against a variety of missile attacks. It is also designed to help engineering managers who must ultimately provide the right designs to satisfy the other decision makers up the chain of command already listed.

It is not the intent of this book to provide more precise data than that already published or to compete with sophisticated modeling and simulation software programs that the detailed analyst must use for any final analysis in particular and specific designs. It *is* designed, however, to place those detailed and sophisticated analyses in context and to *bound* the solutions such that the decision maker can feel comfortable that the design and development (which he is approving) is proceeding along the proper path for the best solution. A strong emphasis in the development of the material in this book has been to "return to the basics" such that decision makers can determine readily, without too much need for sophisticated analysis, the key drivers in setting up the needed defense.

Such basics should also prove useful to the teaching institutions whereby it is hoped that the engineering and management principles can be captured by the material presented in this book.

To add a sense of realism to the management *decision-making methods* developed in this book, data on a variety of missile defense systems have been used where appropriate to indicate general trends in all of the key parameters. All of the data have been obtained from public domain sources, and no attempt has been made to ensure the accuracy of the published programmatic data quoted. Ample references have been supplied as to source material. The user of the *management method* developed here is encouraged to substitute his own data to obtain his own results, or, if desired or if such data are not readily available, he can resort to the "default values" provided in each of the chapters on performance, cost, schedule, and risk and work the example problems and exercises provided at the end of the chapters.

Omissions, corrections, and constructive comments can be sent directly to the author at *mantlep@comcast.net.*

Peter J. Mantle
November 2003

Chapter 1

Making Decisions

That's fine, but I like the red one.
Kathleen Kinney

DIFFICULTY IN MAKING DECISIONS

Making decisions, especially in the defense business, is difficult. Analysts toil constantly to develop their detailed models that will take in all of the considerations needed to make decisions. But, all too often, the person doing the analysis frequently forgets that his or her job is to conduct the analysis *to help* the decision maker and not *to do* the decision making itself. Not all analysts understand the frustration when the decision maker does not properly appreciate the elegance of the analyst's analysis and results. A simple example will illustrate this. Once, while managing a missile defense team, the author was made aware of a particularly sophisticated piece of analysis that purported to reduce a complex set of decisions (choices) between a wide range of weapon systems to a simple algorithm with a closed form and (in the analyst's view) an unassailable solution. In that particular example the analyst showed that many factors contributed to the solution. These factors included the defense weapon's characteristics such as speed, range, guidance control mechanism, and type of warhead ("hit-to-kill" or fragmentation). These factors also included the nature of the threat such as speed, range, type of trajectory (minimum energy, depressed or lofted) of the incoming ballistic missile (BM), etc. Even the political factors (such as treaty implications) and sociological factors (what were the environmental impacts?) were included. In all, the analyst showed that over 256 parameters were included in the decision process—and this method of analysis would take care of all of them!

The method was certainly mathematically elegant. In an attempt to explain the method, a homely example that any decision maker could understand was constructed. Clearly, the thesis was as follows: if the homely example were understood by the decision maker, then ipso facto, the results of the more complex (256 parameter!) problem results would have to be accepted by the same decision maker. The simple and familiar example used to demonstrate the method was *a family decision to buy a car*. The problem statement was, that after touring all of the car lots, the decision was reduced to deciding between three cars: 1) a prestige car (a silver Mercedes) with a high purchase price but good quality construction and good maintenance record; 2) a comfortable familiar Ford sedan at a reasonable price, good quality construction, and excellent maintenance characteristics; and 3) a red sports car again at a good price but limited space (a two seater), excellent condition but a labor-intensive maintenance program. The analyst carefully compiled the key parameters of price, maintenance, prestige, and quality of each of these three cars; applied the weighting factors to each of these parameters as determined through discussions among the family members; and then, after clicking into his computer, showed which was the correct choice for the family based on these agreed-upon factors. To demonstrate how we did things in the office, the author walked through this example with his wife and with a flourish showed the results of the analysis followed by the searching question: "What do you think?"

Her answer was, "That's fine, but I like the *red* one!"

Of course, there was nothing wrong with the method. It was quite elegant and mathematically reproducible. But, an important element was missing. The analysis had done two things. It had taken the decision maker out of the loop (the analysis was done inside the computer) leaving him (her) uncomfortable, and, secondly, it assumed that the decision maker knew *what* questions were important to him (her) to ask in the beginning of the process. The purist would immediately interrupt at this point, of course, to say that the analysis could be repeated but this time adding high weighting to "red;" but the point had been made. Any decision-making tools must make the decision *"feel right"* immediately. And, frequently in the defense business, an opportunity to do the analysis a second time for the decision maker rarely occurs. Returning to the 256-parameter weapon system example, there would be no way the decision maker would know, for example, how the treaty implication parameter influenced the final decision (choice) compared to the influence of, say, the missile trajectory selection. Perhaps his decision would have been different if he had known that.

Further, the results of any analysis must take into account *who* is asking the question. Who is the decision maker? If a congressman or senator is being asked, he might be searching for the weapon system that would be manufactured in his district (corresponding to the "red" one), but he might not be direct in saying that! Such considerations are very real (they provide jobs) but difficult to express in the absence of all of the technical ramifications. If the battlefield commander were the decision maker, he might be looking for the system with the highest lethal capability against today's threat to "protect our boys over there, *now*." If the engineer were the decision maker, it might be the system with the most efficient guidance system that allowed the maximum payload to intercept the target of *tomorrow*. So, what is required is a decision-making method that would accommodate each of the varied and important viewpoints of the many decision makers that would be looking at each of the proposed systems—and at its different stages of development. It must be a simple method that would be both "understandable" and adaptable to questions at hand and adaptable as the circumstances unfold. *It must keep the decision maker in the loop*!

The methods outlined in this book are not meant to replace the many sophisticated modeling and simulation techniques already available on the market for the design of the highly sophisticated defense systems. Neither are they to replace the many design tools used in the final design phases of actual hardware programs. The decision-making methods outlined here are meant to guide the decision makers *early in the selection process* of weapon systems. They are to provide answers as to which are the most likely choices for the most likely situations facing "defenders" in the future. And they must deal with *imprecise data* that frequently exist early in the design process and development of weapon systems. A corollary to the decision methods outlined here is the need to carefully distinguish between *data* and *information*. Oftentimes, immense detail (data) can be provided on irrelevant facts concerning one defense system or another, but the relevant information on a key parameter is missing. For any decision method to provide the proper *aide-de-choix* to the decision maker, this distinction must be recognized.

One of the problems in setting up decision-making tools is the need to recognize the different types of decision making required along the development path of any particular defense system. It could be argued that there are three possible types of decision making, such as the following:

1) The decision maker has already made up his mind, and analysis is unnecessary.
2) The various options are generally understood, but a reliable method is sought to quantify the decisions.
3) There are too many options and variables, and some logical, reproducible "sorting method" is required.

This book cannot do much about the first possibility, but it is aimed at providing the much needed *decision aids* on the last two possibilities to the various management levels to make the path to the best decision easier.

As will become clearer in the following chapters, the method is in succeeding steps with the decision maker apprised of the unfolding results at each stage. There can be more than two stages depending on the complexity of the set of alternatives, but essentially they are as follows:

1) Make selections based on technical, cost, and programmatic issues.
2) Finalize selections based on political and national issues.

It is believed that by sorting the decision process into stages the ramifications of technical, cost, programmatic, political, and national issues are kept clearly in focus. Because of 1) the long development times of most modern defense systems, 2) the rapidity of the changes in the predicted threat, and 3) the delicateness of the changing political and world scene, it is important to keep the reasons for the decisions well understood *as the process unfolds.*

PERFORMANCE, COST, SCHEDULE, AND RISK

Most technical and management decisions on weapon systems revolve around the basic issues of performance, cost, schedule, and risk as shown in the following questions:

1) What is the defense system's purpose? (measured by *performance*)
2) How much will it cost? (measured by *cost*)
3) When will it be available? (measured by *schedule*)
4) Will it work as planned? (measured by *risk*)

Each of these questions usually mean different things to different people, and, further, the answers depend to a very large degree on when (in the development cycle) the questions are being asked. Most of the standard measures used in the defense business, such as *cost effectiveness* and *life-cycle cost* and *reliability and availability* can be derived from the preceding basic parameters as will be shown later.

The actual decision-making method is developed and provided in Chapter 8. In that chapter the method is explained, and the various issues related to the mathematics in the solutions to the missile defense equation are detailed and explained. The method lays out the various types of inputs that are required for the decision-making solutions to be handled. Some of the inputs are quantitative in terms of performance, cost, schedule, and risk; and some are more subjective such as bounds on the problem and qualitative assessments of the parameters involved. Not all decisions can be treated as black and white, and the method in Chapter 8 includes what is called "less precise decision making." In this method, the decision maker has a preference for a particular criteria but is willing to deviate if the characteristics of the candidate defense system do not quite "fit the mold." An example of this would be the decision maker desiring the most cost-effective solution, provided it is not too expensive. This is handled by the method developed in Chapter 8. Of course, in some cases the boundaries of the solution can be quite specific. One example is that the performance must meet a specific value (usually set as a specification requirement). Such specific boundaries of the solution can also be accommodated in the method, but the method allows for the display of the results such that if there are other later-developing criteria (say, political or national interests) which require a change in the specification limit, such influences can be observed easily as deliberate decisions are taken to move away from some "hard limit."

When making decisions, the problem exists of defining all of the terms that are to be understood and agreed upon by both the decision maker and the analyst doing the actual computations. What is meant by performance? What is meant by cost? etc. Accordingly, individual chapters are devoted to each of the key parameters in turn. Chapter 4 develops the many facets of *performance*, similarly for *cost* (Chapter 5), *schedule* (Chapter 6), and *risk* (Chapter 7). It will be shown in each of these chapters that there are many dimensions to each of these main topics, all equally important but needing of a common set of rules.

Although the main purpose of this book is the development of the decision-making method, useful data are provided on the key subjects germane to missile defense on performance, cost, schedule, and risk. The decision-making method can be used with inputs from the user's own database and formulation of requirements. The data, trends, and methods provided in Chapters 4–7 can be used as default values if more detailed data on any specific set of candidates that the decision maker wishes to evaluate are not immediately available. Conversely, the evaluation method can be used to decide on a set of requirements that any new proposed system is required to exhibit.

The main flow of the text is to lead the reader through the development of the method and to present trends in any available data to provide a sense of the state of the art of missile defense as it stands today and is likely to be in any foreseeable future. To avoid the flow being slowed down with unnecessary details, ample use is made of Appendices, where specific information or mathematical derivations can be found. Ample references are provided in each chapter complete with explanatory parenthetical expressions as the text proceeds such that the reader can gain a sense of "context" and historical relevance as the method or data are provided.

All readers are aware of the historical caution: *those who are unaware of history are condemned to repeat it*. Because of this, each chapter begins with the historical background of each subject to provide both a contextual reference and a historical trend in the material at hand.

It is suggested that senior decision makers might wish to read the first few pages of each chapter to gain an appreciation of the subject at hand and that engineering managers keep reading into the main body of each chapter. The student should also delve into the Appendices to see the derivation of much of the material and the inevitable assumptions that have to be made to make any analysis tractable.

Example problems are provided in each chapter. Some of the example problems are worked in the main text to illustrate some point; other exercises are added at the end of the chapters to be worked by those interested readers who wish to explore some variations on the points made in each chapter. A CD-ROM is also provided so that the exercises can be worked using preset spreadsheets for ease of use.

Chapter 2

MISSILE THREAT

The probability of missile attack may be low but the price to be paid for no defense is too high.
Peter J. Mantle

HISTORICAL BACKGROUND

Ever since man began fighting he has sought to retreat further and further from the scene of actual man-to-man combat (for obvious reasons). Some examples might be, first, the Macedonian thrusting spear (*sarissa*) at 3.6-m range (circa 335 B.C.), then the Roman mechanized bow and arrow or *ballista* (a mechanized dart or stone thrower; one of many variants of *katapeltes*, shield piercers) at 0.50-km range (circa 50 B.C.),* then much later the Paris Gun (Kaiser Wilhelm Geschultz) in 1918 at 125-km range,[1] and today with ballistic missiles with ranges up to 10,000 km or more. The long range of today's missile dramatically changes the nature of warfare. The Gulf War in 1991 was the first war where both ballistic and cruise missiles were used as primary weapons of choice.

The missile (or rocket) has been a weapon of choice for many centuries; its use frequently was only tempered by the available technology. Battlefield commanders, in their consideration of the use of missiles, have called for either greater firepower (lethal mass) or accuracy and preferably both. Historians may differ on the true beginnings of the missile threat, but the siege of the Honan Province capital K'ai-feng by the Mongols in 1232 with their "arrows of fire" is often cited as the beginnings of missile attack. After this early Mongol use the story of military battles in the 12th and 13th centuries includes many uses of missiles in warfare. During the 16th century, one could find many examples and designs of multistaged missiles. The work by Conrad Haas, an artillery officer in Sibiu, Romania, documents many such configurations, which were further documented later in 1650 by a Polish artillery expert.[2] In the 19th century the most notable rockets were those designed by Colonel William Congreve[3] of England. These missiles were small (less than 30 kg in launch weight) and short range (1–2 km) but very

---

*Based on a design, built in Jerusalem, by King Uzziah of Judah (Judea) in the eighth century B.C. to launch arrows or stones from the ramparts of his fortifications.

effective in several wars. The Congreve rockets (ballistic missiles) were used in the war of 1812 between Great Britain and the United States, and from that conflict all Americans can associate with Francis Scott Key's "The Star Spangled Banner," which includes the line "... *and the rockets' red glare.*" In the various wars throughout the world over the succeeding centuries, there have been uses of missiles of one form or another as the technology has improved. Depending on the technology for range and accuracy, sometimes the gun was the preferred weapon and sometimes the missile, but eventually, as propulsion and guidance technology improved, the missile (both *ballistic* and *cruise*) became the weapon of choice.

DEFINITION OF BALLISTIC MISSILE AND CRUISE MISSILE

In attempting to analyze the characteristics of missiles (ballistic or cruise), one is confronted with the difficult task of defining what exactly they are. On the face of it, it appears simple enough to say that a *ballistic* missile is one that is launched on a ballistic trajectory into space and that a *cruise* missile is one that depends on aerodynamic lift as it "cruises" through the atmosphere. Such simple characterization, however, immediately breaks down when follow-up questions are asked, such as "*what is the difference between a space launch vehicle and a ballistic missile?*" or "*what is the difference between a cruise missile and an unmanned air vehicle?*" Are they only weapons of war or can they be used for peaceful purposes? Because a sounding rocket for upper-atmosphere research or for meteorological use or for other peaceful uses has all of the same (external) characteristics as a ballistic missile, then some other discriminate must be used. Is the only difference between a cruise missile and an unmanned air vehicle (UAV) whether or not the payload is a warhead or surveillance package? If so, then what is meant by an attack UAV? To further add to the confusion is that in the inventory of weapons worldwide, there are both ballistic missiles and cruise missiles and a *third* category that is really a mixture of both. Several weapons used in the battlefield, for example, are launched ballistically and then use aerodynamic surfaces to zoom in on their targets. Frequently, the real difference between certain types is nothing more than the *intent.*

As both the *ballistic missile* and *cruise missile* have progressed over the last few centuries through development for various military missions, various names have been used to describe them. Unfortunately, names have sometimes been given to reflect particular *types or brands* and sometimes to reflect *generic* forms. The choices have not always been consistent, which has caused confusion in attempts to characterize their performance, cost, mission, and other attributes. It is generally assumed that the ballistic missile is so-called because it spends *most* of its flight time in a ballistic trajectory and, further, is not dependent on aerodynamic forces (generally) for its mission or purpose. [Those ballistic missiles that use reentry vehicles (RV) in the accomplishment of their mission with aerodynamic control or thrust vectoring in the final "flight phase" belie the simple "*only* follow a ballistic trajectory" definition.] The cruise missile does however require aerodynamic lift for it to function. It is also generally agreed that both are for military purposes and that the word "missile" attests to this understanding. (Recognizing that surveillance is a military mission, then substituting a warhead for a surveillance package "converts" a cruise missile to a UAV?!) There are examples however when such an understanding would not be correct for either the ballistic missile or the cruise missile as already mentioned.

The following list collects some of the names used to describe missiles. No attempt has been made to sort out (in this list) which names best apply to ballistic missiles, which to cruise missiles, and which to other forms of missiles that encompass features of each. The purpose is to emphasize that many names have been used, with the normal inconsistencies that have crept into the missile community parlance.

Strategic ballistic missiles (SBM)
Tactical ballistic missiles (TBM)
Theatre ballistic missiles (TBM)
Surface-to-surface missiles (SSM)
Surface-to-air missiles (SAM)
Air-to-surface missiles (ASM)
Guided weapons (GW)
Tactical aerodynamic missiles (TAM)
Aerodynamic missiles, cruise missiles (CM)
Remotely piloted vehicles (RPV)
Unmanned air vehicles (UAV)
Unmanned aerial vehicles (UAV)
Attack unmanned air vehicles (AUAV)
Uninhabited combat air vehicles (UCAV)
Unmanned vehicle system (UVS)
Air-to-air missiles (AAM)
Interceptors
Air vehicles
Reentry vehicles (RV)
Rockets
Space launch vehicles
Guided projectiles
Boost glide vehicles (BGV)
Boost glide missiles (BGM)

Many more names exist that are clearly related more to launch means or are mission related, such as ICBM, GLCM, TASM, SLBM, ALCM, etc. The most common denominator taken from this list is that both ballistic and cruise missiles are *guided weapons*, which could be defined as those weapons where the warhead is delivered by a pilotless guided vehicle. Both the ballistic missile and the cruise missile fall under this definition. Both have warheads, both are guided, and both are unmanned. [It is taken here that a "warhead" can either be one containing an explosive charge or be simply a kinetic warhead for those missiles exhibiting high enough speed to generate sufficient destructive power through kinetic energy. The ballistic missile is guided through the selection of launch parameters (see Appendix C) as well as onboard thrust control, and the cruise missile is guided by either aerodynamic or thrust control.] Accordingly, in the context of this book, it will be taken that a *ballistic missile* is best defined as any pilotless missile that generally follows a ballistic trajectory for most of its flight path to place its warhead on target. (The *American Heritage Dictionary* takes care of the thrust cutoff features and noncontinuous power in a ballistic trajectory in its definition "*a projectile that assumes a free-falling trajectory after an internally guided, self-powered ascent.*")[4] A *cruise missile* is best defined as any pilotless missile that is generally flown under continuous power and aerodynamic lift through the atmosphere to place its warhead on target. This definition paraphrases that in the 1987 INF Treaty, which defines a cruise missile as "*an unmanned self-propelled guided vehicle that sustains flight through aerodynamic lift for most of its flight path and whose primary mission is to place an ordnance or special payload on a target.*"

The main subject of this book is the defense against *missiles*, but it would be remiss not to acknowledge the other classification of air vehicles, frequently called *unmanned air vehicles* (UAV). With the subtle recognition that the word "weapons" is normally restricted to mean those carrying some type of warhead and that the word "vehicles" is normally reserved for military purposes other than an actual weapon, the UAV is normally associated with having 1) no warhead and 2) a return flight feature. Hence, the term UAV is normally associated with those military vehicles used for reconnaissance or surveillance or similar missions. In which case, in the same vein as the preceding definitions, the *unmanned air vehicle* would be defined as any pilotless recoverable vehicle that is generally flown under continuous power through the atmosphere for the conduct of some military mission. [The word "*unmanned*" has become part of the parlance in the defense community although attempts have been made to be more politically correct, using for example the incorrect term "*uninhabited*" (because even manned aircraft are uninhabited). The word "*pilotless*" is suggested as accomplishing both objectives. The collected data, however, will use the originator's designation in listing names of vehicles.]

The words "generally" and "most" in the preceding definitions allow variations of particular designs depending on the mission and the ingenuity of the designer without the need to create a new definition each time a new design appears on the scene. It is also apparent from the preceding definitions that the ballistic missile accomplishes most of its trajectory *outside* the atmosphere and that the cruise missile generally follows its trajectory *inside* the atmosphere. (The popular term "atmosphere" has been used here, although it is recognized that the more correct term is "troposphere" where aerodynamic flight can be sustained.) Such definitions also imply that they are "one-way trips with a killing payload." The word "missile" (or projectile) clearly is intended to connote killing or causing damage. (In some circles, the word "*hittile*" has been used to reflect a specific set of projectiles where the warhead has a contact fuse, reserving the use of "*missile*" to that which uses a proximity fuse, but this is thought to be too restrictive.) As will be shown later, the influence of the atmosphere determines the *shape or geometry* of the ballistic missile and cruise missile. The need to operate efficiently in the atmosphere gives the cruise missile

an *unsymmetrical* geometry (of body, wings, and engine), whereas the dynamics and lack of atmosphere allow the ballistic missile to be *symmetrical* in its body shape. This can be seen in the various designs, but such differences are the result of the different modes of operation between ballistic and cruise missiles and not fundamental to the definition. It does help however in missile and mission recognition.

In the preceding definitions care was taken not to restrict the means of control. For example, the missile (ballistic or cruise) could be guided or controlled either by onboard computer or by remote control. This is a matter of design and not fundamental to the generic features of *ballistic* vs *cruise*. Indeed, both techniques of onboard and remote control have been applied to both types of missile in the world's inventory. Given the general guidelines on the fundamental differences between *ballistic* missiles and *cruise* missiles, it is useful to put this into picture form; Fig. 2.1 provides a general idea of these two missile types that will be the subject of further analysis throughout this book.

A glance at the generic forms of *ballistic* and *cruise* missiles depicted in Fig. 2.1, will immediately bring to mind the possible variants depending on the mission. For example, a short-range, ballistic missile (say, 300-km range) will have an apogee of approximately 60 km. This means that about one-half of its trajectory will be through the atmosphere, and it would be normal to add fins (as shown) for stabilization. For longer-range ballistic missiles (say, 1000 km) the speeds are of the order of 3 km/s or about Mach 10, clearly in

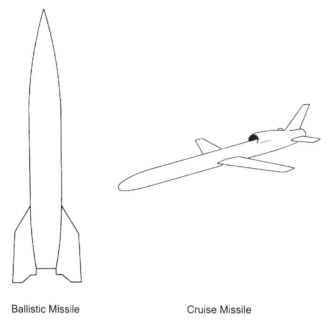

Ballistic Missile   Cruise Missile

Fig. 2.1   Generic forms of ballistic and cruise missiles.

the high hypersonic flow regime, and other shapes such as flared cones would be better used for stabilization through the atmosphere. For ballistic missiles designed to travel distances greater than 3000 km where speeds are of the order of 5 km/s or more (Mach 16$^+$), such features disappear, and the sleek tube-like geometries with no external aerodynamic features, that is, shapes normally associated with *strategic ballistic missiles*, begin to appear in the designs. The differences then are not fundamental but rather the normal application of best geometries for the mission. Similarly, the generic *cruise missile* shape shown in Fig. 2.1 is clearly most applicable to missiles designed for slow speed, long range. As the speeds are increased to supersonic values, then all of the normal shaping familiar to supersonic fighter aircraft would be applied to the cruise missile also. Further, if the cruise missile is to have a low signature [radar cross section (RCS) or infrared (IR)], then again all of the normal design techniques familiar to aircraft designers would be applied as necessary, and the shape of the cruise missile would change accordingly. But there would be no need to restructure the definition.

   An example of the ballistic missile that does not always fly a ballistic trajectory is the THAAD [Theater High-Altitude Area Defense System, currently in development by the U.S. Ballistic Missile Defense Organization (BMDO), now Missile Defense Agency (MDA), with a planned initial

Fig. 2.2   THAAD in flight and energy management maneuver.

operating capability of 2007 (see Appendix A)] interceptor. Figure 2.2 shows both the configuration of THAAD and a typical maneuver that the missile must go through when launched against a short-range threat.

Several key points can be made from Fig. 2.2. Because THAAD is designed to intercept long-range BM over a wide area, its design speed is in the high hypersonic speeds (in excess of Mach 6–7), and it can be seen that it utilizes a sleek slender design with minimal external surfaces and no aerodynamic fins. (It uses a controlled conical flare at the stern for maneuvering and stability in the atmosphere.) As part of the onboard control, one can see the exhaust from the divert propulsion nozzles high on the body. Because it has a single-stage solid rocket propulsion motor, it must use an onboard controlled *energy management system* to burn off the rocket propellant when it is required to engage short-range targets. This energy management system is shown in operation during a flight test (2 August 1999) on the right-hand side of Fig. 2.2. Such a photo is dramatic evidence that a *ballistic missile*, as defined previously, is not always required to follow a ballistic trajectory. As a hint of things to come, an adversary could incorporate such an energy management feature into his BM as a boost phase intercept countermeasure.

Finally, regarding the definition of ballistic and cruise missiles, Fig. 2.3 shows two very different military configurations designed to operate in the atmosphere. The upper photograph in Fig. 2.3 shows the Tomahawk *cruise missile* (containing a warhead). The lower photograph in Fig. 2.3 shows the Global Hawk *unmanned air vehicle* (containing a nonwarhead, military

Designed for High Speed and High Maneuverability

Designed for Long Endurance

Fig. 2.3   Pilotless aerodynamic military vehicles.

mission payload). As can be seen from Fig. 2.3, all of the normal design features common to any flying vehicle (manned or unmanned) have been incorporated as dictated by the aerodynamics and mission requirements. It should be obvious from examining Fig. 2.3 that no fundamental reason exists why either vehicle could be carrying either a warhead or a nonlethal mission payload. Hence, in the collection of data for all missiles in this book both types of military vehicles have been examined to determine the state of the art and possible trends in future missile design.

Part of the problem in the definitions is that in some cases the form of the *trajectory* has been used as the discriminator and in other cases the *mission* has been used as the discriminator. Table 2.1 shows the various classifications of missiles (and military vehicles) that have been used by different groups by mission or trajectory. Such a display emphasizes the many different configurations that are possible.

The basic characteristics of *ballistic* and *cruise* missiles have already been described, but Table 2.1 shows the various possible combinations of features required for particular missions. For example, the design of a missile for a strategic mission might concentrate on maximizing the ballistic portion of the trajectory. In the defense against such a missile, an interceptor can still be predominantly ballistic but would optimize more on the surface-to-air aspects. Many air-launched missiles could optimize, say, on *boost/glide* or *boost/sustain* trajectories, and in each case various features of ballistic and cruise missiles trajectories would be incorporated resulting in various "looks" to the missile.

As modern technology comes on stage, it is of interest to see how such definitions are tested, at least in a "systems sense," where combinations of designs could become the reality in the not too distant future. For example, the Predator started life as UAV[5] searching for Serb missile launchers during the Kosovo crisis, but its 70 kn cruising speed made it vulnerable to easy attack. The U.S. Air Force thus is planning to equip this unmanned air vehicle with its own missiles (the Stinger, which was originally designed as a *surface-to-air* missile but now would be adapted to an *air-to-surface* missile for this new mission, in this instance). Another example of combining missile concepts could be Boeing's concept for the U.S. Air Force[5] to make greater use of unmanned vehicles that can go in harm's way without

Table 2.1   Variations in trajectory and mission classification for missiles

	Type of mission				
		Tactical mission			
Type of trajectory	Strategic mission	Surface to surface (SS)	Surface to air (SA)	Air to air (AA)	Air to surface (AS)
* Ballistic					
* Cruise					
* Boost/glide					
* Boost/sustain					
* Gravity/glide					

endangering human pilots. Designs such as the UCAV shown in Fig. 2.4 offer the possibility of carrying *weapons on weapons*, incorporating the best features of both *ballistic* and *cruise* missiles. (This is the author's opinion and is not necessarily reflective of any official planned use of the UCAV shown.)

The preceding precursor has sought to categorize both ballistic missiles and cruise missiles into a readily definable classification, yet not to exclude the possibility that radically new design concepts of *missiles carrying missiles* can emerge in the not too distant future.

DAWN OF THE MISSILE AGE

Modern-day development and use of missiles can be traced almost directly to the work in Germany in 1923 when Hermann Oberth (a young professor of high-school mathematics in Transylvania) published his famous pamphlet, "*The Rocket into Interplanetary Space.*" [6] In that pamphlet Oberth first put down the basics of mathematics behind the designs. [This is not to discount the early work by Roger Bacon in England (1248) in formulating the equations for black powder (gunpowder) in rocket propulsion.] Professor Oberth's purpose was interplanetary travel, but this work quickly was adopted for military purposes.

BALLISTIC MISSILES

By 1937, Capt. Walter R. Dornberger in the German Army working with a young Wernher von Braun of the German Rocket Society had established a

Fig. 2.4   A possible concept for a missile-carrying missile (copyright Boeing Company).

large group of scientists and engineers at Peenemunde to develop "rockets" for military use. The first *ballistic missile*, as it became known, was the A-4 (Model or Aggregat-4) developed by the Dornberger–Braun team. The name A-4 was changed at the direction of the German Ministry of Propaganda to V-2 for Vergeltungswaffen Zwei (Vengeance Weapon 2). It was first used against Paris on 6 September, 1944 and then again on London on 8 September, 1944. It was a significant weapon, and, although some writers say that the V-2 was not a decisive weapon in WWII, there was virtually no defense against it other than seeking and destroying its launcher, and it was the harbinger of the missile age as we know it today. Figure 2.5 shows the V-2.

The statistics of the damage done by the V-2 are that 4300 were launched at the rate of 20 per day (e.g., see Ref. 7). Of these, 1359 were launched against London, of which 1054 (78%) struck somewhere in England. Further, 517 of these landed inside greater London (an area of 1890 km^2) killing 2480 people. The V-2 was a liquid-propellant, single-stage ballistic missile. Its launch weight was 12,853 kg with a 1000-kg conventional high-explosive (HE) warhead as payload. The propellants were a mixture of liquid oxygen and 75% ethyl alcohol-water combination. The missile motor produced a 25,000-kg thrust up to the end of boost phase (*Brennschluss*) of 60 s. Its burnout velocity V_{bo} was approximately 1.6 km/s, which gave it a (*maximum range, minimum energy*) ballistic range of approximately 360 km and a corresponding apogee during flight of approximately 90 km. (In test flights at

Being Readied for Launch

Early Boost Phase

Fig. 2.5   Ballistic missile V-2 (courtesy Duncan Lennox).

White Sands Proving Grounds on 7 March 1947, the V-2 achieved an altitude of 155 km within a constrained range of 90 km.) It was a large missile by today's standards, with a length overall of 14.3 m and diameter of 1.65 m.

CRUISE MISSILES

The forerunner of today's *cruise missile* could be said to be the V-1, developed by the same team that developed the V-2, also during WWII.* The V-1, known in England at the time as the "doodle bug" or "buzz bomb" because of the staccato noise of its pulse jet engine, was a particularly frightening weapon. The V-1 entered the war before the V-2 on 12 June 1944. Over 9000 were launched,[†] of which 2340 landed in greater London, killing more than 5000 people. In the last days of the war, from June 1944 through June 1945 more than 21,000 V-1 cruise missiles were launched against land targets in Europe. The V-1 (originally designated Fiesler Fi-103) was a self-propelled, automatically steered, pilotless airplane (Although some were piloted during the testing phase and planned for mission use, but it was difficult to get volunteers, because the cockpit as designed was difficult to escape from before final descent!). The unique propulsion motor was a pulsejet that used an explosive fuel-air mixture of oxygen and 80-octane gasoline. It was launched from a piston-operated hydrogen-peroxide-powered ramp (reported at the time by Royal Air Force Intelligence as suspicious ski sites in The Netherlands and France) and accelerated to approximately 0.16 km/s (Mach 0.5). It had a range of between 250 and 290 km. The V-1 measured 8.5 m LOA with a wing span of 5.9 m. It carried a warhead of conventional HE of 1000 kg. Figure 2.6 shows a V-1 cruise missile sitting on one of the ski-like ramps. More details on the V-1 may be found in Ref. 10.

MISSILE DEVELOPMENT IN THE LATTER HALF OF THE 20TH CENTURY

After WWII ended, many nations around the globe continued the development of the *cruise missile* and the *ballistic missile* based on the pioneering V-1 and V-2. Indeed, the profusion of missile types and numbers

---

*An earlier version of such a "flying bomb" was developed during the latter stages of WWI (1918) by the U.S. Army Air Corps. That version, called *The Bug* (a disposable wing biplane designed by Charles F. Kettering), was propelled by an internal combustion engine and propeller. It was mass produced but never used because the war ended (see Ref. 8).

[†]The statistics of the V-1 vary among various authors. One authoritative source is the Hill Report.[9] The Hill Report describes the many techniques used for bringing down the V-1 over England. These techniques included Spitfire aircraft flying alongside the V-1 and tipping it over by the Spitfire's own wing tips causing the V-1 to crash. The Hill report states that 24,491 dwellings were destroyed by V-1 attacks and 52,293 more dwellings made uninhabitable. Some 5864 people were killed; 17,197 suffered major injuries, and a further 23,174 people received minor injuries. Nearly 53% of the V-1 cruise missiles were brought down (by guns, "Blimps," Spitfire tipping, etc.), and 32% reached their targeted areas.

Fig. 2.6   Cruise missile V-1 sitting on launch ramp.

has been almost exponential in scale. Both the United States and the Former Soviet Union were instrumental in missile development, employing engineers from Germany involved in the development of the V-1 and V-2. In the first five years after WWII, the United States developed the Hermes (a copy of the V-2) with the help of the captured Werner von Braun and a salvaged set of enough components to build 100 V-2 ballistic missiles, which were test launched in White Sands between the period of 1946–1951. This led to other variants of the V-2 such as the Redstone (1958) and the Pershing (1962). The Soviet developments also followed the V-1 and V-2 developments and produced the 160-km range SS-1* or Scud-A (1955) and later variants (such as the 300-km Scud-B in 1961 and such missiles as the 600-km range Scud-C in 1965). These spawned new ballistic missiles such as Al Husayn (1988), Al Abbas, and others. Later Soviet ballistic missiles started with the 1200-km range Shyster (a stretched V-2) in 1956. Some 15,000 Scuds were produced and sold around the world to nations such as Egypt, Iran, North Korea, Syria, and many other nations. The Soviet export of Scuds ceased in 1993 to meet the 1987 MTCR[†] restrictions (with a 300-km limit and 500-kg payload weight imposed), but it is known that other nations such as the People's Republic of China and North Korea have continued the development and export of such missiles without regard to such limitations.

DEVELOPMENT OF THE BALLISTIC MISSILE

An indication of the plethora of *ballistic missiles* that have been developed and deployed since the end of WWII is provided by the (partial) list of ballistic missiles by missile name and user nation in Appendix A.[‡] This Appendix is not meant to be an all-inclusive list of all missiles (many have come and gone) but more of an indication of the proliferation of the technology and missile weaponry throughout the world. No attempt has been made to categorize these missiles a "strategic" or "tactical" as has been done in various treaties and agreements between nations, because a key ingredient in such categorization must be the "intent" or "mission" of the user in particular scenarios. What is considered important in the context of this book is that the technology has produced such weapons, and thus all must be considered in any decision making for defense against them.

DEVELOPMENT OF THE CRUISE MISSILE IN THE GULF WAR

In December 1950, after WWII ended, Wernher von Braun at the U.S. Army Redstone Arsenal began development of an 800-km range-guided cruise missile that led to later designs in use today. The use of *cruise missiles*

---

*Sometimes the NATO designation "SS" (for surface to surface) for Soviet missiles is used to describe the series.

[†]Missile Technology Control Regime, which is a group of (now) 28 nations to control the transfer of equipment and technology mainly in the area of ballistic missiles, but now considers cruise missile technology.

[‡]Compiled from various editions of *Jane's Weapon Systems, Jane's Strategic Weapon Systems, Jane's Fighting Ships, Aviation Week and Space Technology, International Defense Review, Ballistic Missile Proliferation, National Defense Industry Association committee reports*, and similar publications.

took on a new dimension with the use of the U.S. cruise missile Tomahawk during the war in the Persian Gulf in 1991. The Tomahawk was first used against Iraqi targets on 17 January 1991, when Operation Desert Shield became Operation Desert Storm, an allied offensive against Iraq, and U.S. ships launched BGM-109 Tomahawk cruise missiles at targets in Iraq and occupied Kuwait. They were used to disable Iraqi command and control (C^2) centers to prepare for a follow-on allied air raid to establish air superiority. Some 264 BGM-109C (unitary warhead) and 27 BFM-109D (cluster bombs) were launched from the Iowa class battleships *Missouri* (BB 63) and *Wisconsin* (BB 64), several cruisers and destroyers, and at least two submarines. Approximately 12 Tomahawks were launched from the nuclear submarines *Pittsburgh* (SSN 720) and *Louisville* (SSN 721). It was reported that 85% of all Tomahawks launched hit their targets.[11]

Appendix B provides a partial listing of modern-day cruise missiles listed by country of use. Such a listing shows the proliferation and extent of the development of the cruise missile as it stands today.

WEAPONS OF MASS DESTRUCTION

As weapons are developed to kill more than just individuals or small numbers of people (civilian or military), they take on a more ominous nature, and the description of weapons of mass destruction (WMD) has crept into defense parlance. Although there is no accepted definition of "mass," it can be said that a single weapon that can kill thousands instead of one, ten, or one hundred is a weapon of mass destruction. Originally, this term was reserved to describe those weapons capable of delivering nuclear warheads of awesome power but now has been expanded to include those weapons capable of delivering biological or chemical warheads. Sometimes "NBC (nuclear, biological, or chemical) weapons" is used as a label for WMD. Frequently, it is assumed that the nuclear weapon requires a certain amount of sophistication to design and build (although this does not apply to those warheads containing nuclear waste material) and thus is only available to "rich" nations and that weapons containing biological or chemical agents can be built relatively cheaply. This has led to the description of biological and chemical weapons as the "poor man's nuclear bombs" because of their ability to inflict mass casualties of the order of nuclear weapons at a considerably lower cost.

CASUALTIES FROM NUCLEAR WEAPONS

To place the meaning of weapons of mass destruction into some perspective, consider the damage inflicted by the *single* 14-kiloton nuclear fission bomb ("Little Boy") dropped by the U.S. Air Force on Hiroshima on 6 August 1945 [followed by a 21-kiloton fission bomb ("Fat Man") dropped on

Nagasaki on 9 August 1945].[12] This particular bomb (burst at a height of approximately 560 m on a windless summer day) was dropped on the city of 276,300 inhabitants, killing 68,000 and seriously injuring 76,000 (Ref. 13). The Hiroshima bomb devastated an area of 11 km^2, and the bomb dropped three days later on Nagasaki devastated an area of 4.7 km^2. (The deaths numbered up to 40,000 from this single bomb with 25,000 injured.) It is common practice to quote nuclear weapon warhead capability in kiloton energy, which is 4.2×10^{19} ergs, which is approximately the amount of energy that would be released by the explosion of 1 kiloton (1000 tons) of the "conventional" HE material trinitrotoluene (TNT). Today, nuclear weaponry has reached the capability where warhead yields are frequently measured in units of megaton energy, which is 4.2×10^{22} ergs or the equivalent amount of energy released by 1,000,000 tons of TNT. It is not the intent of this book to get into a discussion of nuclear physics and the difference between nuclear fission bombs (where the energy is quoted in kilotons and involves the splitting of atoms) and nuclear fusion bombs* (where the energy is more likely to be quoted in megatons because of the fundamental difference in the thermal nuclear process and the melting together of the atoms as in the hydrogen bomb). It is the intent, however, to indicate in some easily measurable way the destructiveness of such weapons and the ease in which they can now be incorporated into ballistic and cruise missile design. One might compare these numbers with the *total* tonnage of conventional bombs dropped on Germany in WWII, estimated at 1.3 million tons of HE with an estimated 305,000 civilians killed. Another comparison might be that a 20-kiloton nuclear bomb has 20 times as much power as that of the total salvo of conventional bombs dropped on Hamburg during WWII, which killed 42,000 people.

Most of the damage done in Hiroshima and Nagasaki was caused by the overpressure generated by the nuclear burst, but other forms of injury and death were caused by nuclear radiation, fallout, and heat (approximately 35% of the energy in a nuclear explosion in the air). The additional damage caused by the nuclear-blast-generated electromagnetic pulse (EMP) and transient radiation effects on electronics (TREE), although not directly injurious to humans, does incapacitate electronic equipment (with the extremely high volts per meter) that can cause indirect deaths and injuries if humans are operating the equipment at the time of the EMP or TREE. Further, "flash blindness" from nuclear explosions is also extremely hazardous to humans. All of these effects are beyond the subject matter of

---

*The nuclear fission bombs that were used on Hiroshima and Nagasaki were (confusingly) called atomic bombs or A-bombs, whereas later improved nuclear weapons that used a self-sustaining thermonuclear chain reaction were called hydrogen bombs or H-bombs. Typically, A-bombs were less than 1 megaton, and H-bombs provide greater than 1 megaton of nuclear energy. This development allowed much lighter weight weapons more suitable to use on ballistic and cruise missiles.

this book but are mentioned to bring attention to WMD effects from nuclear weapons.

Figure 2.7 shows available data on the relationship between the weight of the nuclear warhead and its yield in megatons. There is not a unique and direct relationship because of the different efficiencies in the nuclear process, but there is a discernible trend based on the various design principles used to date, sufficient to provide an indication. In the listed values of actual warhead weights in the various publications, there is the added confusion in the definition of "payload" or warhead weight. In some references the warhead weight is strictly the *lethal mass*, whereas in other cases it includes the shrouds, the release mechanisms, and similar mechanisms. In the SALT II agreements the payload or "throw weight" was defined as the sum of the *reentry vehicle*, the *postboost vehicle*, the *penetration aids*, and the *release mechanism*. With these cautions in mind, Fig. 2.7 provides a rough indication of the lethal power of nuclear weapons as related to the size of warhead.

It should be noted in Fig. 2.7 that both nuclear fission and nuclear fusion bombs and missiles are included in the data. As an indication of the improvement over time in the efficiency of design of nuclear weapons, the nuclear bomb dropped on Hiroshima contained only about 50 kg of uranium-235, but the bomb itself weighed over 5400 kg [Requiring a B-29 bomber (*Enola Gay*) to carry it over 2200 km from the Marianas and return]. Today, the nuclear warheads of ballistic and cruise missiles can be measured in

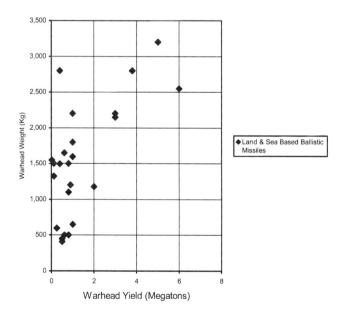

Fig. 2.7   Energy release of nuclear warheads.

hundreds of kilotons of energy also,* but with considerably lower gross weights (see Appendices A and B).† This improved weight reduction coupled with the greatly improved lethality is an indication of the awesome potential for WMD. A subject that is discussed later is the lethality measured in terms of casualties that is directly related to the *population density* in the area being attacked. Hiroshima,‡ with an average population density of 3300 people per km^2 (with peaks to 9700 per km^2 near ground zero) was chosen as a target because it was a port of embarkation, an army headquarters, and a manufacturing center, *not* because it had a high density of population, and so the numbers killed could have been much higher for the same 14-kiloton warhead. This factor of population density makes it difficult to make direct causal relationships between warhead weight, warhead energy, and lethality.

The awfulness of the destructive power of a nuclear weapon and the nature of the human casualties can only be estimated at best. Various studies, both governmental and private, have sought to characterize the effects of such a form of WMD. Table 2.2 has been constructed from several sources[14,15] in an attempt to present the nature of such an attack. The example of a 1-Mton warhead that has exploded over a city in a similar manner to the Hiroshima nuclear attack is used. Unlike a conventional HE warhead, the nuclear warhead produces several damaging and lethal effects. These are light flash, heat, blast (over pressure), electromagnetic pulse, and fallout. Each has devastating effects on people and equipment.

CHEMICAL AND BIOLOGICAL WARFARE CASUALTIES

If it is assumed then that there will be a greater tendency to utilize the poor man's nuclear weapon in the future, that is, biological or chemical weapons, then this should be given further attention. The evidence from recent wars certainly points to this strong possibility. [The use of *mustard gas* first used by German troops on 12 July 1917 in WWI and of chemical agents (a mixture of mustard gas, yellow rain and *tabun*) on Kurdish villages in 1990 are particularly ugly reminders of the use of WMD.] Poisoning one's enemy has been known since biblical times, but usually limited to the killing of a single person. The defense varied depending on one's status in life. In earlier times the king used a hapless food taster, a sort of one-on-one defense. In more recent history attempts have been made to surreptitiously poison one's enemy through poisoning the water supply (done in the American Civil War in 1863 and in many previous military conflicts), but such methods have

---

*The cruise missile *Tomahawk* (BGM-109) with a takeoff weight of 1500 kg has a 200-kiloton yield nuclear warhead.
†It is taken that about 0.50 kg of fissile material generates as much energy as 7.3 million kg of TNT.
‡For comparison, New York City has an average population density of approximately 10,000 per km^2, and Manhattan alone has a population density of much greater than 25,000 per km^2.

Table 2.2 Approximate damage from 1-Mton nuclear warhead

Distance from "ground zero," km	Damage and casualties		
	Flash and heat	Blast	EMP and fallout
0	*Flash* of explosion causes temporary blindness for greater than 150 km from ground zero.	*Heat* travels outward at speed of 300,000 km/s and vaporizes all materials and humans.	—
2–3	*Fireball* forms (2–3 km diameter); air *heats* to 20,000,000°F; radiation (mostly X-rays) lethal to 2–3 km released	*Blast* (overpressure) reaches 30+ times atmospheric pressure. All structures severely damaged or destroyed.	—
2–3	—	—	Radioactive cloud rises to 2 km in 30 s and 3 km in 1 min with wind speeds more than 1000 km/h hurling people and structures into the air
5–6	—	All brick structures completely destroyed to approximately 6 km	EMP generated very dependent on altitude of blast. Could severely damage electronic equipment out to many kilometers
12–14	Most materials ignite spontaneously from the *heat* out to 12–14 km		

(*Continued*)

Table 2.2 *Continued*

Distance from "ground zero," km	Damage and casualties		
	Flash and heat	Blast	EMP and fallout
17–18	Severe burns to those in the open out to 18 km	—	*Fallout* (that can last weeks) extremely hazardous to human organs
30+	First-degree burns to people out at 30+ km from ground zero	—	
40+ km	Extent expected for 1 Mton warhead and minor damage at this distance	—	

uncertain outcome with the result that potential users of biological and chemical weapons give serious consideration to more direct delivery means through the use of *ballistic* or *cruise* missiles.

If one ignores the gruesome early practice of the Tatars (or Tartars) in 1347 of catapulting bodies of bubonic plague victims over the walls in the siege of Caffa,* a Genoese trading post on the Crimean Peninsula, the first deliberately planned use of biological weapons delivered by *ballistic* means would again be the V-2 weapon. Towards the end of WWII, the German Army ordnance sponsored design changes to the V-2 to accommodate either chemical or radiological payloads. The conventional HE nose-mounted warheads of the V-2 were to be replaced with ballast, and then chemical or radiological cylinders were to be mounted circumferentially around the midsection. The radiological payloads were to be formed in the nuclear reactor facility at Watten, Germany. This was an attractive alternative to the German Army to the failed nuclear bomb program that was underway in Germany during the latter days of the war. Fortunately, the war ended before such a weapon could be used and WMD examples from WWII could be added to the list.

A similar set of designs was embarked upon also to use the *cruise* missile V-1 with a similar set of circumferentially mounted chemical weapons, but fortunately this also was not pursued to deployment. The chemical agent that was planned to be used in the canisters was the nerve agent *tabun*. (*Tabun* was selected at the time because it was known that the gas masks used by the British did not contain filters, only charcoal particles, which were ineffective against *tabun*.) Although *tabun* was a particularly unpleasant nerve agent (a hundred-fold more deadly than the *chlorine gas* used by the Germans on British, French, and Canadian troops on 22 April 1915[†] at Ypres Salient in WWI that caused thousands of casualties[‡]), *its* toxicity is dwarfed by "improvements" in chemical and biological agents being considered today. If *tabun* is given a relative toxicity rating of 1 for lethal dosage, then compare this value with that of *bacillus anthracis* with a lethal dosage amount of 10^{-5}, which is a significant increase in lethality. In its pulmonary form *anthrax* has a mortality rate approximately equal to 100% unless treated immediately with appropriate medical procedures. In tests that went awry on Gruinard Island off Scotland by the British Ministry of Defense in 1941 with dispersal tests of *anthrax-B* when over a hundred trillion spores were

---

*Some authorities quote this as the source of the "black death" in Europe, a disease that eventually killed 25 million—more than one-third of the approximately 60 million people of Europe at that time. Today, the population of the United Kingdom *alone* is approximately 60 million.
[†]On 19 October 1915 in Champagne, the German Army added *phosgene* to the chlorine to produce a more powerful and suffocating gas that killed 815 French troops and seriously incapacitated 4000 more.
[‡]Approximately 168 tons of chlorine from 4000 cylinders was released in five minutes along a four-mile front in Flanders fields.

released, the toxicity of the contaminated soil caused that island to be declared off limits for over 40 years into the mid-1980s. It required over 280 tons of formaldehyde and 2000 tons of seawater to return the island to some degree of normalcy. Such chemical and biological agents have lethality characteristics that approach the deadliness of the black death that changed the world after the Middle Ages. The microbes that cause bubonic plague (or black death) are a type of bacteria known as *yersinia pestis* (after the Swiss scientist Alexandre Yersin, who in 1894 discovered the cause of the bubonic plague) or *Y. pestis* for short, which has a relative toxicity of 10^{-6}, which is ten-fold more deadly than *anthrax-B*. Biological agents are generally more deadly than chemical agents, although more difficult to deliver effectively being "live" agents (and thus susceptible to temperature, ultraviolet light, and other difficult to control environmental factors). As pathogens, bio-logical weapons introduce a fear or terror among the intended victims, especially during the incubation period (from 1–12 days depending on the actual biological agent used) as the victims await the unknown results of the attack. The biological agent *pasteurella tularensis* (after the plague-like disease discovered in squirrels in Tulare County, California, in the early 1940s) is 10 times more toxic than anthrax, and its effects, if not deadly, can last for years in the victim. Table 2.3 displays these chemical and biological agents against an approximate scale of degree of deadliness.[12]

For clarity, not all known agents are displayed in Table 2.3. There are also other important subdivisions. For example, the chemical agents are often further divided into *nerve agents* (*sarin,* * *tabun*, *soman*, and *VX*) and *mustard agents* (*sulfur*, *mustard*, *nitrogen*, and *Lewisite*) for medical treatment categorization. Some of these agents have no vaccine, yet can result in death. Several of the agents listed (especially the biological agents) occur in nature and have had devastating effects on the populations of the world in past times (e.g., the black death) and have not as yet been "weaponized," which would require sophisticated processing for the correct particle size. These could include *botulism* (listed) and *smallpox*** and in modern times the *filovirus* [†] (family name of the *ebola Zaire* and *marburg* viruses), *salmonella*,[‡] and the *Crimea–Congo hemorrhagic fever* virus.[§]

If it is assumed that the Black Death (e.g., see Ref. 16) was indeed started by the Tatars *ballistic* launches of diseased corpses, some measure of the awesomeness of the possibilities of WMD using chemical or biological agents can be appreciated by the casualties caused by this event for which

---

*Used in Tokyo subway station in 1995 by Aum Shinri Kyo cult, killing 12 and injuring 5500. Also used by Iraq in attack on Kurdish village of Birjinni in 1988. *Sarin* kills within 2–15 min of exposure.
**Highly contagious and devastating viral disease that killed 500 million before being eradicated in 1977.
†Introduced into the United States (Washington, D.C. area) in 1980 from Kenya.
‡Used by the Rajneesh cult to poison (no deaths) 751 people in The Dalles, Oregon, in 1984.
§A deadly virus that has appeared in refugees fleeing Afghanistan into Pakistan in September 2001.

Table 2.3   Approximate toxicity of chemical and biological agents

Chemical agent	Lethal dose, mg/person	Biological agent
Chlorine	10^2	—
Phosgene	10	—
Tabun	1	Strychnine
	10^{-1}	Ricin
	10^{-2}	Saxitoxin
	10^{-3}	Botulinum toxin
	10^{-4}	—
	10^{-5}	Bacillus anthracis (anthrax)
	10^{-6}	Yersinia pestis

there was no cure at the time. The spread of the disease from Caffa on the northern coast of the Black Sea by the Genoese traders into Europe lasted from 1347–1351 but with most of the deaths occurring in the first *six months*! (There were numerous recurrences of the disease throughout Europe because of poor medical care and lack of knowledge of the "antidote" in 1361–1363, 1369–1371, 1374–1375,1390, and 1400.) This incredible death toll of more than 25 million (approximately *one-third* of the entire population of Europe) was greater than the *combined total* number of military deaths from both WWI and WWII [see Table 2.4, the numbers vary among authoritative sources but the *military* deaths for WWI are often quoted at 8–9 million and 15 million for WWII. Part of the problem in quoting casualties is that many of the deaths were attributed (especially in WWI) to indirectly related causes such as inadequate medical care and illnesses (the U.S. Army lost 62,000 soldiers in WWI because of influenza — more than lost in combat!)]. Although the spread (into Europe) of the effect of the Tatar launches might not have been intentional, the impact of future WMD in a more populated Europe could be devastating.

Both the *ballistic* missile and the *cruise* missile offer attractive possibilities (to the attacker) in delivering WMD. The ballistic missile provides the means to deliver WMD quickly and over large distances with difficult defense means to combat them. The cruise missile offers the ability to deliver WMD in more concentrated locations but at greater risk to delivery. One could imagine a version of the US AGM-109 containing its large number (58) of tactical airfield attack munitions (TAAM) spewing chemical contents over concentrated areas as it cruises along prescribed paths at low altitude.

A casual reading of historical records of previous wars will convince the reader that the use of chemical and biological warfare has been used by many nations. Some examples already quoted from WWI, where German forces

Table 2.4 Casualties from some armed conflicts

Conflict	Combat deaths	Other deaths	Wounded	Note
French Revolution and Napoleonic Wars (1792–1815)	4.4 million		n.a.	Data sketchy and conflicting
War of 1812–1815	2260	n.a.	4505	American Forces only
Civil War (1861–65)				
Union	140,414	224,097	281,881	Plus 25,000–30,000$^+$ prisoners died in prison
Confederate[a]	74,524	59,297	n.a.	
Total	**214,938**	**283,394**		
WWI (1914–1918)				
Allies	5.4 million[b]		7.0 million	Does not include 4.8 million Prisoners of war
Central Powers	4.0 million		8.4 million	Does not include 3.1 million Prisoners of war
Total	**9.4 million**		**15.4 million**	
WWII (1939–1945)	**(Military)**	**(Civilian)**		
Allies	8.8 million	n.a.	17.2 million	Does not include the deaths in the concentration camps
The Axis	6.2 million	n.a.	9.3 million	
Total	**15 million**	**35 million**	**26.5 million**	
Korean War (1950–1953)	4.4 million		n.a.	Data sketchy
Vietnam War (1955–1975)	2.5 million		n.a.	Includes Cambodia, data sketchy
Iran–Iraq War (1980–1988)	1.0 million		n.a.	Data limited

[a]No authoritative data issued for Confederate forces. These are unofficial estimates taken from Ref. 17.
[b]As noted earlier, many deaths were from noncombat reasons, such as the 62,000 U.S. forces lost to influenza, which is more than the 53,400 lost in combat.

used gas at different times in 1915, 1916, 1917, and 1918, are not the only examples. Under Lord Kitchener's orders the British troops used 150 tons of *chlorine gas* at Loos on 25 September 1915, killing 600 German soldiers. French troops used *phosgene gas* shells on 22 February 1916 in defense of Verdun. Bulgarians used *phosgene gas* shells in March 1917 on the Salonica Front. There are many other examples of chemical and biological warfare used sporadically throughout succeeding wars. Chemical warfare has seen more than incidental use since WWI up to the war in the Gulf in 1991, and hence it is important to understand its use and how it can be defended against. The new concern is that technology exists today to efficiently deliver WMD by ballistic or cruise missile means. During the United Nations inspections after the Gulf War in 1991, it was discovered that Iraq had produced four Al-Hussein missile warheads that contained *anthrax* bacteria. Iraq had also stated that it had produced five Al-Hussein warheads with *botulinum toxin*, but fortunately these had been stored at room temperature, which rapidly destroys their effectiveness.

CASUALTIES OF WAR

It is necessary, when discussing the survivability during any armed conflict, to understand the meaning of "casualties." This is necessary whether the subject is weapons of mass destruction or use of conventional weapons. As might be expected, the records are not clear on this subject. When deaths or casualties are listed, it is not always clear whether the numbers quoted are for *both* sides of the conflict and which are from direct combat and which are from unrelated causes. Table 2.4 shows the data from several reputable sources, but even then the data frequently conflict because of incomplete data or differing definitions of "casualties."

As can be seen from Table 2.4, care must be taken when noting casualties from any armed conflict when attempting to compare numbers and, as will be discussed later, in the measurement of the effectiveness of defense systems. Further, some authorities define casualties as the sum of killed and wounded, whereas other authorities record only deaths. Some authorities include the missing, but most do not. Indeed, not all nations have reported their casualties even in recent modern wars. As one observation on "winning," it is noted that the winning side frequently has a considerably *larger* number of deaths from the conflict than the losing side. It is also clear from Table 2.4 that the casualties, however measured, vary almost logarithmically from conflict to conflict.

Because of the divergent characteristics of chemical and biological agents and the extreme sensitivity to environmental factors, it is difficult to give casualty estimates from such weapons in the same manner as for conventional HE or even nuclear weapons (see Table 2.2). Some data do exist, however, on the lethal possibilities of *anthrax*, which are provided here as illustration. It is generally taken that for biological agents to be effective in a weaponized form (as part of a missile warhead) they must be suspended as an aerosol with a particle size of 1–10 microns. The Office of Technology Assessment (OTA)[18] concluded, for example, that a missile containing 30 kg of anthrax spores in a unitary warhead landing on an unprotected city could kill over an area of 5–25 km^2. The report hastens to add that this assumes no protection (passive or medical) and that the weather conditions are such that the aerosol of spores remains suspended for human inhalation. As discussed in Chapter 3 when deciding on the nature of defense, the number of people killed depends on the population density over the city area, but could easily be measured in tens of thousands. It is known that anthrax spores are very resilient and can last for very long times. The spores can last for *days* when suspended in air or *years* when buried in soil (recall the reference to the 1941 experiments in Scotland) or contained in animal skins. The difference in delivery systems (*ballistic* or *cruise* missile) has already been studied, and the Rumsfeld Report[19] pointed out that if the same amount of anthrax as just discussed in a ballistic missile is delivered instead by an aircraft or a cruise missile the number of deaths could be 10 times as high. This is because of the local environmental conditions and dispersal rates.

COMMENT ON STRATEGIC AND TACTICAL CATEGORIZATION

In the political process, in the attempt to control the proliferation of such weapons, there have been several treaties and agreements. In the supporting documents for such treaties and agreements, it has been necessary to categorize the many different types so that nations can continue to develop weapons for their own national defense but be limited in their development on the international front. For example, it has been found to be convenient to categorize the *ballistic* missiles by range,[20] as shown in Table 2.5.

Table 2.5   Categorization of ballistic missiles

Category	Range grouping
SRBM (short-range ballistic missiles)	150[a]–799 km
MRBM (medium-range ballistic missiles)	800–2399 km
IRBM (intermediate-range ballistic missiles)[b]	2400–5499 km
ICBM (intercontinental ballistic missiles)	5500 km and above

[a]Below 150-km range the missile is considered a battlefield weapon and not subject to strategic limitations on development and use.
[b]In the late 1980s this was shortened to IBM.

According to the SALT/START agreements,* a "strategic" missile is one that is nuclear armed and has a range of 5500 km (the distance between the western boundary of the FSU and the Eastern coast of the United States) or more. Further, according to the terms of the INF Treaty (*Intermediate Range Nuclear Forces Treaty* of 1987), all missiles with ranges between 500 and 5500 km are banned. Unfortunately, many nations (included in those listed in Appendix A and B) are not bound by these treaties. Further, as the threat of WMD appears on the scene, where not only can the warheads be nuclear, they can also contain chemical or biological material with a potentially more deadly and strategic impact even with shorter ranges. The 1972 Biological Weapons Convention (with signatures of 110 nations) also cannot boast too great a success in limiting the proliferation of such weapons around the world. Under these circumstances the categorization of "strategic" and "tactical" becomes less clear; hence, such categorization has been dropped in Appendix A, and all ranges of possible ballistic missiles have been included to show that the technology already exists to produce and field such weapons.

GROWTH IN BALLISTIC MISSILE RANGE

Figure 2.8 shows the rapid increase in "reach" of the ballistic missile. Since WWII, the range of ballistic missiles has increased at a rate of (10:1) every decade. This development rate starts with the United States, FSU, France, and the People's Republic of China. With only a generation delay and with "leaky" treaty constraints, this has progressed to nations that are considered as threat or risk nations to the West. Figure 2.8 clearly shows the missile range far outstripping that of the best guns (land or sea based). It also shows the large number of different types of missiles that have capability beyond the START limits.† Figure 2.8 also shows the continued development of longer-range missiles from nations such as the Democratic People's Republic of Korea and Iran.

Because the circumference of the Earth is approximately 40,000 km at its largest (equatorial) circle, one could argue that there is no need to develop missile ranges beyond about 20,000-km range. Figure 2.8 demonstrates that

---

*Actually, there were no agreements on range and speed as performance criteria, but the Interim Agreement of SALT I/SALT II did define the ICBM as a lower limit of 5500-km range. As shown in the Appendix, this infers a burnout speed of about 6.7 km/s.

†A general reference to the SALT/START (circa 1972 and later) talks between the United States and (at that time) the Soviet Union, which attempted to limit the use of antiballistic missile (ABM) systems. These talks are still in the political process and still contain confusing and conflicting definitions of "strategic" and "tactical" missile systems. Part of the confusion comes from attempting to define a limit on range, speed, and type of warhead. The many combinations already in existence belie such a concise definition. However, for reference purposes in this book, a 3000-km range is kept as an approximate milepost on the various charts and figures as a guide to the "boundary" between strategic and tactical missiles. Note that 3000 km represents the distance across most geographical corners of NATO Europe and the expected "battlefields" of the future.

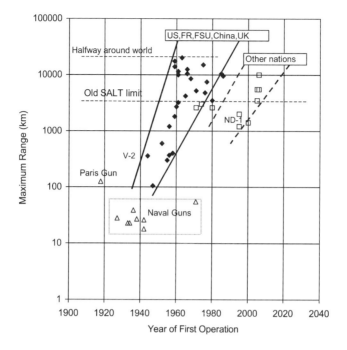

Fig. 2.8   Ballistic missile range: year of first operation.

the technology has already reached this practical limit of any needed reach for this planet.

GROWTH IN BALLISTIC MISSILE PAYLOAD

In addition to the development of missile range, the increase in both launch weight and payload (warhead) has also increased steadily in all ballistic missile developments, as shown in Fig. 2.9.

IMPROVEMENT IN MISSILE ACCURACY

As propulsion and guidance technology improves over time, the accuracy of missiles, that is, the ability to arrive at the aim point within a small margin of error, has also improved by approximately an order of magnitude (10:1) every decade since WWII. If the circle area of probability (CEP) is used as a measure of this accuracy, it can be seen from Fig. 2.10 that dramatic improvements have occurred in all missiles. [Because of security classification issues, it is doubtful that nations will publish the exact value of CEP, but Fig. 2.10 does show the general magnitude and the trend in accuracy (CEP) of missiles taken from the available unclassified literature.]

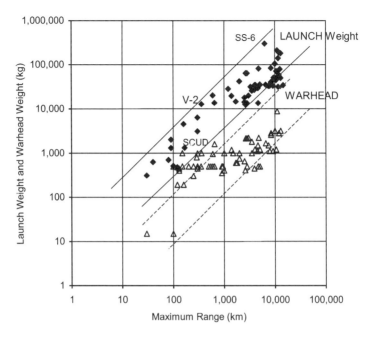

Fig. 2.9   Growth in ballistic missile weight.

The CEP is defined as a circle with radius around a mean point of projected impact within which 50% of the missiles (warheads) will land. This has a definable meaning in the case of ballistic missiles that are unguided after burnout. It is less definable in the case of cruise missiles where frequent correction in flight is normal, but it can be seen from Fig. 2.10 that the cruise missile is becoming extremely accurate.* The ballistic missile is also becoming extremely accurate as improved seekers are integrated into the various designs. As shown on Fig. 2.10, Russian designers have improved the accuracy of India's 250-km range Prithvi and 2500-km range Agni ballistic missiles using upgraded Scud B seekers that reduce the CEP from 900–1000 m to 20–40 m (Ref. 21).

CRUISE MISSILE WEIGHT AND SPEED TRENDS

In contradistinction to the ballistic missile that spends most of its time outside the atmosphere, the cruise missile depends on aerodynamic lift for its

---

*Most readers will recall the January 1991 TV pictures during the Gulf War that showed a Tomahawk "cruising" down a street in Baghdad on its way to its target. Also other pictures showed missiles seeking out specific windows of targeted buildings and flying through them. With the various versions of the Tomahawk equipped with either the terrain-contour matching (TERCOM) or the digital scene matching area correlation (DSMAC), the accuracy of today's cruise missile can be measured in meters or less. This is especially true with the ready availability of a global positioning system (GPS) receiver.

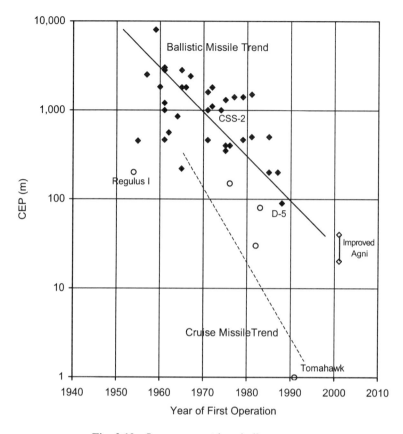

Fig. 2.10   Improvement in missile accuracy.

support and is thus affected dramatically by aerodynamic resistance. This effect (dependent on the square of the speed) causes a distinctly different characteristic in weight vs range performance from that shown for the ballistic missile (see Fig. 2.9) for example. Using the data collected in Appendix B, Fig. 2.11 shows the state-of-the-art trend in cruise missile weight. The data shown are for all types of propulsion mechanisms used in the various designs (turbofan, turbojet, solid rockets, etc).

Note from Fig. 2.11, that, except for a few special cases, the preponderance of cruise missile designs are for ranges of 1000 km or less because of the state of the art of propulsion systems and the problem of overcoming aerodynamic resistance. Some more insight can be gained on the characteristics of cruise missile performance capability by "zooming in" on those cruise missiles with less than 1000-km range, that is, the left-hand part of Fig. 2.11. This allows a focus to be made on the types of propulsion systems used. Figure 2.12 shows such an expansion of the data, and the state-of-the-art, AD 2000 boundary from Fig. 2.11 has been superimposed on

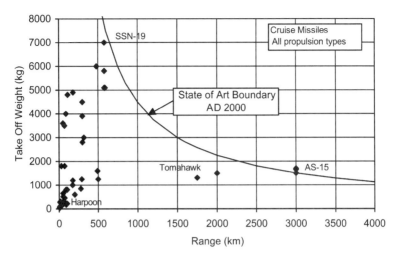

Fig. 2.11   Trend in cruise missile weight and range.

Fig. 2.12 for reference. Because of the wide variety of missions used for the cruise missiles and the variety of geometries, there is no definite trend for each type of propulsion system used, although the expected characteristics can be discerned. Of the airbreathing engines, for example, the *turbofan* is known for its low specific fuel consumption and high propulsive efficiency and is thus used for the long-range (strategic) cruise missile such as the

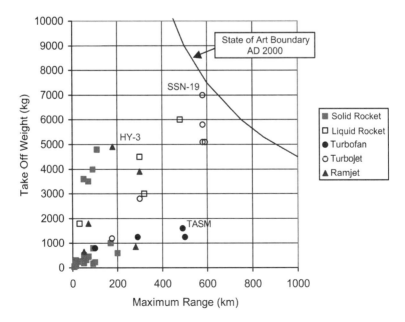

Fig. 2.12   Cruise missile weight trends for less than 1000-km range.

Tomahawk Land Attack Missile (TLAM) and Tomahawk Antiship Missile (TASM). Such a choice, however, limits the takeoff weight capability as shown. The *turbojet* (also an airbreathing engine) has comparable specific fuel consumption as the turbofan but only at lower speeds (around Mach 2.5) but has a thrust capability approaching that of the rocket and thus is capable of boosting higher weight cruise missiles as shown but over shorter distances. Examples of this type are Harpoon and Sea Eagle.

Many cruise missiles use the *solid-rocket* motor because of its simplicity of handling in the field and its ability to generate high specific thrust (speed) over short distances. This can be seen in Fig. 2.12, where the highest weight, shortest-range cruise missiles use this form of propulsion. Some examples of the solid fuel rocket are the Phoenix, Sea Skua, and Exocet. A quick glance at Fig. 2.12 will show that the ballistic missile and the cruise missile have comparable performance in this range. The problem with solid-rocket propulsion, of course, is the difficulty in controlling the thrust level on demand and hence the use of *liquid-rocket* motors, which have been used in some cruise missiles over the same missions but with better (controlled) performance. In recent years there has been renewed interest in the *ramjet*, which offers high thrust and low specific fuel consumption but suffers from the need for a booster motor (usually a solid rocket) to get it to the speeds where the ramjet is most efficient (approximately Mach 3 and above). This complexity has limited the number of cruise missiles that use this form of propulsion. An example of this type is the Air Sol Moyenne Portée.

The takeoff weight of cruise missiles is a key parameter, and the trend in weight has been shown in Figs. 2.11 and 2.12. The *payload* is also an important parameter, but is less reliable in trend setting because of the obvious tradeoff that must be made in weight, range, and speed in any given mission design. Figure 2.13 does show, however, the state-of-the-art design pressure that tends to keep cruise missiles at low values of payload and range.

For reference, the Missile Technology Control Regime (MTCR) limits are shown on Fig. 2.13, which encompass most designs. Those cruise missiles that are outside the MTCR limits are usually older (heavier designs). The similarity between the early Russian cruise missiles and the V-1 is noted. The speed of the cruise missile is also part of the design trade, and the state-of-the-art trend line can be seen from Fig. 2.14.

Because of the need to satisfy the demands of aerodynamic flight, it should not be surprising to see that the cruise missile must either be designed for high speed, short range or low speed, long range. This is represented by the state-of-the-art, AD 2000 boundary superimposed on Fig. 2.14. The majority of designs have been subsonic with a few special purpose supersonic designs. The familiar Tomahawk cruise missile (version AGM-86B shown), an

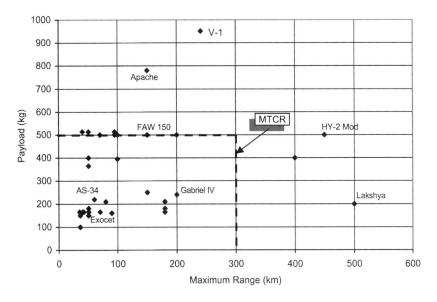

Fig. 2.13   Cruise missile payload trends.

example of a cruise missile designed for long range already discussed, for example, has speeds of Mach 0.73 to higher (classified) speeds. At sea level this corresponds to 0.25 km/s. Some of the cruise missiles designed for supersonic speeds and high altitude (of the order of 20–25 km) have achieved speeds in excess of Mach 4 (i.e., at speeds in excess of 1.0^+ km/s).

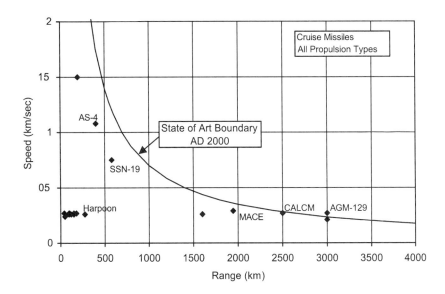

Fig. 2.14   Speed of cruise missiles.

MACH NUMBER AND SPEED

The main discussion in this book is centered on the ballistic missile and the cruise missile, each designed by two different aerospace communities, and in compiling the data for the charts it was frequently necessary to convert units into a common format. Because the ballistic missile spends the majority of its flight outside the atmosphere, it is common to quote speeds in absolute units such as kilometers/second. On the other hand, because the cruise missile must contend with aerodynamic forces, speed is most often quoted in relative terms to the speed of sound, that is, through the use of Mach number (the ratio of speed to the speed of sound). The speed of sound varies with absolute temperature and thus the altitude at which the cruise missile flies. This deceptively simple difference masks a major difference in the performance regimes of each type of missile.

Figure 2.15 is a direct comparison between actual speed (in kilometers/second) and Mach number for a range of altitudes most likely to be used by the cruise missile.

It has been assumed in Fig. 2.15 that the values for a standard day as defined by NACA (now NASA) are generally applicable[22,23]. For such conditions the speed of sound at sea level is approximately 0.34 km/s and *decreases* (and thus Mach number *increases*) as altitude increases to a value of approximately 0.30 km/s at the tropopause. The tropopause varies from an altitude of 8.54 km at the Earth's poles to 16.4 km at the Earth's equator.

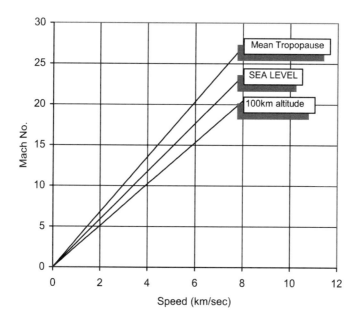

Fig. 2.15 Mach number and speed relationship.

For simplicity, the tropopause in Fig. 2.15 has been taken at an average altitude of 10 km. Above the tropopause the speed of sound remains fairly *constant* up to about 35 km altitude and *increases* above that altitude (see Appendix D on Standard Atmosphere for more information). There are other decrease/increase variations in the speed of sound at much higher elevations, but these occur at altitudes above those considered of value for aerodynamic flight of cruise missiles (see Appendix D for discussion of the Standard Atmosphere). The value of Mach number for an arbitrary altitude of 100 km has been included in Fig. 2.15 to show the "reversal" in Mach number caused by the effect of the variation of the speed of sound with altitude. (The air density at 100-km altitude is approximately one millionth that of the air density at sea level and is not likely to sustain aerodynamic lift.)

Most cruise missiles built to date operate subsonically at low altitude (less than 100 m) with a few designs capable of cruising at higher altitudes (approximately 20–25 km) with the capability of diving supersonically onto their targets. (An altitude of 25 km is approaching the limit of sustainable aerodynamic flight.) From such a simple comparison of speed and Mach number, it is immediately apparent that the ballistic missile and the cruise missile operate in vastly different "environments." For reference, Table 2.6 lists typical values of speed and Mach number (taken from Fig. 2.15) that are applicable for both cruise missiles and ballistic missiles.

SPEED OF BALLISTIC MISSILES

Unlike the *cruise missile*, the speed of the *ballistic missile* after burnout has a direct correlation with its range. It is useful to show the various estimates of ballistic missile speeds at burnout, which initiates the ballistic trajectory. Figure 2.16 shows the speeds of missiles of interest. The curves in Fig. 2.16 come from 1) the derivation of ballistic trajectories based on the

Table 2.6   Mach number and speed

Speed, km/s	Mach number (at 10-km altitude)
0.3	1.0
1	3.32
2	6.64
3	9.96
4	13.28
5	16.61
6	19.92
7	23.33
8	26.67

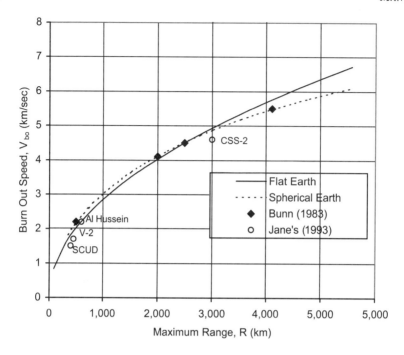

Fig. 2.16   Burnout speed of ballistic missiles.

principles of orbital mechanics of projectiles moving around a spherical Earth and 2) the assumption of a flat Earth (see Appendix C).

Included in Fig. 2.16 are a few examples (e.g., see Refs. 24 and 25) of the burnout velocities of ballistic missiles plotted against the advertised range. As can be seen, ignoring the curvature and rotation of the Earth, the assumption of a flat Earth provides a good approximation to ballistic missile speeds. (This applies to those missiles taken as tactical ballistic missiles with ranges equal to or less than 3000 km. As the range approaches strategic values, that is, where the range begins to be a more significant fraction of the Earth's circumference, this assumption begins to break down.) Appendix C provides a detailed derivation of the relationship between a ballistic missile's burnout velocity V_{bo}, angle at burnout θ_{bo}, and the ballistic range R. For the ranges of interest, this relationship reduces to a simple equation:

$$R = \frac{V_{bo}^2 \cdot \sin 2\theta}{g} \tag{2.1}$$

Upon substitution of an average value (see Appendix C on choice of burnout angle in range selection) for θ_{bo} of 42 deg and the value of g at sea level, it is easy to show that the speed of ballistic missiles can be expressed quite simply as

$$V_{bo} = 0.09\sqrt{R} \tag{2.2}$$

[**Rule of Thumb:** *Divide the square root of the range (in kilometers) by 10 to get the burnout speed of a ballistic missile (in kilometers/second).*]

CAPSULE COMPARISON OF BALLISTIC AND CRUISE MISSILES

The preceding sections have described the available evidence on the many types of *ballistic missiles* and *cruise missiles* and their basic characteristics. For ease of reference and to highlight the basic differences, Figs. 2.17–2.19 have been compiled using the trend lines from the earlier figures to show the relative "domains" for both types of missiles for launch weight, payload, and speed. Figure 2.17 shows that for launch weights of 1000–2000 kg and ranges of 100–200 km both types of missiles have similar capability. As range is increased beyond 300 km, the ballistic missile has the capability to carry larger weights. The state of the art boundaries of the cruise missiles for take-off weight and payload shown in Figs. 2.17 and 2.18 are taken from Figs. 2.12 and 2.13. The payload or warhead or throw weight values for ballistic and cruise missiles taken from Fig. 2.9 (for the ballistic missile) and

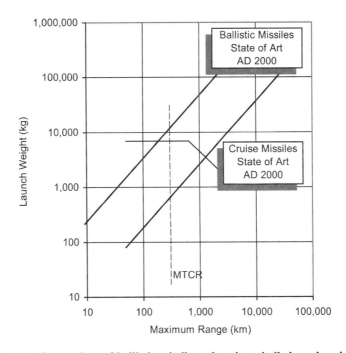

Fig. 2.17 Comparison of ballistic missile and cruise missile launch weights.

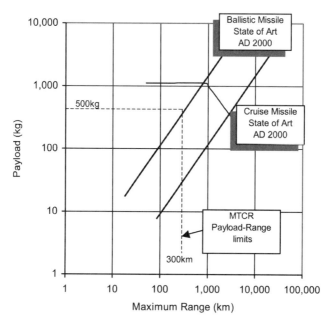

Fig. 2.18   Comparison of ballistic missile and cruise missile payloads.

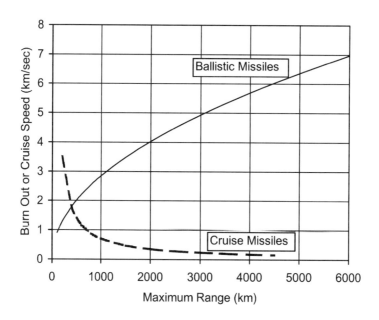

Fig. 2.19   Comparison of cruise missile and ballistic missile speeds.

the cruise missile data from Appendix B have been compiled into Fig. 2.17. The payload–range (500 kg–300 km) limit from MTCR for cruise missiles has been superimposed on Fig. 2.18 to show the relative positions of most of the available data. Finally, in Fig. 2.19, the speeds of ballistic missiles and cruise missiles have been compared. In the case of the ballistic missile, the burnout velocity has been shown for the corresponding range. For the cruise missile the state-of-the-art AD 2000 boundary from Fig. 2.11 has been shown as an indication of cruise missile capability.

It is of interest to compare the trajectories of ballistic missiles and cruise missiles as one way of showing their relative domains. Figure 2.20 shows the expected trajectories of a ballistic missile launched at the same speed as a cruise missile. The choice of 1.3 km/s (Mach 4^+) was made to be representative of the highest speed cruise missile likely in the near future. This corresponds roughly to the crossover point in Fig. 2.19.

Typically, the cruise missile designed for high (supersonic) speeds will cruise at the high altitude (say between 20–25-km altitude) and then dive on its target in the last few kilometers before reaching the target. During the descent, it would be indistinguishable from a ballistic missile of the same speed. The shape of the ballistic missile trajectory shown here is somewhat idealized as is that of the cruise missile because the trajectory of an actual descent will depend on the aerodynamic resistance. This is discussed in more detail in Appendix C, but the essential features are as shown in Fig. 2.20.

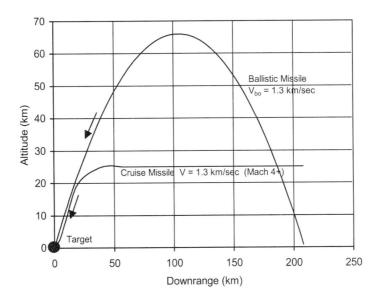

Fig. 2.20   Comparison of cruise missile and ballistic missile trajectories.

THREAT DISTANCES

To put the preceding missile threat into a frame of reference, it is useful to tabulate the distances of some of Europe's capitals from existing known threat areas. Libya (which launched a ballistic missile against NATO territory in 1986) and Iraq (which launched SCUD missiles into Israel and Saudi Arabia during the 1991 Gulf War and into Kuwait during the Iraq War that started on 19 March 2003) will be used as examples. Table 2.7 shows the threat distances.

Table 2.7 shows that ballistic missiles with ranges from 1450–4550 km can threaten most of the capitals within Europe. Further, these missiles can arrive *some 9–15 minutes after launch*! Additionally, ballistic missiles with much shorter ranges (*less than 1000 km and 7 minutes of flight time*) can threaten many strategic targets within the European continent from possible threat or risk areas. This means that any defense system must react and intercept within these short time windows. Land-based systems must be in place and ready. Satellite early warning systems must be coming into view on their orbits, detect the launch, process the data, and provide coordinates to the defense interceptor system in that time. Sea-based systems must be on station and operational and conduct the BMD mission even if engaged in some other mission. Airborne defense systems must be either "on combat air patrol" and headed toward the BM launch site or be scrambled from their air bases, get airborne, and be cued to launch their interceptors within these time windows. Additionally, in Chapter 4 the times for boost phase are given, where in that case, if any intercept is expected, then the intercept window is reduced to *1–2 minutes.*

CAUTIONARY NOTE ON THREAT DISTANCES

There is a danger in noting the distances from likely threat areas. Indeed, some writings have indicated that friendly states might or might not be at risk from the missile threat depending on whether the state of the art of ballistic

Table 2.7   Distances and flight times from Sabha and Baghdad to Europe

European capital	From Sabha, Libya		From Baghdad, Iraq	
	Distance, km	Flight time, min	Distance, km	Flight time, min
London	2900	$12\frac{1}{2}$	4350	$15\frac{1}{2}$
Berlin	2750	$12\frac{1}{4}$	3500	$13\frac{3}{4}$
Paris	2550	$12\frac{3}{4}$	4150	15
Madrid	2200	11	4550	$15\frac{3}{4}$
Ankara	2200	11	1450	9
Rome	1650	$9\frac{1}{2}$	3150	13
Athens	1500	9	2150	$10\frac{3}{4}$

missile or cruise missile design has matured enough to reach the long geographic distances from a possible threat area to the nation under consideration. Unfortunately, there have been a sufficient number of instances in history where such an evaluation could cause an undue sense of complacency from the missile threat. For example, during WWII, sometime in 1943, specific plans and designs were put forward by Nazi Germany to attack the United States (New York was specifically identified) by the V-2 ballistic missile (e.g., see Ref. 26). The plan was for a Type XXI class U-boat to tow a watertight canister containing the V-2 behind the submarine until it was within about 150 km off the coast and still in international waters. The canister would then rotate, through suitable ballast control, into a vertical position, at which time the V-2 would be launched toward its target. The canister contained the necessary tanks for the storage of the V-2 liquid oxygen; and the towing cable from the submarine included the necessary electrical power for launch actuation. The project was called *Projekt Lifevest*. Fortunately, this project did not come to fruition, but one of the canisters was completed at the time of VE Day on 8 May 1945. Figure 2.21 shows the basic features of the design and operational concept.

Such a scheme would have effectively reduced the distance from Peenemunde, Germany to New York City (a distance of about 6300 km) to 150 km and within the range of an existing (at that time) missile, the V-2.

German Type XXI Submarine towing V-2 in tug toward target area

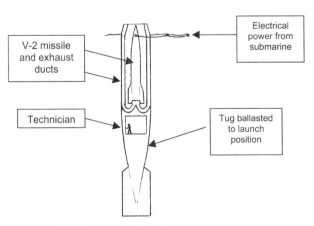

V-2 in launch position in target area

Fig. 2.21   Planned V-2 attack on New York in U.S. (in WWII).

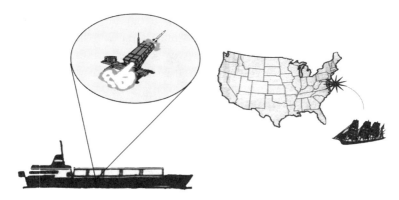

Fig. 2.22   Missile threat may not always be as predicted. (The author originally put in the pictorial of a sailing vessel launching missiles somewhat tongue-in-cheek and was considering replacing it with a 'more suitable' pictorial until, on 24 March 2003 during the Iraq War, the coalition forces intercepted a private dhow in the Persian Gulf that was actually a minelayer. An even more insidious example is the intercept by Indian custom agents in the port of Kandi (near Calcutta) of a North Korean freighter on 13 August 2003, with final destination Libya, that had a hidden assembly line for ballistic missiles in progress on board! Despite international law and conventions, history confirms some aggressors will continue to use combatants disguised as noncombatants in warfare.)

There is essentially nothing to stop a potential adversary from using a similar technique today. Even the use of a standard land-based launcher bolted to the deck of (say) a container ship (see Fig. 2.22) and cruising international waters, sufficiently camouflaged to avoid detection, would constitute a similar threat to that planned by the V-2—the forerunner of today's SCUD.

Converting surface ships or submarines to carry missiles is not an expensive proposition and can provide a threat of significant proportions. The concept of submarines carrying surface-to-surface missiles was expanded by the U.S. Navy after WWII in the late 1950s with the *Halibut, Growler*, and *Grayback* guided-missile submarines. The *Halibut* was the first in 1956. These submarines went through a series of modifications and changes and were capable of carrying the transonic *Regulus I and II* missile in shelters on the forward deck. The *Regulus II* was capable of carrying a nuclear warhead over 1500 km. The Regulus I guided missile on the *Grayback* saw operational duty in the Pacific Fleet in 1958 before being replaced by the Polaris program.

Figure 2.23 shows a U.S. Navy photo of the *Grayback* with a *Regulus I* missile that has been extracted from its shelter on the forward deck and is being readied for launch.

A third and disturbing example from today was the "missile" attack in New York City and Washington, D.C., on 11 September 2001. Here, the

US Navy photo

Fig. 2.23 *USS Grayback* **with one of four Regulus I guided missiles.**

missiles were civilian aircraft, and the "warheads" were full tanks of aircraft fuel with sufficient thermal energy once ignited to destroy the buildings. The threat "nation" was al-Qaeda who had effectively converted the "distance" of approximately 11,000 km from Kabul to Washington and New York to an effective 300–600 km from Boston to these same cities. Such statistics show the danger of thinking that the threat range is a simple geographical measure of distance using a standard map.

A corollary of this recognition of the possible forms of missile threat is that any decision to provide defense against missile threats should be aware of the danger of setting up early warning schemes that might be "looking" in the wrong direction and placing the defense in a possible Maginot Line* situation.

Any decision maker would be wise to recognize that potential adversaries can be quite imaginative. Although range will often be used in the following chapters as a way of characterizing the threat, this reminder from history should caution against using distance as the sole criterion of the state of theart of the missile threat.

---

*The Maginot Line was a fixed line of fortification built on the eastern border of France prior to the start of WWII with guns locked into facing the presumed direction of the German threat. The German army simply went around this fortification, effectively negating the defense. Although the Maginot Line is often quoted to illustrate this problem, history is replete with such examples by many nations. The warning is still valid in this missile age.

Chapter 3

DECIDING ON THE DEFENSE

Qui desiderat pacem, praeparet bellum.*
Flavius Vegetius Renatus, c. AD 375

The legitimate object of war is a more perfect peace.
William Tecumseh Sherman
Speech in St. Louis, 20 July 1865

This book is about missile defense, but it is worth pausing for a moment to consider the subject in the contextual nature of protecting the peace. Given that some form of military defense will always be required, it is not immediately obvious which approach will ensure the greatest safety to the nation or nations being defended. The ideal would be some form of defensive shield that would protect all areas and all key assets within that area to such an extent that no missiles (ballistic or cruise) of any kind could penetrate. That is, what is desired is *a zero-leakage, all-expansive defensive shield.* Unfortunately, today and in the near future, technology will not support such an ideal and compromises must be made.

One of the compromises that must be made is to decide on the nature of the threat. It is generally agreed that the likely forms of missile threat are as shown in Table 3.1.

Without being too prescriptive it is generally taken that the Western world has moved from being under threat of attack from *thousands of missiles, that is, a massive attack from a known direction*, to facing the possibility (probability?) of attack from *hundreds of missiles from unknown directions* (so-called limited attacks). Conversely, although it is hoped that the end of the Cold War has lowered the probability of nuclear attacks, the new world order has given rise to concern about the possibility of attacks from missiles containing chemical or biological warheads (see Chapter 2). Whether the agents are nuclear or chemical or biological, the use of weapons of mass destruction is still a high probability. It is debatable whether the use of massive or limited use of WMD is any less devastating to the nation or

---

*"Let him who desires peace prepare for war."

Table 3.1   Spectrum of likely threats

Intensity of attack	Conventional	Weapons of mass destruction		
		Nuclear	Chemical	Biological
Massive attacks				
Limited attacks				

nations seeking defense. Some of the options in broad scope are summarized in Table 3.2.

Table 3.2 summarizes the broad scope of issues facing decision makers in the general context of providing missile defense. As discussed in Chapter 2, the warhead can be delivered by a ballistic or cruise missile. Also, the nature of the defense can have ramifications to the policy maker (e.g., a good *active defense* capability could in the eyes of a potential aggressor be a sufficient

Table 3.2   Options and capabilities sought in missile defense

Defense possibilities	Intent	Comment
Treaty	Everybody agrees not to fight in exchange for peaceful coexistence.	Obviously the preferred approach but with a few exceptions the history of mankind does not support a high degree of success with this approach
Deterrence	Walk softly but carry a big stick.	Three selected examples suggest that this is a workable defense: US/USSR Cold War, Formation of NATO, France's "*Force de Frappe*".
Extended air defense[a]	If a war does start (or is imminent), be prepared to win it.	Sufficiently good defensive capability has the additional benefit of being a *deterrent*.
Active defense	Actions taken to destroy or mitigate the effectiveness of an enemy attack by intercepting enemy threats in flight.	See footnote.[b]

(*Continued*)

Table 3.2 *Continued*

Defense possibilities	Intent	Comment
Passive defense	Protective measures that can reduce the effectiveness of an attack by degrading an enemy's target acquisition capability, reducing the vulnerability of critical forces, and improving their potential to survive. These measures include deception, dispersion, and use of protective construction. Measures also include timely dissemination of early warning information to civil defense organizations.	See footnote.[b]
Conventional counterforce (CCF)	Offensive counterair option to prevent launch of missiles by neutralizing essential elements of the opponents' attack capability.	See footnote.[b]
Battle management, command, control, communication and intelligence (BM/C^3I)	Comprises the capabilities, processes, procedures, and information for coordinating and synchronizing both offensive and defensive measures during peace, crisis, and war.	See footnote.[b]

[a]This book is about missile defense and thus will discuss only extended air defense (EAD) here and will not discuss the equally important aspects of warfare such as antisubmarine warfare, antisurface warfare, marine amphibious forces, army and navy forces other than those necessary for missile defense, etc.
[b]Although no universally agreed-upon set of definitions exists, these following definitions have been extracted from various unclassified NATO and other documents to provide a basis for understanding of the various aspects of missile defense.

deterrent). An incorporation of the correct form of *passive defense* can alleviate the amount of *active defense* required. Further, the proper *intelligence* (as part of BM/C^3I) can alert the defense such that *conventional counterforce*, *passive defense*, and *active defense* can be more efficient. A discussion of the various aspects of a defense posture for the NATO Alliance based on the new world order can be found in the (new) Alliance's Strategic Concept of 1991 and related documents.[27]

Understandably, there are no iron-clad definitions in missile defense because of the continuously evolving nature of this form of defense and because technology is rapidly expanding such that new forms of both offensive and defensive capability are appearing on the scene in the world's inventory at an ever increasing frequency. Chapter 2 provides indications of the rate of increases of technology advancement.

ROLE OF ADVANCING TECHNOLOGY

The "leakiness" of treaties and changing alliances make control of the technology a very difficult issue. The MTCR (U.S. export control Missile Technology Control Regime) has served to slow down the technology transfer but not to stop it completely. Historically, warfare planners have sought to either match the *numbers* of systems of the potential adversaries in their annual defense budget requests (i.e., if they have 100 guns, then we want $100+$ guns) or *improved performance* of the systems (i.e., if their guns can shoot 50 km, then we must be able to shoot $50+$ km). There have been some unfortunate instances in history in which those with the superior (advanced technology) weapons have not always won the day. Over the last 25 years, a new feature has appeared on the scene called *information warfare*, that makes it possible that even with less advanced technology systems in the adversary's inventory, he can win the day.

As the reach of today's and tomorrow's weapons expands beyond "eyeball-to-eyeball" fighting, this feature of *information warfare* takes on significant importance in any defense strategy. A further complication in deciding on advanced technology systems is 1) the length of time it takes to develop such systems and 2) the current trend to form transnational alliances in defense systems development on a global scale.

MILITARY PRIORITIES AND CIVILIAN PRIORITIES

The nature of warfare changes with the nature of the advancing technology. For example, until WWI the fighting took place "on the ground." When the new technology of the airplane was introduced, it was seen as a means to fly over the battlefield to provide a "bird's eye view" to *aid* the fighting on the ground. It was not until WWII that the phrase *"air superiority"* was created, and the nature of warfare changed again. It is now taken that the strategy is to first obtain air superiority and then to finish the battle on the ground, because in the final analysis that is where the war must be won. Such a distinction tends to influence the military viewpoint of what to protect first (usually forward-based military assets) and then to protect the civilian assets further behind the "line in the sand." It is quite possible that in a war that uses predominantly *ballistic missiles*, a change in warfare strategy will be required. It will be recalled from

Chapter 2 that even with the relatively simple technology involved in the V-2 there was virtually no defense against these weapons. If the war had not ended when it did, there would have been a severe strain on the part of the Western Alliance technological capability to defend itself against such weapons. Some 50+ years later in the Persian Gulf War (1990–1991), the defense against ballistic missiles was similarly severely strained.

Hence, whether to protect forward-based military assets or civilian assets back in the homeland area is a question that must be answered in deciding on the nature of the defense.

MAKEUP OF MISSILE DEFENSE SYSTEMS

With the preceding caveats and questions regarding the nature of technology advancements, military and civilian needs, etc., in mind, it is necessary to at least define the nature of defense systems to consider. The definitions summarized in Table 3.2 represent the general agreement today among NATO allies and are sufficiently robust to guide the defensive solutions considered here. In *active defense* (which will receive the greater part of attention in this book) there are three other important subdivisions that will be discussed. These subdivisions are 1) early warning systems, 2) interceptor systems, and 3) BM/C^3I systems.

There are necessary overlaps in these systems and in the definitions. For example, various sensors (space, air, or surface borne) can serve both as fire control sensors and early warning. The BM/C^3I elements of active defense can serve in other mission areas as shown in Table 3.2. One can be too rigorous and attempt to redefine all elements or accept that multicapabilities are possible and indeed frequently desirable. As such overlaps occur, they will be identified as necessary in the analysis. It is normally found, that this active defense must be arranged in *layers*.

Figure 3.1 illustrates the basic features of a *layered active defense* system with most of the major components displayed. As described later, it is generally agreed that it is not practicable or feasible with today's technology or that projected into any foreseeable future to provide total defense with a simple defense system arranged around the key asset to be protected. Some form of combinations of defense systems will be required.

In very broad terms there are essentially three main areas of defense, as shown in Fig. 3.1. They can be described as 1) a *lower-layer system* to provide point defense around key assets, 2) an *upper-layer system* to defend a wide area encompassing the key assets, and 3) A *forward-base system* to intercept the threat early in its flight.

Depending on the technology and capability of specific systems, these three basic areas of defense can be combined or expanded as necessary. The

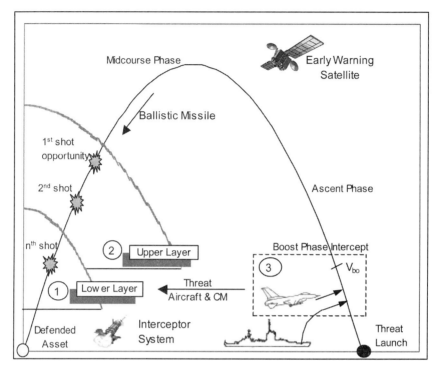

Fig. 3.1   **General concept of layered active defense.**

upper- and lower-layer systems, for example, can be subdivided as necessary to encompass different portions of the ballistic missile trajectory (*boost phase*, *ascent phase*, *midcourse*, or *terminal phase*).

As drawn, Fig. 3.1 shows the *boost phase intercept* possibilities (by aircraft or ship), but these same systems could just as well intercept during the *ascent phase*. Also, the airbreathing threat (aircraft and *cruise missile*) is normally thought of as being intercepted by the static defense (made up of lower- and upper-layer defense systems) arranged around the key assets to be defended. The upper layer is sometimes referred to as the first layer and can provide multiple and successive "shot" opportunities (see later) before "handing over" the defense to the final layer (usually the lower layer) as shown. In any complete battlefield arrangement, there are additional layers embodied in short-range air defense systems (SHORAD), but these will not be discussed in great detail in this book except to note where such systems can be of benefit in the broad subject of *ballistic* and *cruise* missile defense. The lower altitude limits of lower- and upper-layer systems shown in Fig. 3.1 are determined by mission needs and technology limits.

Also, to keep Fig. 3.1 from being overcrowded, the various possibilities for surface-based radars for both early warning and interceptor fire control purposes have only been indicated. The possibility of providing *early warning* by satellite is shown, however. Of the three basic subdivisions of *early warning, interceptor systems*, and the connecting BM/C^3I *systems*, the complex network of BM/C^3I has been omitted from Fig. 3.1 and is discussed separately later in this chapter.

The geographic scenario will determine which system (land, sea, or air) will be the dominant line of defense. This decision will also depend on the technological maturity of the various systems. For example, less capable systems will have to be placed further forward than more capable systems, and this will affect the degree that any particular defense arrangement is in "harms way." This will, in turn, influence the choice of systems and their associated cost.

In certain scenarios the land and sea geography may be such that the *air* and *ship* systems shown, in Fig. 3.1 in the forward-base mode, could instead (or as well) provide the defense around the key assets and provide the same functions of the lower- and upper-layer defense systems shown. This is simply a matter of geography and not any fundamental characteristic of air-, land-, or sea-based defense systems. (Such variations of system components are considered in Chapter 8 to indicate the possibilities, especially as they are driven by the availability or projected technology of the defense.)

AIM OF THE DEFENSE

The ultimate aim of the defense is to protect the nation, its *population*, and its *key assets*. In doing so, the military forces have two aims: 1) to protect itself as it conducts its military operations throughout the entire process from peace, through to crisis, in actual combat (war), and finally safely guiding the situation back to stability, and 2) to protect the civilian population and national assets.

In any event, whether the goal is to protect civilians and property or to protect military forces and equipment, the nature of defense essentially can be expressed as protection of a given *area* to a certain *level of protection* and protection of key assets within that same area to some required level of protection. As already stated, the ideal defense would provide 0% leakage of incoming missiles over 100% of the nation's geographical area.

PROTECT THE AREA, KEY ASSETS, AND POPULATION

To provide a broad guideline of what needs to be protected, it is helpful to summarize the statistics of the nations of the Alliance. The geographical areas of the 19 nations within NATO, together with their population estimates, are shown in Table 3.3. The nations are listed in descending order of geographical area.

Table 3.3   Statistics of nations within NATO[28]

Nation	Area km^2	Estimated population	Population per km^2	Radius of circle of same area, km
Canada	9,976,139	29,123,194	3	1780.28
United States	9,372,610	267,954,767	29	1765.79
Turkey	780,580	63,528,225	81	497.76
France (excludes overseas depts.)	547,030	58,470,421	107	417.00
Spain	504,750	39,244,195	78	400.56
Norway	324,220	4,404,456	14	350.73
Germany	356,991	84,068,216	235	336.85
Poland	312,678	38,700,291	124	315.27
Italy	301,230	57,534,088	191	309.46
United Kingdom	244,820	58,610,182	240	278.59
Greece	131,940	10,583,126	80	204.81
Iceland	103,000	272,550	3	180.94
Hungary	93,030	9,935,774	107	171.97
Portugal	92,080	9,867,654	107	171.00
Czech Republic	78,703	10,318,958	131	158.33
Denmark	43,069	5,268,775	122	117.04
The Netherlands	37,330	15,653,091	419	115.36
Belgium	30,510	10,203,683	334	98.49
Luxembourg	2587	422,474	163	28.67

For reference purposes the last column in Table 3.3 lists the radius of circles that have the same area as that nation's geographical area. Although clearly each nation has different areas, some groupings help place the requirements in perspective; these are shown in Table 3.4. Figure 3.2 shows the radius of circles of the equivalent areas of the NATO nations, but, except for United States and Canada, the size of the European nations can be grouped into about five major categories (Table 3.4).

Table 3.4   General sizing of European NATO nations

Nations	Radius (circle of same area), km
Turkey, France	400–500
Spain, Norway, Germany, Poland, Italy	300–400
United Kingdom, Greece	200–300
Iceland, Hungary, Portugal, Czech Republic, Denmark, The Netherlands	100–200
Belgium, Luxembourg	20–100

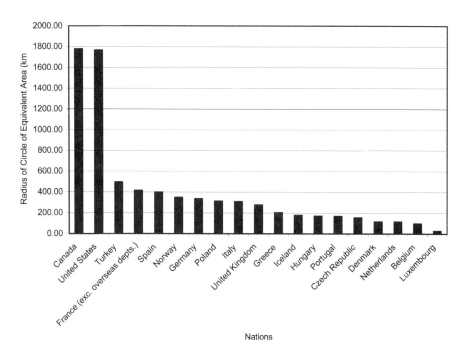

Fig. 3.2   Radius of circles of same areas as NATO nations.

Because of the actual geographical layout of the nations, the *circumscribed* circles provide additional information. Figure 3.3 is a map of Europe with just a few circumscribed circles added (for clarity). As shown in Table 3.5, for example, United Kingdom, France, Spain, and Italy lie within circumscribed circles of approximately 500-km radius. Germany is circumscribed within a 350-km radius and Greece within a 300-km radius circle. Turkey is unique because of its land mass arrangement, in that it lies within two adjacent circles, each with a radius of approximately 350 km. If the intention is to defend the immediate land mass, then the radii of circles of the same area as the nation is close to the required reach of the system. If, however, as in most European nations bordering on bodies of water, there is a need to reach beyond the land mass area, then the circumscribed circles are more representative of the required reach.

Clearly any detailed analysis of particular nations can achieve more precision on the area coverage needs, but the values quoted here are sufficient for early decision-making and sizing studies. As will be shown later, the actual area coverage by the defense systems varies greatly depending on many defense system parameters including the choice of scenario and direction of the threat axis.

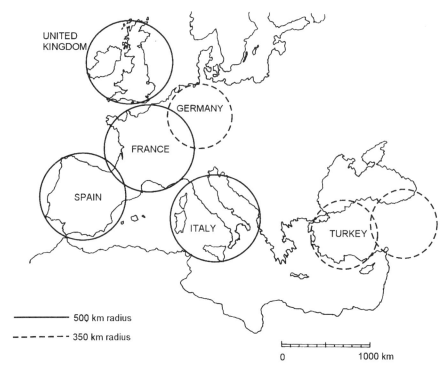

Fig. 3.3 Examples of circumscribing circles of European nations.

A second key factor is the population to be defended. Figure 3.4 shows the data for the estimated population for each of the nations of the Alliance (plotted from the data shown in Table 3.3), listed in descending order of population estimates. The population estimates are plotted in Fig. 3.4 in a descending order of size. The order of the nations is different than the areas

Table 3.5 Radius of circles for representative nations

Nation	Radius of circle of same area as nation, km	Approximate radius of circumscribing circle of nation, km
United Kingdom	278.59	500
France	417.00	500
Spain	400.56	500
Italy	309.46	500
Germany	336.85	350
Greece	204.81	300
Turkey	497.76	2×350

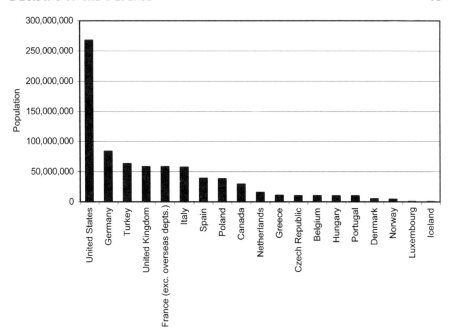

Fig. 3.4   Population of Alliance nations.

plotted in Fig. 3.2. Although the spread of the population figures varies quite markedly from the more than 250 million people in the United States to the less than half a million people in Luxembourg, there is a level value of between approximately 40 and 80 million people per nation in the countries of Western Europe. The total land area of NATO Europe is 4,050,172 km^2 (out of a total Europe area of 9,900,000 km^2, including the land area that extends to the Urals). The total population (2000 census) of NATO is approximately 774 million people, of which approximately 477 million are in NATO Europe.

A second factor of significance in determining defense needs is *population density*. Using the same data as for Figs. 3.3 and 3.4, Fig. 3.5 shows the population density (measured in people per square kilometer) distributed among the Alliance nations. As can be seen from both Table 3.3 and Fig. 3.5, there is a significant variation in the population distribution or density across the Alliance. The numbers vary between *374 people/km^2* in the most densely populated nation of Netherlands to approximately *3 people/km^2* in the least populated nations of Iceland and Canada. This shows a factor of 120 : 1 variation across the nations.

A third factor in the population makeup is how the population is distributed within each nation across its land area. There is a common trend in all nations that the population is not evenly distributed and that most of the people live in

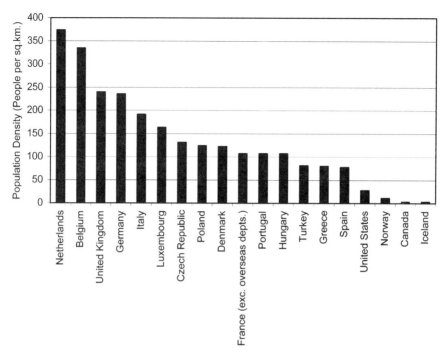

Fig. 3.5 Population density of the Alliance nations.

a small percentage of the available land area. These population centers tend to group around natural geographical and trade nodes and around military assets. Figure 3.6 collects the available data on the population distribution of the NATO Europe nations. For any particular nation and a more detailed analysis it would be necessary to compile the specific (current) data for the nation under analysis. The general trend however is as shown in Fig. 3.6: approximately 50% of the population live in 10% of the area, and 90% live in 50% of the nation's area.

Two examples, by way of illustration, are Germany, where approximately 48% of the population live in 10% of the area and 90% live in 52% of the area, and Italy, where approximately 56% live in 10% of the area and 90% live in 50% of the area. Although there is a significant spread of data for all nations as seen from Fig. 3.6, it is helpful to emphasize that a large percentage of the population in all nations lives in only a small fraction of the area. Very roughly (from Fig. 3.6), for the "typical" nation 50% of the population live in 10% of the area and 90% of the population live in 60% of the area.

An additional factor in determining where the population resides is the distribution of the population around the city centers themselves. Again, as

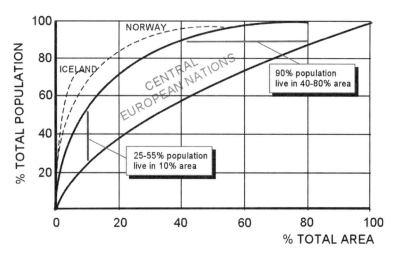

Fig. 3.6   Population distribution in Alliance nations.

for the population distribution data in the nation, only scant data are available for distributions around specific cities. Figure 3.7 shows a possible distribution compiled from U.S. data (constructed from data from Ref. 29). This figure shows the predicted square-root function distribution of the population expressed per lateral size of the city. Some (limited) actual data have been included.[29]

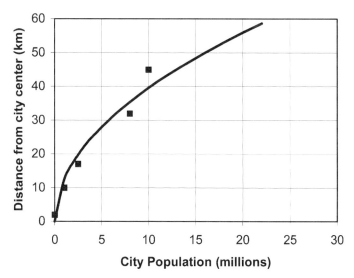

Fig. 3.7   Typical distribution of population in a city.

NUMBER OF SYSTEMS REQUIRED

In any detailed analysis for particular situations, more complete analysis needs to be conducted to determine the number of systems required. However, the general sizing can be determined quite readily by showing how many circles of equivalent areas to the nation to be defended are required depending on their coverage capability. Assuming that the nation (or nations) to be defended has an area expressed in equivalent circles of area πR^2, and that the capability of the selected defense system has a defended area capability of πr^2, then the number of systems n required to cover the complete area to be defended is approximately expressed by $n = (R/r)^2$. Figure 3.8 shows the number of defense systems required for a range of national defended areas of interest.

Table 3.5 shows the general sizing of the equivalent circles required to either equal the nation (or nations) area or to circumscribe the desired area. Figure 3.8 shows the number of systems required to cover those areas if they have different area coverage capability. To illustrate the relative

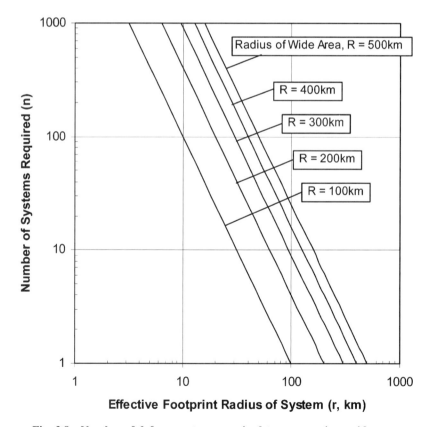

Fig. 3.8   Number of defense systems required to cover a given wide area.

Table 3.6 **Number of defense systems required to cover wide area**

Defense system area capability (radius, km)	Number of defense systems required		
	Wide area $R = 300$ km	Wide area $R = 350$ km	Wide area $R = 500$ km
25	144	196	400
50	36	49	100
100	9	12.25	25
150	4	5.44	11.11
200	2.25	3.06	6.25
250	1.44	1.96	4
300	1	1.36	2.78

magnitudes of possible systems, Table 3.6 summarizes numerical values taken from Fig. 3.8 that apply to the areas of interest for those nations listed in Table 3.5.

In Table 3.6 the noninteger values would have to be rounded up to ensure complete coverage but have been left in their exact numerical form for completeness. It is immediately seen in either Fig. 3.8 or more dramatically in Table 3.6 that the number of defense systems become impractical very quickly as the defended area capability reduces much below 50–100-km radius capability. Because the simple calculation used is to equate areas, it is recognized that these areas will naturally overlap. Figure 3.9 shows this feature for three selected defense systems of radius 100-km, 200-km, and 300-km radius capability designed to cover a 500-km radius wide area. In any particular situation, overlap of systems would be normal to provide additional protection and also to ensure a certain level of redundancy.

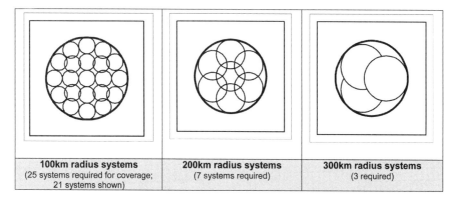

100km radius systems (25 systems required for coverage; 21 systems shown)	**200km radius systems** (7 systems required)	**300km radius systems** (3 required)

Fig. 3.9 **Defense systems areas within 500-km radius wide area.**

PROTECT THE POPULATION TO A SPECIFIED LEVEL OF PROTECTION

One of the basic questions facing the decision maker is *what level of protection is deemed acceptable* (*against conventional weapons*) *over the areas to be protected*? As stated at the beginning of this chapter, the ideal defense system would be one that protects the civilian population or the military personnel and equipment to a level that provides no casualties or zero leakage, that is, no missiles get through. This is not possible, of course, and realistic compromises must be made.

Historically, it has been common practice to express the level of protection in terms of *probabilities*. Specifically, discussions usually center around the *probability of survival* P_{surv} when discussing the end result of use of any defense system, or the *probability of kill* P_{kill} when the discussion centers on the effectiveness of the defense system. Fortunately, one probability can always be determined from the other by subtracting from unity. The more difficult question is the understanding of the difference between a "hit" and a "kill" for any given interceptor.

Before the introduction of WMD, it was normal practice to equate a hit with a kill, and the analysis became relatively straightforward. This was the case for defense against conventional weapons, and this case will be treated first.

LEVEL OF PROTECTION FROM CONVENTIONAL WEAPONS

Unfortunately, there are no universally agreed on definitions related to the probabilities of kill. If the *kill* (of the incoming missile) is defined in terms of *probabilities*, one can write

$$P_k = P_{kill/damage} \times P_{damage/hit} \times P_{hit/launch}$$

$$\times P_{launch/detection} \times P_{detection} \qquad (3.1)$$

Equation (3.1) is the mathematical way of saying that the *probability of kill* depends on a string of *probabilities*, such as

P_k, probability of kill of incoming missile;

$P_{kill/damage}$, probability of kill when incoming missile is damaged;

$P_{damage/hit}$, probability of damage if incoming missile is hit;

$P_{hit/launch}$, probability incoming missile is hit when interceptor is launched;

$P_{launch/detection}$, probability interceptor is launched when missile is detected; and

$P_{detection}$, probability that incoming missile is detected in flight.

It is clear that in any battlefield scenario there are other probabilities to consider, such as malfunctioning, false alarms, personnel-induced delays, etc. The string of probabilities must include considerations of all subsystems in the defense architecture (radars, support vehicles, BM/C^{3}I equipment, etc.) as well as just the interceptor. Such a string of realities makes it difficult to generalize for any specific set of equipment, hence the difficulty in establishing general rules for defense communities. The aforementioned serves mainly to warn the reader to be wary when comparing quoted values for P_k from various sources.

Another complication in determining the level of protection is that in some scenarios it is acceptable to have a *"mission kill"* instead of a *"weapon kill"* (as just discussed). In such circumstances, if the incoming missile is hit and deflected away from the prime target, this would be termed a "mission kill." As will be discussed later for WMD, such a definition of "kill" would not be acceptable given the nature of the spread of the lethal contents on either the battlefield or over civilian population centers. Except in special circumstances, such a deflection of kill is not acceptable in the case of missile attack (ballistic or cruise) and thus is not treated further here.

As already stated, although not a completely universal rule, it is normally assumed that with *conventional* warhead missiles, a "hit" is equated to a "kill." This simplifies the mathematics, and it becomes possible to write the survivability of the asset being defended in relatively simple terms once the form of interceptor launch has been selected. It is assumed in what follows that a *"layered defense"* is used. That is, the asset to be defended (city, military site, naval battle group, etc.) is defended by weapons that have different reach and capability. The assumption of layered defense is the most likely form of defense, as it provides a backup defense philosophy that has been in military circles since the early battles in military history. Variations of this approach bring in such phrases as "defense in depth" and similar descriptions. Given the layered-defense posture, the probability of survival now depends on the nature of interceptor launch. The interceptors can be launched in *salvos* or in a *shoot-look-shoot* firing doctrine. Launching interceptors in salvos is self-explanatory. The *shoot-look-shoot* firing doctrine means that after the first shot (which might be a salvo) the defense system determines whether the incoming missile was hit or not before it decides to launch a second (salvo) or single interceptor. Such a technique allows the defender to determine whether a second shot is needed or not and thus to economize on inventory of interceptors.

LEVEL OF PROTECTION WITH LAYERED DEFENSE AND SALVO FIRING

A more complete treatment can be found in a work by Larson and Kent,[30] which covers the mathematical probabilities of survival of defense assets when there are 1) multiple warheads incoming to a single asset, 2) multiple defense layers each with differing defense capabilities, and 3) dependent and independent events and other complicating factors related to the determination of the survival of any particular asset from missile attack. (The usual equation of a hit with a kill might not always be correct, and the probabilities of survival change if the second shot has to contend with a hit-but-not-killed intercept.) What follows is a simplified treatment of the probability of defense asset survival P_{surv} from missile attacks where there are multiple layers (each with their own SSPK and salvo n assumed against a single incoming warhead.

In addition, in what follows no attempt has been made to ascertain the effects of the (likely) use of *countermeasures* by the threat. Such countermeasures can take a variety of forms that could have a significant impact on the final value of P_{surv} for any interceptor defense system and architecture considered. Hence, the values used in the following analysis could be degraded significantly, and more stringent defense measures used. However, it is informative to at least determine the basic numbers of possible survival rates using classical probability theory.

For *single or salvo shots* and *multiple layers* the survivability of the asset can be expressed by the following equation.

$$\text{Survivability of asset } (P_{surv}) = 1 - (1 - \text{SSPK}_1)^{n1} \times (1 - \text{SSPK}_2)^{n2}$$

$$\times (1 - \text{SSPK}_3)^{n3} \tag{3.2}$$

In Eq. (3.2) the string of multiple terms relate to a layered defense around the asset to be defended. The layered-defense approach shown in Eq. (3.2) includes the single-shot kill probability ($SSPK_1$) of the interceptor used in the first (outer) layer and the single-shot kill probability ($SSPK_2$) of the interceptor used in the second layer, and so on. The number of interceptors used in layer 1, layer 2, layer 3, etc., are denoted by "$n1, n2, n3$, etc."

The single-shot kill probability (SSPK) of the interceptor would be determined from Eq. (3.1) for P_k. By way of example, for a single-layer defense in which the SSPK of all interceptors equals 0.70 and in which a salvo of $n = 2$ is used, then the survivability of the asset would be predicted as 0.91. For a two-layer defense system in which the SSPK of all missiles (in both layers) also equals 0.70 and in which a salvo-firing doctrine of $n = 2$ is used, then the survivability of the defended asset would be predicted as 0.9919. Another way of stating these results is that if 100 missiles are launched at the asset, then nine missiles reach their target with a single-defense-layer protection, and one missile reaches its target for a two-layer defense system.

Using the salvo-firing doctrine, it is useful to see the possible levels of protection that can be achieved for various values of single-interceptor probability of kill (SSPK). Table 3.7 shows these values for a single-layer defense system. Although it is not practical or economical to have salvos of

Table 3.7   Survivability level for a single-layer defense system (salvo-firing doctrine)

Salvo	Survivability level			
	SSPK = 0.90	SSPK = 0.80	SSPK = 0.70	SSPK = 0.50
1	90%	80%	70%	50%
2	99%	96%	91%	75%
3	—	99%	97%	87%
4	—	—	99%	94%
5	—	—	—	97%
6	—	—	—	98%
7	—	—	—	99%

greater than 2 or 3, the values are shown for each combination of SSPK and salvo to achieve a protection level of at least 99%.

The value of a two-layer defense system can be seen in Table 3.8, where for convenience and clarity the same SSPK has been assumed for the interceptors in each layer. Such tables emphasize the need to develop interceptors with high values of SSPK.

Such calculations show the desirability of the use of a two-layer defense system so as to achieve survivability levels of protection above 99% against conventional warhead incoming missiles. As can be seen from Table 3.8 (and Table 3.7), there is a tradeoff between developing interceptors with a high value of SSPK and low salvo values and using a firing doctrine of more salvos of less capable (lower SSPK) interceptors. For example, it would take a salvo of four interceptors with an SSPK of 0.70 to achieve the same survivability of the defended asset as a salvo of three interceptors with an SSPK of 0.80. Would the cost of developing such a higher SSPK interceptor offset the inventory cost of four vs three interceptors per attacking missile?

Table 3.8   Survivability level for a two-layer defense system (salvo-firing doctrine)

Salvo	Survivability level			
	SSPK = 0.90	SSPK = 0.80	SSPK = 0.70	SSPK = 0.50
1	99%	96%	91%	75%
2	99.99%	99.84%	99.19%	93.75%
3	—	99.9936%	99.927%	98.438%
4	—	—	99.9934%	99.609%
5	—	—	—	99.902%
6	—	—	—	99.976%
7	—	—	—	99.994%

LEVEL OF PROTECTION WITH LAYERED DEFENSE AND SALVO-LOOK-SALVO FIRING

With the high cost of interceptors (see Chapter 5), it is reasonable to consider the alternative firing doctrine of assessing the kill (or hit) before launching the second and later shots. In this case, after the first shot the defense system determines whether additional shots are necessary and the inventory of interceptors can be controlled to some lower value. The number of interceptors NR required in this case is shown in Equation 3.3,

$$NR = k\left[1 + \sum_{i=2}^{n} (1 - \text{SSPK})^{(i-1)k}\right] \tag{3.3}$$

where k is the number of rounds in a salvo, n the number of salvos, and $i \leq n$. The potential savings in inventory of interceptors can be seen from the calculation for typical values of rounds and salvos for typical values of single interceptor P_k (or SSPK). Table 3.9 shows these values for a single-layer defense system.

The degree of savings by using S-L-S firing doctrine instead of straight S (single-shot) firing will depend on the total inventory. If, for example, the war requires defense against 500 incoming missiles and the defense has interceptors with SSPK $= 0.70$, then all else being equal, the defense would require 1000 interceptors using a salvo-firing doctrine ($k = 2$) or 650 interceptors using a S-L-S firing doctrine. (This simple illustration ignores consideration of interceptors in the depots and pipeline, logistic requirements, and other planning factors.)

Table 3.9   Number of interceptors used in salvo-look-salvo firing

	Numbers of interceptors required per incoming missile			
Firing sequence	SSPK $= 0.90$	SSPK $= 0.80$	SSPK $= 0.70$	SSPK $= 0.50$
S[a]	1	1	1	1
S-L-S[b]	1.1	1.2	1.3	1.5
S-L-S-L-S[c]	1.11	1.24	1.39	1.72
SS[d]	2	2	2	2
SS-L-SS[e]	2.02	2.08	2.18	2.5
SS-L-SS-L-SS[f]	2.02	2.08	2.20	2.63

[a]S = Single shot or shoot.   [b]S-L-S = Shoot-look-shoot.
[c]S-L-S-L-S = Shoot-look-shoot-look-shoot.
[d]SS = Salvo of two simultaneous shots.   [e]S-L-SS = Salvo (2)-look-salvo (2).
[f]SS-L-SS-L-SS = Six conditionally launched interceptors with two intermediate "looks" or "kill assessment" before successive launching.

LEVEL OF PROTECTION FROM WEAPONS OF MASS DESTRUCTION

The concept of destroying missiles or more particularly warheads or submunitions of incoming missiles that contain lethal contents that can inflict damage commensurate with mass destruction is much more difficult to analyze. The preceding equations and treatment of the *probability of kill* P_k of the incoming warhead do not ordinarily apply to WMD, because even if hit, the "residue" from such weapons can retain lethality capability much beyond conventional weapons that contain HE material. The atmospheric effects also become more significant in the lethality equation than conventional HE weapons. Some comments follow on the *probability of kill* P_k as it applies to WMD, especially as it relates to the equation

$$P_{\text{kill}} = P_{\text{kill/hit}} \times (P_{\text{hit}}) \tag{3.4}$$

which expresses the fact that the probability of kill of the incoming missile by the defense interceptor depends on 1) the probability of hitting the missile and then 2) the probability of killing the warhead given that it has been hit.

PROBABILITY OF KILL OF NUCLEAR WARHEADS

A difference between intercepting a missile with conventional warhead and a missile with a nuclear warhead is that with the latter there is the additional complication of the possibility of generation of electromagnetic pulse (EMP) or a transient radiation effect on electronics (TREE), even if the basic missile is hit and destroyed. Both the EMP and TREE can incapacitate the defense (which usually has some form of electronics in its function). Hence, the question of survival of the defense becomes debatable.

If the incoming missile is hit at a high enough altitude, then in all probability

$$\Gamma_{\text{kill}} = P_{\text{hit}} \tag{3.5}$$

But, if the intercept occurs at a low altitude, then in all probability

$$P_{\text{kill}} \neq P_{\text{hit}} \tag{3.6}$$

because of the effect of EMP and TREE on the defense assets. Hence, the effectiveness of the defense against nuclear warheads depends on many other factors than the dynamics of the intercept itself. Chapter 2 (specifically Table 2.2) expands on the forms of damage from a nuclear warhead.

PROBABILITY OF KILL OF CHEMICAL AND BIOLOGICAL WARHEADS*

The makeup of chemical and biological weapons is an extremely complex subject that is not easy to cover in a few short pages. Chapter 2 summarizes many (but by no means all) of the types of weapons and their encapsulated agents that have been either used in war or at least tested in the thousands of laboratories across the world in many nations. A further category of *entomological* weapons is not discussed here. [This particularly offensive type of weapon applies to the use of delivering in a container in the warhead a collection of, say, infected mosquitoes that would then, once the container breaks open on the ground, proceed to infest the target area and beyond. A recent United Nations weapons inspectors report stated that Saddam Hussein was planning (during the 1991 Persian Gulf War) on launching 25 Al Hussein missiles armed with deadly microbes against Allied forces, if Iraq were attacked by nuclear weapons.]

There is a continuum between chemical and biological agents without a clear line of demarcation between them. There are *toxic* agents and *infective* agents. Some are lethal by percutaneous means, but some (most) are lethal through respiratory means. Some are effective over large areas, whereas others are only effective over localized areas. Because of these and many other factors, it is difficult to summarize generalized values of the probability of kill, but some general comments can be made.

PROBABILITY OF KILL OF BULK BIOLOGICAL WARHEADS

The comments provided here apply mostly to the defense against those agents that are respiratory in nature, that is, the agent (in some form to be discussed further) must be ingested into the victim's lungs to either incapacitate or kill. The issue, from an attacker's viewpoint, is to weaponize the biological agent in such a way that 1) the particle size is of the correct size to enter the victim's lungs and 2) the living microorganism in the agent is alive and can survive the missile launch and dispersal to the ground. Hence, from a defense viewpoint, if the intercept can be made at a high enough altitude then the biological contents are either destroyed on contact, or they are dispersed by the local meteorological conditions. If on intercept the biological agent breaks up into small particle sizes, say about $5-10$ μm, which is the maximum size for ingestion into the lungs, they will be dispersed by the winds. If the droplets are larger than about 10 μm, then they might fall to the ground but would be too large for ingestion.

---

*Most of the technical characteristics of chemical and biological warheads including particle sizes, dosage rates, and concentration levels have been extracted from the work done in the 1970s by the Stockholm International Peace Research Institute (SIPRI).[31]

Given the preceding facts, it appears reasonable to assume that the defense system can reasonably control the intercepts so that it can be postulated that

$$P_{\text{kill}} = P_{\text{hit}} \qquad (3.7)$$

PROBABILITY OF CHEMICAL AND BIOLOGICAL SUBMUNITIONS AND BULK CHEMICAL WARHEADS

Within these three classes of weapons, that is, 1) chemical submunitions, 2) biological submunitions, and 3) bulk chemical warheads, the problem of the definition of probability of kill becomes much more difficult to generalize. Here, a return is made to the basic definition, which is written as

$$P_{\text{kill}} = (P_{\text{kill/hit}}) \times (P_{\text{hit}}) \qquad (3.8)$$

in which now the two terms in Eq. (3.8) can be written as

$$P_{\text{hit}} = \text{Function (SSPK, salvo)} \qquad (3.9)$$

in which the values of the SSPK and the choice of firing doctrine involving various combinations of salvos are as given earlier for conventional warheads. It is the first term $P_{\text{kill/hit}}$ where the difficulty occurs. This term depends on many factors, most of which depend on scenario and local conditions to determine the lethality of the incoming missile and the form of defense.

The term $P_{\text{kill/hit}}$, that is, the probability of kill given a hit (intercept), has been tackled by many authors ever since Professor Fritz Haber first postulated his Haber product in 1914. (Professor Fritz Haber at the Kaiser Wilhelm Insitut für Physikalische Chemie, Berlin, was in charge of the German chemical warfare effort during WWI.) This key parameter can be written as

agent concentration × exposure time

in which the agent concentration is typically measured in milligrams per cubic meter and the exposure time in *minutes* to give a dosage measure in milligram-minute per cubic meter. This deceptively simple product brings with it many complex factors. Different military terms of importance from a weaponization viewpoint need to be taken into account. The level of concentration can vary from an *effective* concentration level (i.e., that level of concentration that will have some measurable effect) to a *lethal* level. Some military authorities refer to a *military significant level of incapacitation.* These terms are usually applied to a statistical percentage of the exposed target population. Once the particular form of concentration has been agreed to, then the Haber product is expressed in a form such as LCt_{50}. This particular form would be interpreted as *that dosage that has the probability of killing half of the target population exposed to the agent.* The dosage that would have some effect would be labeled ECt_{50}, and that which would have a militarily significant incapacitating capability would be called ICt_{50}. There have been many instances in which military forces have sought merely to incapacitate the enemy rather than kill. The concept of nonlethal weapons explores this different philosophy of warfare. Unfortunately, all of the parameters that determine these levels of dosages are not precise, and small changes in conditions can turn an intended *incapacitating* weapon into a *lethal* weapon.

If the intent (of the attacker) is to kill the targeted population (civilian or military), then the capability of the defender's interceptor would be expressed as

$$P_{\text{kill/hit}} = \text{Function } (LCt_{50}) \tag{3.10}$$

The Haber product LCt_{50} is a function of several scenario-related parameters, such as 1) density of population in target area (see earlier in Chapter 3); 2) meteorological conditions over target area (to high altitude); 3) toxicity of the chemical or biological agent (see Chapter 2); 4) exposure time to the agent; 5) breathing rate of the targeted population; and 6) protection given to the targeted population. The last two items are quite critical. Clearly, if the targeted population is wearing appropriate gas masks or protective suits, the effects of the chemical or biological agents can be either minimized or alleviated. The breathing rate is also a factor. A man at rest typically breathes at a rate of about 10 liters per minute, but when in strenuous activity (or fearful) this breathing rate can increase to 70 liters per minute.[31] Such a change in breathing rate dramatically changes how much chemical or biological agent can be ingested into the respiratory tract, which in turn dramatically changes the lethality as expressed by LCt_{50}.

Whether or not the chemical or biological agents can get to the ground and remain at the required particle size before dissipation by local wind conditions depends on the nature of how these agents are "packaged." Several missile developments have tackled this problem, and the submunitions that were transported by the U.S. Army *Honest John* missile in the early 1960s (and subsequent missiles *Sergeant* and *Lance*) provide illustrations of how this might be done. This is best illustrated by the features of the U.S. Army *Honest John* missile, which began its life as a battlefield missile in the early 1950s but then was modified to carry either nuclear or chemical or biological submunitions in the early 1960s. Figure 3.10 shows the key features of this short-range (38-km) missile.

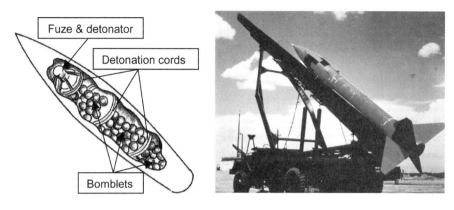

Honest John Warhead
containing bomblets

Honest John Missile
on Launcher

Fig. 3.10   1960s U.S. Army Honest John missile.

In the M190 version (as shown), the *Honest John* warhead contained about 368 M139 bomblets. The total payload weight of all bomblets was about 217 kg. Each bomblet was about the size of a cricket ball. These bomblets had curved ridges built onto the outside to provide spin sufficient that the Magnus effect would provide stability and an accurate trajectory of the bomblet toward the ground. The design was to launch the *Honest John*, with the warhead remaining as a unit (with the bomblets inside) until the missile neared its intended target. At the appropriate time, detonation cords (see Fig. 3.10) would activate and cause the warhead to breakup and disperse the bomblets. The design area over which the bomblets were expected to fall was approximately 1 km^2. Because the chemical agents were contained inside the bomblets, they were not subjected to dispersal by the wind or turbulence or other atmospheric effects, but immediately fell to ground. Inside each bomblet was an explosive charge, which caused the bomblet to break open upon reaching the ground to disperse its deadly contents of either GB or VX.

The design problem for the defender was to design the interceptor such that it would completely destroy *all* 368 submunitions. If even *one* bomblet were not destroyed during intercept (with hit-to-kill or fragmentation warhead interceptors), then that would have been equivalent to a 99.73% kill probability of the interceptor system using either a single or salvo-firing doctrine. From the tables shown earlier, this would require either a very high performance interceptor SSPK ($=0.9973$) or a very large salvo (at least $3-4$ interceptors if each interceptor had an individual SSPK $= 0.70$). The current statistics of missile intercepts, even under test conditions, suggests that this might be difficult to achieve with significant ramifications on the defense of the asset(s).* (In recognition of this problem, the new PAC-3 missile that is designed to be hit-to-kill also has a feature that includes releasing 24 tungsten fragments as a defense against any submunitions that are not destroyed by the direct impact.)

The lethality of each bomblet depended on the many parameters listed previously. If the contents were toxic agents, then if the recipient were exposed for a matter of minutes and wore no protective masks, death could occur anywhere from *minutes* to *hours* depending on the agent used. If the agent were of the infective kind, then death could occur within *days*. The specifics of whether each bomblet (with less than 1 kg of chemical or biological agent) when broken open could kill *thousands* or *tens of thousands* or even more would depend on the many factors already cited, and no general rules can be given at this stage of development. The most logical choice would be to devise means of stopping the launch (by preemptive strike) or to intercept during the boost phase or certainly before any release of the bomblets or submunitions.

Because of these many variables, it is clearly not possible to give typical values of P_k of the defense systems against such weapons of mass destruction.

Such considerations emphasize why a significant amount of research, development, test and evaluation dollars are being spent on boost-phase intercept (BPI) today such that any intercept would cause the attacking missile (with its deadly load of bomblets or submunitions) to fall back on the attacker's territory and his own troops. This approach to defense against WMD might require a paradigm shift in missile defense philosophy to BPI as such an approach sidesteps the problem of the defender having to intercept all submunitions with the associated insufficient P_k against WMD—and transfers the problem back to the attacker!

QUESTION OF THE CORRECT MEASURE OF DEFENSE

As missile defense evolves in the modern era of warfare, it is not clear what *is* the correct measure of defense, especially in the case of the use of "terror weapons" that might or might not be equated with WMD. Two observations from the results of previous wars suggest that the correct measure has not yet been found.

The first concerns the preoccupation with the measure SSPK. Chapter 2 has already shown that for many previous wars (Civil War, WWI, and WWII

---

*See remarks by Undersecretary of Defense Pete Aldridge to Congress on 17 March 2003,[32] in which the planned interim missile defense system to be installed in 2004 in Alaska to protect the United States from missile attack is expected to have a capability that will provide a 90% success rate against incoming ballistic missiles. This represents the state of the art of missile defense, pending more sophisticated systems that are coming out of development within the decade.

and perhaps others) the winning side has suffered more casualties than the losing side; hence, from the losing side viewpoint simply killing more people did not seem to work.

The second observation comes from the 1991 Gulf War, in which it was reported that one of the main effects from the SCUDs landing in Israel (whether they exploded or not) was not the accuracy or the lethality of the missile so much as the fear or terror factor that kept the civilian population inside their homes. This meant that they did not go to work, and the statistics showed a distinct drop in the nation's gross domestic product. If the objective of the attack was to reduce Israel to a nonfunctioning nation, then perhaps that was a good measure (from the attacker's viewpoint). In the same vein, some writers have put forward the idea that the V-1 and V-2 (the forerunner of the SCUD) attacks during WWII were not very effective (using the SSPK philosophy). However, the war ended before these weapons could show their true effect, which could have been the same result as for the SCUDs during the Gulf War. There was no defense against them at the time.

The analyses presented in this book continue the approach of determining the probability of survival P_{surv}, or the inverse $1 - P_{surv}$, or leakage, but the preceding questions are intended to stimulate thought on choice of defense systems that will eliminate the preceding concerns and perhaps provide a more correct overall measure of defense.

Chapter 4

PERFORMANCE

Never promise more than you can perform.
Publilius Syrus, 1st century B.C.

According to most dictionaries, *"performance"* is defined as *"the taking of action in accordance with the requirements . . . to fulfill some function."* In the context of missile defense, as discussed in Chapter 1 and again in Chapter 3, the function or the intent of *defense* is to ensure the safety and survivability of the defender and his assets against any attack. This is quite distinct from the role that is taken in *offense*, in which the function is to win over some territory or the subjugation of other parties, which is not the subject of this book. (The reader should consider this difference carefully even with such well-worn quotes as *"the best defense is a good offense."*) The quote in Chapter 3, by General Sherman, provides a reminder that the legitimate object of war is a more perfect peace. This subtlety was recognized in 1947 when the *U.S. Department of War* changed its name to *U.S. Department of Defense*.

President Reagan on 23 March 1983 put forward a noble objective in missile defense with his new initiative to devise a defensive shield over the United States against *any* missile threat. In those early days of the Strategic Defense Initiative (SDI), the analysts toiled with many technological approaches to achieve that aim. As each technology was examined and discarded, it became clear that the analysts were challenged and caught between two approaches: a *defense shield* or a *defense weapon* (Fig. 4.1).

The United States quickly learned in 1983 and in the intervening years up to today that technology does not support, at least in the foreseeable future, a *defense shield* capable of deflecting or destroying any incoming missile. The era of "Star Trek" and *energy shields* is not yet here. Accordingly, analysts were forced to consider the concept of *defense weapons* with their obvious limitations. A simple arithmetic calculation will illustrate the problem. Suppose, as was done in 1983, that the missile threat could be anticipated to be of the order of thousands of missiles in a raid. Assume 10,000 missiles for this example. Further assume that the best defense weapon system had a 90% probability of intercepting (and destroying) such a

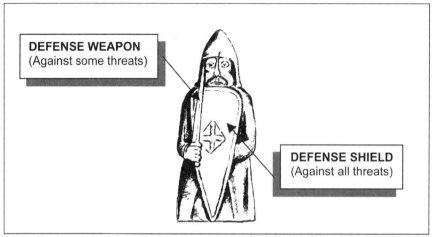

Author's sketch of Isle of Lewis chess piece

Fig. 4.1   Defense shield and defense weapon.

raid. Even with such a high-performing defense system, where 1000 missiles still get through to their targets, which if they were nuclear tipped (as was predicted at that time), would mean that 1000 cities would be annihilated or severely damaged. (See discussion of the effects of nuclear war in Chapter 2.) Clearly, this would be called an ineffective defense. If the capability (or performance) of the defense weapon systems could somehow be raised to 99%—a value not yet achieved in the past 20 years of intense development— then 100 cities would still be destroyed. Going further, a defense system with a performance capability of 99.99% would still mean *one* major city could be destroyed, which again would be considered an unacceptable defense. Such a realization quickly brought about the need to study other means to defend the nation that are beyond the scope of this book. They include political measures to dismantle world inventories of offensive weapons, a restatement of the meaning of deterrence, and the postulation of different technological solutions and operational tactics (which are discussed in this book).

As the specter of nuclear warfare has somewhat diminished since the days of the Cold War, the other forms of weapons of mass destruction (WMD), namely, chemical and biological weapons, have begun to be considered and even used in recent wars. (See Chapter 2 for further discussion on this possibility.) In the 1990s this has led to new initiatives missile defense, including

1) Political actions to reduce the number and sale of offensive weapons;
2) Development of more accurate interceptors;
3) Introduction of directed-energy interceptors;

  4) Boost-phase and ascent-phase intercept; and
  5) Preemptive strike.
The first and the last of these have incurred intense debate around the globe
for obvious reasons. (Up until 1990, many nations and NATO itself forbade
the consideration of preemptive strike in their military analyses, but as the
implications of the technically advanced weapons came on the scene it
became a necessity to at least study the ramifications of such defense
methods. The difficulty remains, of course, of differentiating *defense* and
offense if such tactics are used.) The other innovations for *defense weapons*
will be discussed and analyzed in this book. In all cases the objective is to
increase either or both 1) the performance effectiveness and 2) the defended
area around the assets (cities, troops, etc), that is, to keep the attackers and
the effects of their weapons as far away as possible.

DEFENSE IN DEPTH

  As events have unfolded over the last 20 years, so have the defense
strategies and terminology used. It is useful to point out some of the key
terms that appear frequently in the analysis to follow. The concept of
"*defense in depth*" has been in military terminology since the battle tactics of
the Roman Army in the 3rd century B.C. and was used even earlier by
Alexander the Great. Figure 4.2 shows a modern-day version of *defense in
depth*, which captures the terminology in modern-day warfare. The "battle
area" expands from *self-defense* nominally out to about 10–20-km radius,
then *area defense* out to about 150-km radius, the *outer-area defense* to 500–
1000 km, and finally the *long-range surveillance* out to 1500 km or more.
  In Fig. 4.2, each of the rings around the fighting force (or asset) to be
defended indicates a different defense system to be used depending on
threat and technological advances available. In this example (Navy antiair
warfare), guns and Phalanx systems would be used in the self-defense ring,
then surface-to-air missiles (SAM) and strike aircraft in the area defense and
outer-area defense zones, and finally satellites and long-range early-warning
aircraft would be used in the outer ring of long-range surveillance.

LAYERED DEFENSE

  As the missile defense era began to take shape in the early 1980s, the
concept of *layered defense* began to be used to distinguish between the
various performance capabilities of the defense systems. To ensure no
confusion would arise, sometimes the term *active layered defense* was used
to make a distinction between those capabilities desired from a defense
weapon system and the complementary *passive defense* system(s) (see
Chapter 3). The basic terminology introduced is as shown in Fig. 4.3.

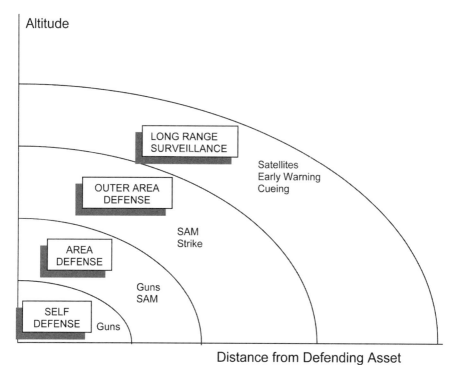

Fig. 4.2 Defense in depth.

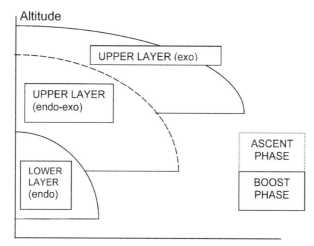

Fig. 4.3 Layered defense.

There are several important differences between Fig. 4.2, which illustrates *defense in depth*, and Fig. 4.3, which illustrates *layered defense*. The first important difference is that the system shown in Fig. 4.2 applies for all aspects of extended air defense (i.e., it includes defense against airbreathing threats as well as missile threats). It is an all-encompassing description of the battle area, whereas the system shown in Fig. 4.3 was structured for ballistic missile threats only. In this regard Fig. 4.3 does not show areas (or layers) below some altitude, which "belong" to other systems such as SHORAD or VSHORAD. It also neglects description of the domain of the cruise missile and how defense against it should be handled. It has become common practice to assume these requirements and treat them separately without too much rigor on defining areas or layers of treatment. This omission is returned later in the analysis of the cruise missile (CM) threat.

Also, it has become common practice to assume that the boost phase might have to be extended into the ascent phase if the defense system is unable to intercept during the boost phase. In Fig. 4.3 a dotted rectangle above boost phase is intended to remind the reader (and analyst) of the existence of this very real mode of defending against ballistic missiles. As in the description of *layered defense* shown in Fig. 4.3, it has become common practice to divide the upper layer into two categories. This division is based on the differing abilities of potential defense systems to operate for any appreciable amount of time in the atmosphere, aerodynamic heating can cause radical changes in the design of the system. The terms "endo" and "exo" to indicate where the layers are in relation to the tropopause (at approximately 30-km altitude where the atmospheric density begins to have a strong effect on the aerodynamic characteristics).

Defense Segments for Missile Defense

The concept of layered defense is still the main form of description for ballistic missile defense, but it is useful to mention the terminology introduced by the U.S. Missile Defense Agency (MDA) in January 2002, when it was decided to group all programs and activities into the four main categories of *terminal segment*, *midcourse segment*, *boost-phase segment*, and the *ballistic missile defense system and sensor segment*. Although this was done primarily to emphasize program priorities, it does serve to help describe the main phases of ballistic missile defense. This form of description is given in Fig. 4.4.

Again, as for layered defense in Fig. 4.3, the MDA description is primarily for ballistic missile defense, and one must adjust the segments or layers if the discussion is to include defense against the airbreathing threat, cruise missiles, and (very) short-range air defense systems. Similarly, although

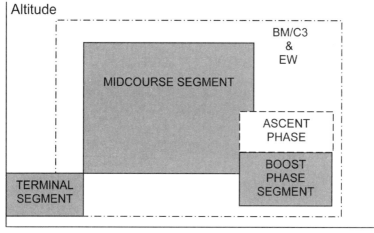

Fig. 4.4   Segmented defense categories.

pertinent programs can be grouped in the boost-phase segment, they might actually be ascent-phase systems. Finally, the MDA description collects the essential elements of early warning and battle management command, control and communication (BM/C^3) in the fourth segment called *BMDS and sensor segment*. The solution-sensitive delineation of the upper-layer defense systems in "exo" and "endo-exo" have been subsumed into the *midcourse segment* and *terminal segment*. [As just one example, the THAAD system, originally conceived as an upper-layer system (in Fig. 4.3), has now been grouped programmatically into the terminal segment in MDA's grouping (Fig. 4.4).]

As the analysis of the performance of the various possibilities of systems design and decision making in missile defense proceeds, it will be necessary from time to time to refer to one or the other of these various ways of describing the regions of interest of the missile defense systems as laid down in Figs. 4.2–4.4.

PERFORMANCE REQUIREMENTS

Given the preceding description of the main elements, it is clear that there are two main objectives to be met by the defense system.* These objectives are to maximize the area around the asset to be protected and maximize the

---

*In this chapter the reference to "defense system" usually has the broad connotation of all of the elements of the defense, which will include 1) the interceptor system with any supporting systems, 2) any required early-warning system, and 3) the necessary BM/C^3. If the reference is to just one of these elements, the text will be added as needed.

level of protection or survival level desired within that area. The other performance characteristics of the defense system are to ensure that objectives 1) and 2) are achieved. Because the main objective of *defense* is the attainment of these objectives, then the decision maker can be content with what is described in military circles as *mission kill*, that is, it might be acceptable to divert or otherwise deter the attacking missile from reaching its target without actually having to destroy it. This becomes a matter of operations analysis with due consideration of collateral damage (to neighboring allies for example). This could be the case in boost-phase intercept (BPI), where certain benefits can be obtained if the attacking ballistic missile (BM) can be either destroyed immediately after launch or simply "knocked off course" so that it lands on the attacker instead of the attacked. One can postulate several scenarios that would have a *deterrent* effect if the attacker knew that the deadly WMD agents that he was to launch into the intended target nation were diverted onto his own territory and troops.

CHARACTERISTICS OF THE BALLISTIC MISSILE THREAT

To determine the performance requirements for the defense system, it is pertinent to summarize the main features of the missile threat as they would influence the design of any potential defense system. In Appendix C, the equations have been developed for all three phases of flight as depicted in Fig. 4.4. Essentially, the ballistic missile trajectory can be characterized in this way and shown in Fig. 4.5. The characteristics of the typical ballistic missile trajectory as shown in Fig. 4.5 are in agreement with those shown in Fig. 4.4. Several important factors need to be highlighted. They are as follows:

1) The *boost phase* has a distinct beginning and end. It begins with liftoff and ends with thrust termination. It can include *energy management* to consume fuel during this phase.

2) The *midcourse phase* of the trajectory has a distinct beginning with thrust cutoff or burnout, but might not always be in "freefall" acted on by gravity alone. It might receive flight adjustment through midcourse guidance. The end of this phase is also less distinct in that as it reenters the atmosphere additional forces (lift and drag) begin to take effect.

3) The *terminal phase* is essentially a continuation of the midcourse phase except that the effect of the atmosphere can have a significant effect on both its speed and trajectory.

The equations of motion for each of these phases are provided in Appendix C, where it is shown that various effects can be accounted for with different treatments such as the use of orbital mechanics to account for the

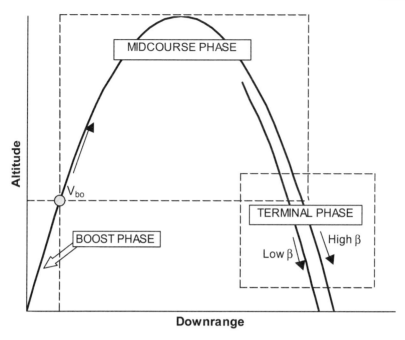

Fig. 4.5   Three main phases of ballistic-missile trajectories.

rotation of the Earth; the use of steering laws to properly align the flight trajectory angle for best performance (such as seeking maximum range); and the effects of burnout altitude on the final range and other factors.

EXAMPLES OF BALLISTIC MISSILE TRAJECTORIES

Both the threat BM and the interceptor exhibit similar characteristics in their performance parameters, and to aid the discussion of how to intercept the missiles in flight sets of trajectories are provided, first with a "typical" set of parameters and then with variations of key parameters shown to indicate the various means of treating defense against them. A set of ballistic missile trajectories is provided for five classes of trajectories for ranges of 100, 500, 1000, 2000, and 3000 km. These have been called "classes" because the actual ranges in each case do not have precise values of final range. They have been determined through a set of initial parameters and the trajectories have been allowed to follow the equations of motion to their termination on the ground. The collection of trajectories is shown in Figs. 4.6–4.10, grouped by the three main phases of *boost phase, midcourse*, and *terminal phase*.

The trajectories are shown on the left-hand side of each figure, and the speeds in each phase are shown on the right-hand side. The missile is shown

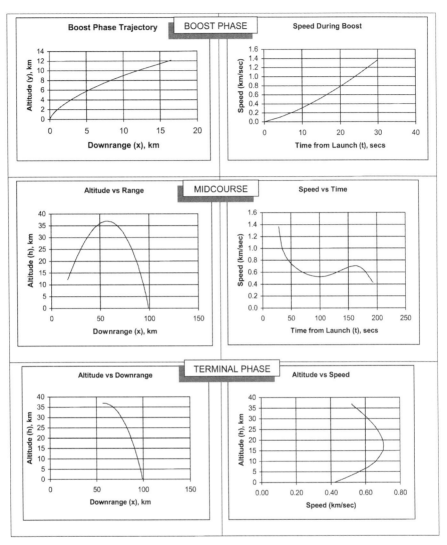

Fig. 4.6   Three main phases of typical 100-km-class ballistic missile.

to accelerate through the *boost phase* up to the burnout speed V_{bo}; then, in the *midcourse phase* it is seen to decelerate up to apogee through the effects of gravity and then accelerate again in free fall, postapogee. As it reenters the troposphere, the aerodynamic resistance effects begin to slow the missile down again. In the *terminal phase* it can be seen that the trajectory itself does not change shape significantly, but the speed slows appreciably as it approaches the ground. It is shown later under the discussion of defended footprints that these characteristics directly influence the size and shape of the defended area.

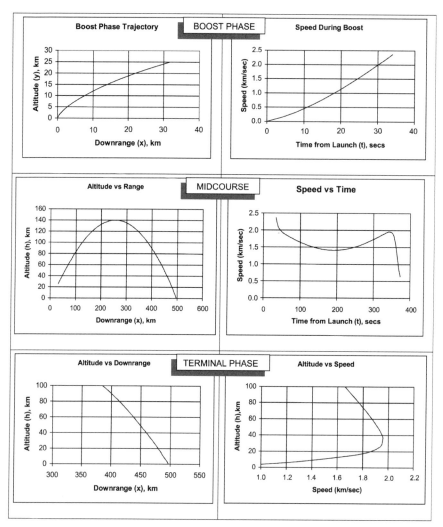

Fig. 4.7   Three phases of typical 500-km-class ballistic missile.

Figures 4.6–4.10 are presented as typical BM trajectories, but it is important to recognize the basic assumptions in their generation. They can have significant impact on the performance of any one of these trajectories and must be considered when designing any defense against them. The assumptions are best presented for each phase.

BOOST-PHASE ASSUMPTIONS.   All missiles are launched vertically ($\theta = 90$ deg) and follow an exponential steering law to turn the missile over to achieve the required angle for maximum range (minimum energy) at the end of the boost

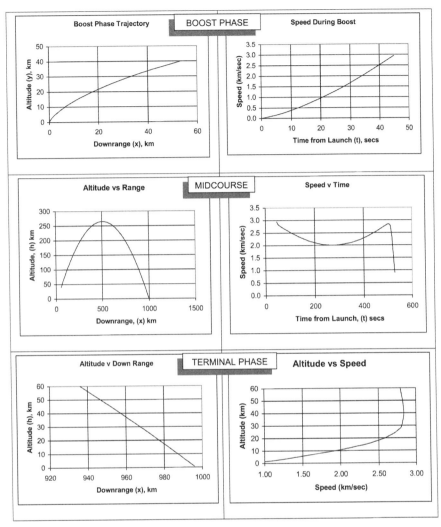

Fig. 4.8   Three phases of typical 1000-km-class ballistic missile.

phase. The burn times t_{bo} have been taken to be the average values of the statistical sampling of data shown in Appendix C for each appropriate range. That means that these typical trajectories are neither the slowest burning nor the fastest burning missiles in existence or in development today. Although each case is slightly different, the 1000-km class BM values for the steering law and boost-phase acceleration are shown in Fig. 4.11 to show the nature of these parameters. The acceleration throughout boost is determined from the need to meet the required V_{bo} in the stated burn time. The burn rate matches those of typical propellants.

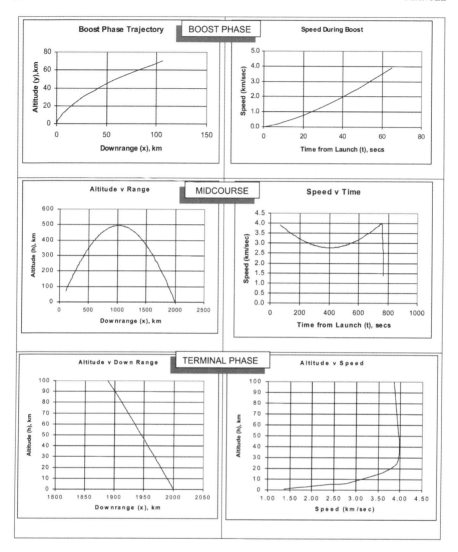

Fig. 4.9 Three phases of typical 2000-km-class ballistic missile.

Figures 4.6–4.10 show, for each typical BM trajectory, the point (x_{bo}, h_{bo}) in space that represents the end of boost and the start of the ballistic trajectory into the midcourse phase. These points in space affect the apogee and range of the missile in a distinct manner and are important to understand when examining, for example, the quoted values of missile performance in the literature. As can be seen from Figs. 4.6–4.10, these boost-phase endpoints vary from more than 30% of the missile apogee and 15% of the range for the 100-km class BM to about 15% of the apogee and 5% of the range for the 3000-km class BM.

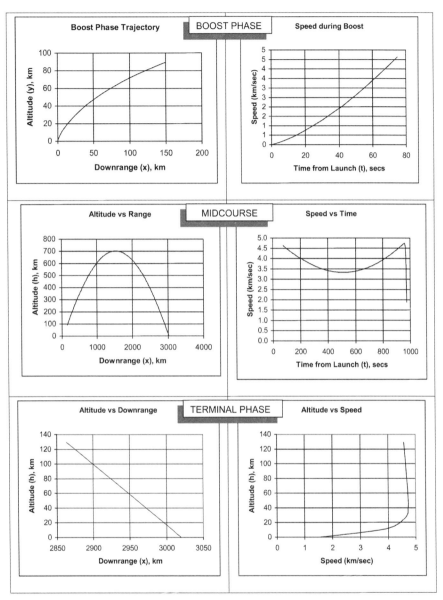

Fig. 4.10   Three phases of typical 3000-km-class ballistic missile.

These values can change even more significantly for different values of the missile ballistic coefficient β as discussed later.

The ballistic coefficient β has its greatest effect during the terminal phase but also helps to determine the basic range of the missile through its impact during the boost phase. The aerodynamic resistance or drag D and the thrust term T combine during the boost phase to produce the

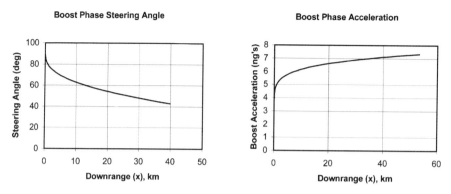

Fig. 4.11   Boost-phase steering angle and acceleration (1000-km-class BM).

acceleration (ng) that can reach high values for many BM. As can be seen for the case of the 1000-km class BM shown in Fig. 4.11, this acceleration begins at about 2.58 g and increases to about 7.34 g (as the propellant is used up and the mass decreases). This gives an average value of acceleration during boost of 5.65 g, which can be compared with the statistical database in Appendix C.

In these typical BM trajectories it has been assumed that the missile has been launched vertically, to clear the launch area, but is then steered exponentially to the angle required for maximum range. In the case of the 1000-km class BM, this was shown to be (in Appendix C) approximately 43 deg. The appropriate values of θ for maximum range were used in the other trajectories shown in Figs. 4.6–4.10. Because the boost phase is of short duration, if the alternate means of launching the missile directly at the required launch angle for maximum range is used a slight, but not significant variation, in the performance characteristics would occur.

A graphic illustration of the launch angle and the subsequent steering of the trajectory toward the required angle for the desired range can be seen in Fig. 4.12, which shows a launcher of the *Sol-Air Moyen Portée/Terrestre* (SAMP/T, ground-to-air medium-range/land) missile system (left-hand side) and a test firing of the SAMP/T missile, the ASTER (on right-hand side). The stop-action photos of the ASTER immediately after launch clearly show the rapid turnover after launch to the desired angle for the range selected. The track of the angle can be compared to the exponential steering angle plot shown in Fig. 4.11.

RELATIONSHIP OF BOOST PHASE TRAJECTORIES TO MISSILE DESIGN. The typical ballistic missile (TBM) trajectories shown in Figs. 4.6–4.10 use the acceleration profiles as just discussed. Such profiles can be related to actual

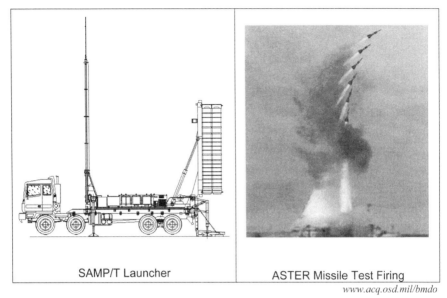

| SAMP/T Launcher | ASTER Missile Test Firing |

Fig. 4.12   Launcher and launch of SAMP/T missile.

missile designs through Professor Oberth's (see Chapter 2) original rocket equation where the thrust T of the missile at liftoff is given by

$$T = V_e \frac{dM_p}{dt} \tag{4.1}$$

where V_e is the exhaust velocity of the missile and M_p is the mass of the exhaust propellant. The exhaust velocity is often expressed in terms of the *specific impulse* I_{sp}, which has the units in seconds. In the SI system the thrust T would then be expressed in newtons, the exhaust velocity in meters/second, and the propellant mass in kilograms to give I_{sp} in seconds. The exhaust velocity V_e is related to the specific impulse by the equation

$$V_e = gI_{sp} \tag{4.2}$$

(See Appendix E for a more complete derivation of the velocity terms for multistage rockets.)

If it is assumed that the propellant mass is consumed totally by the time of burnout (i.e., at time t_{bo}), then the thrust can be expressed (approximately) by

$$T = gI_{sp} \frac{M_p}{t_{bo}} \tag{4.3}$$

By way of example, consider the characteristics of the single-stage SCUD B missile with a takeoff weight of 6370 kg (see Appendix A). The SCUD uses a liquid propellant (a mixture of nitric acid and dimethylhydrazine) that has a specific impulse I_{sp} of approximately 240 s. (The formulation for specific impulse I_{sp} is a cumbersome formulation of the combustion process, type of engine, and the specific heat of the exhaust mixture. Typically, values of I_{sp} range from 1000 to 4000 s for jet engines in aircraft, down to about 200 to 300 s for rocket propulsion. Appendix E provides more detail on the expected values for I_{sp} in missile design.) With a propellant mass of 4000 kg and a t_{bo} of 70 s, this combination of characteristics produces a thrust of approximately 135 kN. This gives an acceleration at liftoff of $n = 2.20 \, g$. As can be seen from the set of typical trajectories, this is the expected value.

The shape of the acceleration curve (Fig. 4.11b) after the initial acceleration is a little more difficult to reduce to simple algorithms because of the nature of the burning mechanism and whether it is a liquid- or a solid-propellant design. In practically all cases the desired design feature is to burn rapidly during the initial stages. As the propellant is consumed the burn rate and the missile rate of increase in acceleration (see Fig. 4.11b) are gradually reduced to the desired values at burnout t_{bo}. Appendix E provides notes on relating missile design characteristics to meet the performance parameters used in the typical trajectories shown in Figs. 4.6–4.10.

MIDCOURSE-PHASE ASSUMPTIONS.   First, it is important to note that in all of the analysis presented it has been assumed for all three phases that the trajectories are over a spherical Earth which is nonrotating. The effect of the Earth's rotation is treated separately later in this chapter.

In Figs. 4.6–4.10, the trajectories and speeds are shown for the midcourse and terminal phases of the flight. The "starting point" in each case is the condition at the end of the boost phase. Specifically, the start of the trajectory is shown to begin at the point in space (x_{bo}, h_{bo}) and with a burnout speed of V_{bo}. Three critical assumptions are made in the simulations, which are: 1) The representation of the acceleration caused by gravity; 2) The representation of the density of the atmosphere and how it varies with altitude and the air temperature; and 3) The representation of the effect of the shape and size of the missile on its aerodynamic resistance.

GRAVITY

In Appendix C it is assumed that the Earth is spherical and that gravity can be represented by

$$g = g_o \left(\frac{r_e}{r_e + h} \right)^2 \tag{4.4}$$

in which acceleration caused by gravity at the Earth's surface is 9.806 m/s/s and the radius of the Earth is taken to be the average value of 6378.14 km. The altitude h is measured from the Earth's surface and includes the burnout altitude h_{bo}.

ATMOSPHERIC DENSITY

The characteristics of the Earth's atmosphere vary significantly around the globe and also with altitude. Appendix D summarizes these different characteristics in terms of pressure, temperature, and air density. The majority of the trajectories used in this book, unless specifically identified otherwise, assume a standard atmosphere. (See Appendix D for summaries of the various treatments of "standard atmosphere.") The assumption of a common standard to describe the atmosphere is of course necessary for any logical comparison of results from different sources and for consistent analysis. It should be recognized that despite the need to establish a standard atmosphere, it is very rare that such a standard atmosphere occurs, and it is also necessary to evaluate the effect on missile trajectories on a nonstandard day. Appendix D shows that the atmosphere can be considered as being made up of seven layers, but the three layers (that includes the troposphere) closest to the Earth's surface have the greatest effect on the shape and characteristics of the trajectories. The air density in these three layers can be represented by the exponential relationship

$$\rho = \rho_1 \cdot \exp\left(\frac{-h}{H}\right) \tag{4.5}$$

in which the constant ρ_1 varies in each layer. Because the atmosphere does not neatly follow the common exponential formulation through all layers, it is usual to adjust the scale height value of H also in Eq. (4.5) to various values in each of the layers. Appendix D shows that the following equations for air density give a good approximation (within 1–2%) of the standard atmosphere. (In most of the work, the 1976 U.S. Standard Atmosphere values are used.)

Layer 1 (troposphere); altitude limits (sea level to 11 km or tropopause):

$$\rho = \rho_0 \cdot \exp\left(\frac{-h}{9.15}\right) \tag{4.6}$$

in which ρ_0 is the air density at the Earth's surface with a value of 1.225 kg/m^3 on a standard day of 15°C.

Layer 2; altitude limits (>11 to 20 km)

$$\rho = 2.000 \cdot \exp\left(\frac{-h}{6.45}\right) \tag{4.7}$$

Layer 3; altitude limits (>20 to 32 km)

$$\rho = 2.550 \cdot \exp\left(\frac{-h}{6.00}\right) \qquad (4.8)$$

Above layer 3 the air density is so low (approximately 1/100th of the sea-level value at 32-km altitude), declining exponentially to less than 1/15,000th of the sea-level value at the outer reaches of the discernible atmosphere (greater than 70-km altitude). Because of this, the layer 3 approximation has been used for all altitudes above 32 km (in the stratosphere). A quick examination of the typical trajectories shown in Figs. 4.6–4.10 shows that the missile spends varying amounts of time in the atmosphere depending on its range. The amount of the trajectory in the discernible atmosphere is shown in Table 4.1.

Hence, the exponential representations used here are seen to be reasonably accurate especially when it is recognized that the accepted values of the standard atmosphere are approximations in themselves.

Later in this chapter the effects of a nonstandard atmosphere, especially on a hot day, are shown using data collected by the U.S. Air Force in the desert regions of Iran and Iraq (see Appendix D and Ref. 33).

BALLISTIC COEFFICIENT

In all of the trajectories shown in Figs. 4.6–4.10, it is assumed that the missile (or the incoming warhead) is a nonlifting body ($L = 0$) and that the only additional force acting on the missile is the aerodynamic drag D. In addition to the speed V effect, aerodynamic effects show up through the effect of the air density ρ just discussed and the effects of mass, size, and shape.

Table 4.1   Amount of trajectory in discernible atmosphere

BM range, km	Part of trajectory below 32 km altitude, %
100	Approximately 100
500	23
1000	12
2000	6
3000	4

These latter parameters are conveniently combined into a single parameter, the ballistic coefficient β, defined by

$$\beta = \frac{W}{C_D \cdot S} \qquad (4.9)$$

The ballistic coefficient β is a way of conveniently combining several effects in the equations of motion. It combines the inertia effect of the missile mass W, the size of the missile's surface area S, and the shape of the missile through the drag coefficient C_D. (In this treatment the usual practice of using the base area of the missile as the reference area is used.) For all of the trajectories shown in Figs. 4.6–4.10, a typical value of $\beta = 5000\,\text{kg/m}^2$ has been assumed. Such a value corresponds to missiles that have a relatively clean shape (aerodynamically) and do not tumble on reentry. Appendix C provides an appreciation of the likely values of the ballistic coefficient based on existing missiles. A discussion of the various possible values of β is presented later.

As can be seen on examination of Figs. 4.6–4.10, the characteristic shape of the speed vs time plot during the midcourse phase changes significantly from that of the 100-km class BM that essentially stays within the atmosphere to 3000-km class BM that spends only a limited amount of time in the atmosphere. For the short-range (100-km) missile it is seen (Fig. 4.6) that the speed cannot build up fast enough during the descent from apogee before the atmospheric effects (through both ρ and β) slow the missile down. This speed effect rapidly becomes of less importance as the range increases from 100 to 3000 km.

TERMINAL-PHASE ASSUMPTIONS. The terminal phase is simply a continuation of the midcourse phase except that the effects of the atmosphere already discussed become more pronounced. It has been assumed in the terminal phase that there are no lift effects ($L = 0$) on the reentering missile or its warhead. It is highly likely that the missiles could indeed have such shaping to give lift. This would be an effective decoy mechanism in the latter stages of the BM approach to target. If the attacking missile is designed to release chemical or biological agents where specific target points are not important, then this lift effect could be quite effective as an intercept avoidance technique. Although important, because such features are highly system or design specific, they have been omitted from these typical BM trajectories.

COLLECTED PARAMETERS FOR TYPICAL BM TRAJECTORIES

For ease of discussion, some of the pertinent characteristics of the five typical BM trajectories (100-km class to 3000-km class) are summarized in Table 4.2.

Table 4.2 Summary of characteristics of typical ballistic missile trajectories

Phases	Parameters of typical BM trajectories (ballistic coefficient, $\beta = 5000\ \text{kg/m}^2$)				
	100-km class	500-km class	1000-km class	2000-km class	3000-km class
Boost phase					
Burn time, s	30	34	45	65	75
Initial θ, deg	90	90	90	90	90
End of boost, θ, deg	45	44	43	40.5	38
Initial acceleration, ng	2.5	2.72	2.50	2.55	2.53
Final acceleration, ng	5.07	7.33	7.20	6.74	6.86
Average: acceleration, ng	4.13	5.69	5.52	5.28	5.36
V_{bo}, km/s	1.30	2.22	2.89	3.91	4.63
x_{bo}, km	15.83	29.87	52.25	106.21	149.09
h_{bo}, km	11.42	23.18	38.80	69.99	88.90
Midcourse phase					
Apogee, km	35.59	137.94	264.71	493.82	697.18
Speed at apogee, km/s	0.55	1.42	2.01	2.77	3.32
Time to apogee, s	95	194	275	400	510
Terminal phase					
Speed at 10-km altitude	0.67	1.70	2.20	3.34	3.85
Reentry angle (@10 km), θ_{re}, deg	56	48	46	43	41
Time of flight, s	191	369	527	770	972
Time of flight, mins	3.18	6.15	8.78	12.83	16.20
Range at the surface, km	101.16	497.44	999.10	1997.78	2997.56

APOGEES OF TYPICAL BM TRAJECTORIES.   The apogees of the BM trajectories shown in Table 4.2 and in Figs. 4.6–4.10 are shown in Fig. 4.13. Figure 4.13 shows that despite the variations of key parameters in the boost phase and midcourse phase they are close numerically to the values that would be calculated by the flat-Earth assumption, which, as presented in Appendix C, is given by

$$\text{Apogee} = \frac{R \tan \theta_{bo}}{4} + h_{bo} \qquad (4.10)$$

Because θ_{bo} is close to 45 deg for most cases and because h_{bo} is a small percentage of the total height of the trajectory (as seen in Fig. 4.13), this led to the rule of thumb that the *apogee (in kilometers) of most BM trajectories is determined by dividing the range (in kilometers) by 4.* This simple rule of thumb applies to the minimum energy trajectories and for ranges less than about 4000 km. Appendix C provides similar calculations for lofted and depressed BM trajectories.

TIME OF FLIGHT.   It is also of interest to determine the time of flight (TOF) using simple formulas. The time of flight for the typical BM trajectories from Table 4.2 is shown in Fig. 4.14.

In Fig. 4.14 the curve for the time of flight for typical BM trajectories is taken from the complete set of calculations in which the effects of gravity

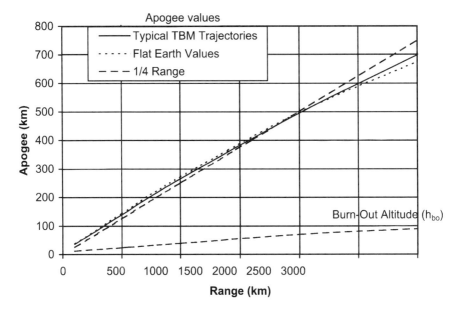

Fig. 4.13   Apogees of typical ballistic-missile trajectories.

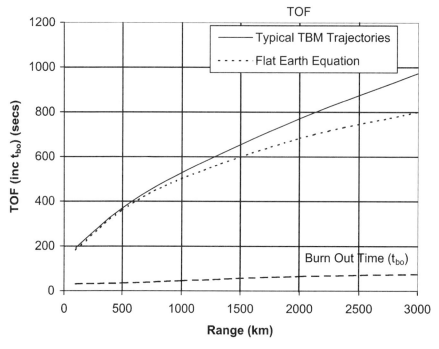

Fig. 4.14   Time of flight of typical ballistic-missile trajectories.

and a nonrotating, spherical Earth are included with the effects of aerodynamic resistance when the BM is passing through the troposphere. The actual values are shown in Table 4.2 and displayed in Figs. 4.6–4.10. The predominant reason for the gradually increasing departure of the actual time of flight from the flat-Earth assumption results is the effects of gravity that decrease with altitude (which increases the time of flight).

Without the correction for gravity variation, the difference between the flat-Earth results and those of the more accurate spherical, nonrotating Earth is about 2% at the 500-km class BM, 5% for the 1000-km class BM, 11% for the 2000-km class BM and about 18% for the 3000-km class BM. From Appendix C the TOF over a flat Earth is given by

$$\text{TOF} = \sqrt{\frac{f(h_{\text{bo}}) \tan \theta_{\text{bo}}}{g}} \sqrt{R} + t_{\text{bo}} \qquad (4.11)$$

in which the function $f(h_{\text{bo}})$ brings in the effect of "starting" the ballistic (zero thrust) part of the midcourse trajectory at altitude h_{bo}. In Appendix C the function $f(h_{bo})$, essentially a potential energy term, is shown to be

$$f(h_{bo}) = \sqrt{1 + \frac{4gh_{bo}}{V_{bo}^2 \sin^2 \theta_{bo}}} \qquad (4.12)$$

The numerical value of the second term in Eq. (4.12) is close to unity for most trajectories of interest here (less than 3000-km range missiles) such that the time of flight can be expressed more simply as

$$\text{TOF} = \sqrt{\frac{2 \tan \theta_{bo}}{g}} \sqrt{R} + t_{bo} \qquad (4.13)$$

In the case of simple flat-Earth calculations, it is common to use the value of g for the sea-level value of 9.806 m/s/s and, further, acts perpendicular to the ground. However, the apogee of the 3000-km range BM is shown to be approximately 700 km (see Fig. 4.10). Here the acceleration caused by gravity has dropped to about 82% of that at sea level. Such a drop accounts for the difference in the time of flight obtained using the more correct calculations shown in Figs. 4.6–4.10 and using the correct gravity vector (magnitude and direction) over a nonrotating Earth. For quick calculations, ignoring this effect, Appendix C show the rule of thumb for the *time of flight could be determined by simply multiplying the square root of the missile range (in kilometers) by 14 to get the time of flight (in seconds)*. The effect of gravity could be included by a ratio of the gravity terms as a further refinement. As before, these quick rules of thumb are provided for minimum energy trajectories, and Appendix C provides other expressions for the cases of depressed and lofted trajectories. Such rules of thumb are intended only to give a quick appreciation of the magnitude of the various parameters. The equations provided in the text allow for calculation of the more accurate set of values.

EFFECT OF BALLISTIC COEFFICIENT.   The ballistic coefficient β has the largest effect in the terminal phase. In the calculations of typical BM trajectories shown in Figs. 4.6–4.10, a value of the ballistic coefficient $\beta = 5000 \, \text{kg/m}^2$ was used. This value corresponds to those BM that are designed to be aerodynamically shaped and follow reasonable design standards on surface irregularities and roughness. The value also changes in direct proportion to the mass of the BM (or its released warhead). Using the same trajectory simulations as shown in Figs. 4.6–4.10 but varying the ballistic coefficient, one can see the immediate effect. Figure 4.15 shows the effect of the ballistic coefficient on three selected cases: the 100-km class, the 500-km class, and the 1000-km class BM trajectories.

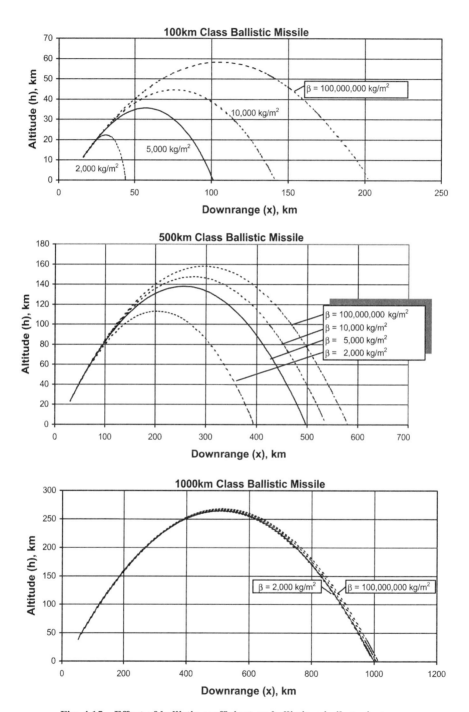

Fig. 4.15   Effect of ballistic coefficient on ballistic-missile trajectory.

As can be seen from Fig. 4.15, the effect of the ballistic coefficient is most pronounced for the shorter range BM (100-km class), but the effect rapidly decreases in terms of the shape of the trajectory as range increases. For BM ranges of about 1000 km and above, the effect of β on the trajectory shape is negligible. The ranges of β selected vary from $\beta = 2000$ to 100,000,000 kg/m^2. This last value was selected arbitrarily to represent the trajectory essentially in a vacuum. The values of the trajectory parameters used in Figs. 4.6–4.10 and summarized in Table 4.2 were selected to give BM ranges of the selected cases for the common value of $\beta = 5000$ kg/m^2. The ranges then seem to vary both above and below the nominal values of range as the ballistic coefficient is varied from $\beta = 2000$ to 100,000,000 kg/m^2.

Figure 4.15 shows that the ballistic coefficient has a marked effect on the shape of the trajectory for those missiles with ranges less than 1000 km and little effect for those missiles with ranges above 1000 km. However, a more significant impact can be seen in the deceleration effects on the missile's speed as it renters the atmosphere. This effect is significant for all ranges of BM. Figure 4.16 shows this effect for the same range of BM classes as before and with the common value of $\beta = 5000$ kg/m^2.

Figure 4.16 shows that β is seen to begin its influence at about the "top" of the third layer of the atmosphere (i.e., at about 32-km altitude). The effect is progressively more significant as the missile speed increases, that is, as the

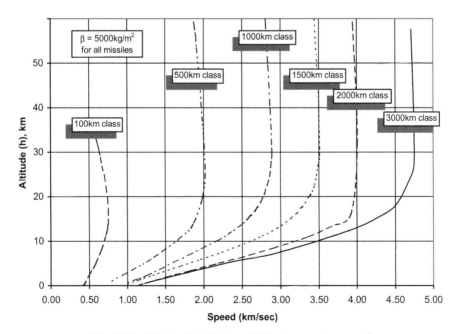

Fig. 4.16   Effect of ballistic coefficient on reentry speed.

nominal range of the BM increases. The effect is more noticeable as the missile approaches the ground. If the intercept altitude is set at a 10-km altitude, the effect is not so pronounced.

If now a particular BM trajectory is selected and the ballistic coefficient varied for that missile design, the effect of such a design parameter can be immediately determined. Taking the 1500-km class BM from the set of trajectories shown in Fig. 4.16, it is possible to show the effect of varying β on both the shape of the trajectory (in the reentry or terminal phase) and on the speed. This has been done in Fig. 4.17.

The upper set of curves in Fig. 4.17 show the shape of the reentry phase of the trajectory. The scale has been greatly expanded to show the very small difference in the shape of the trajectory at these speeds and ranges. The spread of the endpoint in the range is of the order of 5 km (about $\frac{1}{3}$ of 1% of the nominal range), well within the calculation accuracy in the trajectory simulation. The lower set of curves shows the more significant effect, which is the deceleration of the missile caused by β. The curve for the extremely high β (100,000,000 kg/m^2) provides an indication of how the missile would perform in the vacuum of space. As the ballistic coefficient is reduced (i.e., the drag increased), the deceleration increases causing a significant fall off in speed. It is the result of the assumption in the simulation of zero lift ($L = 0$), that is, the assumption of a nonlifting body that results in essentially no or minimal change in the trajectory shape but a large change in the speed, which is decelerating along the flight path.

EFFECT OF A NONSTANDARD DAY.   The conditions for a standard day do not always occur and it useful to see the effects of a nonstandard day, especially those of a hot day. Miller[33] has compiled the density altitude plots for all of the seasons in Iran and Iraq. The plots show temperatures up to 30°C can occur on regular intervals in the month of July. If such a temperature is taken as an indication of what could occur on a hot day, then adjustments can be made to the typical trajectories assuming that the temperature affects the air density according to the gas laws. If this is done, it is expected that the air density would be reduced to 95% of the value on a standard day. The impact of such a change on missile trajectories is quite small, with the greatest effect occurring on those missiles that spend the majority of their flight time inside the atmosphere.

For missile ranges around 100 km, the effect of a 30°C hot day would be to increase the range by about 3%. This effect will decrease as the range is increased and the missile spends proportionately less time in the atmosphere (see Table 4.2). For a missile range of 500 km where the amount of time spent moving through the atmosphere is less than a quarter of the time, the range increases through the effects of a lower air density to less than 1%. Normally, such differences are insignificant in determining the range of a missile especially in the light of the original assumptions of the treatment of

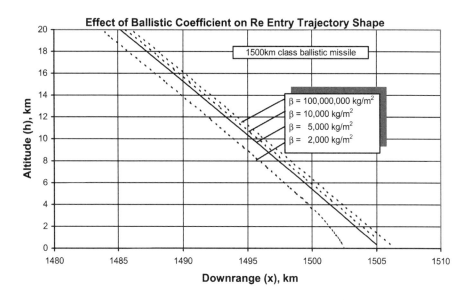

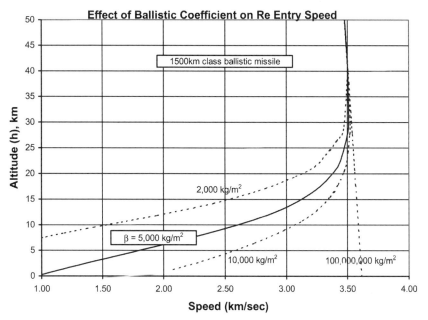

Fig. 4.17 Effect of varying ballistic coefficient on 1500-km-class BM.

the layers in the atmosphere over a spherical Earth with constant altitudes around the globe of all layers.

EFFECT OF THE EARTH'S ROTATION. Most of the analyses of missile trajectories considered here have assumed a nonrotating Earth. To include such an effect introduces much more complex equations of motion and treatment of the trajectories that will now depend on the specific locations of the launch points and azimuth selection for the launch. The characteristics will now depend on the Earth's latitude selected for the launch point.

To include the effect of the Earth's rotation, it is necessary to switch to the treatment of the equations of motion in their vector form, so that all three components (x, y, z) of a point in space can be properly accounted for as the missile moves over the three-dimensional rotating globe. In polar coordinates, if the vector $r(x, y, z)$ represents the position of the missile measured from the center of the Earth, then the acceleration of that missile can be represented by

$$\frac{d^2\mathbf{r}}{dt^2} = \mathbf{g} - \left(\frac{C_D \rho S V}{2W}\right)\mathbf{V} - 2\boldsymbol{\omega}_\wedge \mathbf{V} \tag{4.14}$$

in which the rotation of the Earth $\boldsymbol{\omega}$ is 7.3×10^{-5} radians/s (i.e., 360 deg in a day, West to East, or anticlockwise when looking down on the North Pole). The terms g, C_D, ρ, S, and V are as used earlier. The last term in Eq. (4.14) is the familiar Coriolis acceleration. This additional term, which accounts for the effects of the rotation of the Earth, is a cumbersome expression in terms of the missile coordinates and the Earth's rotation. The format of the expression in terms of the Earth's latitude depends on the choice of convention of the coordinate system.* If it is taken that the z axis is vertical (and away from the center of the Earth), the x axis is toward the East, and the y axis points North, then the Coriolis acceleration can be written as[†]

$$\boldsymbol{\omega}_\wedge \mathbf{V} = \omega(V_z \cos \Omega - V_y \sin \Omega)\frac{dx}{dt} + \omega V_x \sin \Omega \frac{dy}{dt}$$
$$+ \omega V_x \cos \Omega \frac{dz}{dt} \tag{4.15}$$

---

*There is an example in British naval history when, during World War I, the Royal Navy in its first engagement in the southern hemisphere forgot that the effect of the Coriolis acceleration is reversed from that in the northern hemisphere, and in its initial engagement with the German cruisers consistently missed until the correction in the launch coordinates was made. Then the British won the sea battle!

[†]Various texts such as *Fundamentals of Astrodynamics* [34] provide more detail on the equations pertaining to the inclusion of the Earth's rotation in orbital mechanics.

in which Ω is the latitude on the Earth. The speed of the missile now has three components (V_x, V_y, V_z) such that $V = sqrt(V_x^2 + V_y^2 + V_z^2)$. The notation (V_x, V_y, V_z) represents the components of the missile speed $(\mathrm{d}x/\mathrm{d}t, \mathrm{d}y/\mathrm{d}t, \mathrm{d}z/\mathrm{d}t)$, respectively.

It is difficult to provide general conclusions on the effect of the Earth's rotation because the effect varies depending on where the launch point of the missile is on the globe and also on the missile range. If the missile range is short compared to the circumference of the Earth, then the effect is small. If the missile range is long, and especially if it reaches across the poles, the effect can be quite significant.

The significance of the inclusion of the Earth's rotation is best shown by example. Consider the case of a series of missiles with different nominal length trajectories being launched due North, due East, due South, and due West from (as an example) Baghdad. The coordinates of Baghdad are (Latitude 33° 20′ N; Longitude 44° 25′ E). Figure 4.18 shows the results of including such an effect. It is seen that the tendency is for the missile trajectories if launched in a northerly direction to "drift" toward the East; if launched in a southerly direction, the trajectory drifts towards the West. If the missile is launched in either an East or West direction, the effect of the Coriolis acceleration is small—unless the range is large (say 10,000 km or more), in which case sufficiently long flight times have passed such that the

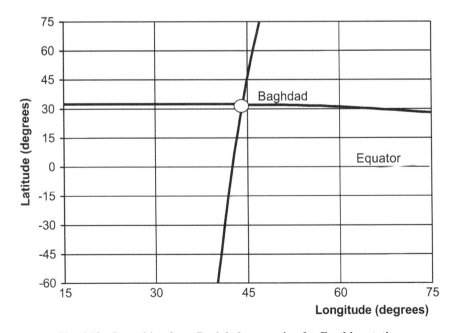

Fig. 4.18   Launching from Baghdad accounting for Earth's rotation.

Earth has rotated a significant amount beneath the missile flight path. When viewed from above the North Pole, this drift effect is opposite to the direction of Earth's rotation, that is, the missile trajectory will tend to drift toward the East when launched in a northerly direction, and drift toward the West if launched in a southerly direction. All of these characteristics of missile trajectories will reverse if the launch point is in the southern hemisphere (as quickly learned by the British sailors in World War I!).

Compared to the missile trajectory results given thus far, the effects of the Earth's rotation tend to either increase or decrease the actual range and flight times (TOF) depending on the conditions of launch. When launched North or South, the missile range and flight time are approximately equal to those values already derived for the nonrotating Earth. If the missile is launched toward the East, the range and flight time will be longer than the nonrotating Earth results. If launched toward the West, the range and flight times will be less than the nonrotating Earth values. These effects are very small for short range missiles but can be significant for long-range missiles. Table 4.3 gives some approximate results for a set of missiles launched from Baghdad with various ranges.

DEFENDED FOOTPRINT

The concepts of *defense in depth* and *keep-out range* and other similar terms used in analysis have one thing in common and that is to determine the "safe boundary" around the asset to be defended. This safe boundary has become known as either the *defended footprint* or simply the *footprint* of the defense system against some required or perceived threat. Definitions of this *footprint* vary among authors, making it difficult to compare results of different analyses or marketing brochures, but the intent is the same. It is generally agreed that this footprint is that which is projected onto the ground so that it becomes a tangible measure of the defense. As used here, the

Table 4.3 Approximate effect of Earth's rotation on missile flights (when launched from midlatitudes in Northern Hemisphere)

Nominal missile range (over a nonrotating Earth), km	Approximate increase in range and TOF if launched eastward, %	Approximate decrease in range and TOF if launched westward, %
300	1–3	1
1000	1–3	1–3
1500	1–3	1–3
5000	6–9	4–7
10,000	~25	~15

footprint is the net result of all defense elements working as a coherent whole defense system. In the literature it is common to also use the word "footprint" to describe the detection boundary of, say, the early- warning satellite, also projected onto the ground. As the satellite circles the globe, it will mark out a swath of detection area on the ground that is frequently referred to as the satellite footprint. In this analysis such areas contribute to the performance of the defense system, but the description *defended footprint* or simply *footprint* will be reserved for the safe boundary as just discussed of the total defense system.

BATTLESPACE AND DEFENDED FOOTPRINT

At the beginning of this chapter, Figs. 4.2 and 4.3 highlighted the concept of "defense in depth" and the "region of applicability of missile intercept." Any particular system composed of early-warning and interceptor and BM/ C³ capability when operating in unison will produce a battlespace with boundaries that will be discussed later. It is the projection of the outer limits of this *battlespace* onto the ground that defines the *defended footprint*. This is illustrated in Fig. 4.19.

From Fig. 4.19, a definition of the (defended) footprint can be stated as *that area projected along the threat trajectories onto the ground of the outer edges of the battlespace of the defense system*. If Fig. 4.19 is taken as a section in the plane of the incoming threat axis, then the most forward edge of the footprint is that of the earliest intercept possible at the intercept

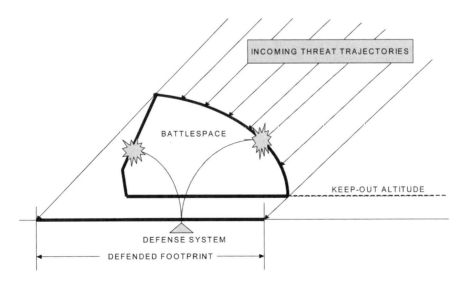

Fig. 4.19 Basic concept of battlespace and defended footprint.

altitude projected along the threat flight path onto the ground. Similarly, the back edge of the footprint is a similar projection along the threat flight path onto the ground of the latest intercept possible by the system. This definition is applicable to both BM and CM threats. The limits of the battlespace will be discussed later. One limit shown in Fig. 4.19 is that of the "keep-out altitude." This is the lowest altitude desired to intercept the threat to prevent the fall-out from any intercept from having a secondary damaging effect on the assets to be defended. Such a requirement is particularly important in the case of intercepting nuclear, chemical, or biological warheads.

If, now, a plan view is taken of the same defended footprint, a shape similar to that shown in Fig. 4.20 is seen. The actual shape in both profile (Fig. 4.19) and plan of the defended footprint (Fig. 4.20) will vary quite widely from system to system and also according to the criteria set for the battlespace boundaries.

The defense system, as shown in Figs. 4.19 and 4.20, is a collocated missile interceptor launcher and radar. It will be shown later that the launcher and radar can be set apart from each other (in a system of remote launchers and sensors) in some systems to provide a larger footprint. The combination of launcher and radar also can be placed in different locations but always in a manner such that the asset to be defended is somewhere within the footprint. The threat axis in Fig. 4.20 is shown coming in from the right of the diagram. If that threat axis rotates to some other angle, the defended footprint will also rotate into a

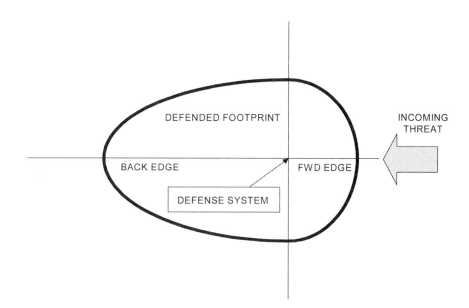

Fig. 4.20   Plan view of defended footprint.

different position. It has become common parlance in the missile defense community to use different names for different size footprints. For those systems that generate footprints of approximately 10–50 km in diameter, the defense is often referred to as *point defense*. ("Diameter" and "radius" are used here as a general sizing term, because footprints are rarely circular in shape.) For larger diameters the term *area defense* (say 50–100 km diam) and *wide-area defense* (greater than 100 km diam) are frequently used. There is no rigid definition of such terms (nor need there be) and often times the wide-area defense is made up of several systems acting in unison to provide protection over a wide area, say the size of a nation. (In Chapter 3 it is shown that most nations in Europe can be circumscribed with circles of approximately 500-km radius.) This is simply a matter of design and tradeoff of several parameters that are discussed in more detail in Chapter 9.

DEFINING THE EDGES OF DEFENDED FOOTPRINTS

Although the general concept of defended footprints is accepted by most in the missile defense community, the actual criteria used to define each of the edges are far from a common set. This is because of the many different types of defense systems in use today and projected for use in the future and also by the lack of common standards. Despite the long history of the use of missiles in warfare (see Chapter 2), it is only in recent history over the last decade that the various governments have sought different schemes to reduce the analyses to a common systems engineering approach. As will be shown later, even the basic treatment of the incoming missile trajectories has been treated differently from the early days of strategic missiles (in the 1980s) to today's treatment of tactical missiles.

It is not possible to develop a closed-form analytical solution for the defended footprint because of all of the disparate functions of intercepting trajectories, radar search volume shapes, differing criteria for the boundaries, and many other factors. Indeed, computer simulations and detailed step-by-step integration processes are required in order to properly assess the impact of the many variables in order to make reliable predictions on the geometry of defended footprints in any particular case. Once the general understanding of the problem at hand has been determined, it is always preferable to use the complete solutions with all key parameters included. However, in the early stages of decision making large computer simulations can hinder the understanding of the problem because it is not intuitively obvious, which are the driving factors. Accordingly, it is useful to make certain simplifying assumptions and to consider the main features that set the general sizing of defended footprints. In what follows, techniques are derived that can be used to "size" the defended footprint initially, as determined by the radius dimensions of front edge, back edge, and the side edge of the footprint. By the use of simplifying assumptions, it is possible to reduce the major parameters to analytical treatment that can provide sufficient insight as to the requirements of any particular system to provide the needed defense. Figure 4.21 shows some of the more prevalent conditions that define the

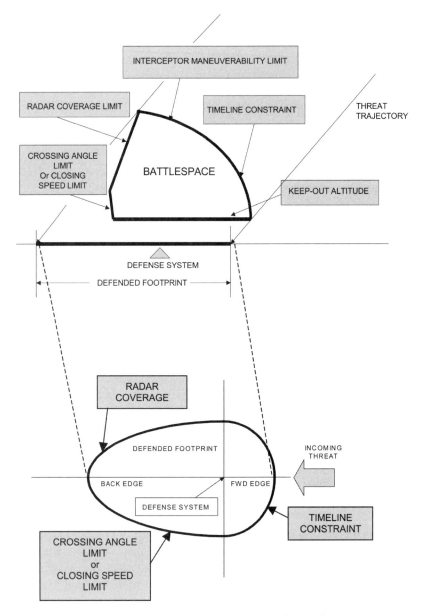

Fig. 4.21   Typical boundary conditions on defended footprint.

edges of most footprints. There are others, as will be discussed, but the main driving limits are as shown.

A brief synopsis of the boundary conditions shown in Fig. 4.21 is as follows:

1) *The keep-out altitude* is the minimum intercept altitude set by operational requirements and the need to nullify the effects of WMD.

2) *The timeline constraint* is the earliest possible intercept as determined by the kinematics of the threat, interceptor, and radar detection range.

3) *The interceptor maneuverability limit* is the lack of sufficient aero-dynamic forces for control of the interceptor (usually a lower-layer inter-ceptor limit).

4) *The radar coverage limit* is determined by physical elevation limits of radar and also by the need to keep the intercept in field of view of the radar (for midcourse guidance and kill assessment).

5) *The crossing angle and closing velocity limit* is the limit set by the lethality parameters of the interceptor. There are many other conditions that can and are set in particular cases, but the preceding broad conditions show the type of constraints that are applied in the determination of any particular footprint of any defense system operating against a set of threat trajectories. As will be discussed later in this chapter, the additional constraints of ensuring that a specific set of survivability levels (see Chapter 3) is achieved over the desired defended footprint will also determine the final size. It is the combination of 1) the area defended and 2) the level to which that area is protected that determines the final characterization of the system defense. Although there are several possible constraints that will determine the specific shape of the defended footprint in any particular instance, the forward edge R_{FE} and back edge R_{BE} (see Fig. 4.21) can be generally determined by the kinematic constraints of the intercept and thus are more amenable to straightforward analysis. These two dimensions of the defended footprint thus serve as a convenient measure of the size and are often used to indicate the defended area of any given system. These two dimensions are often referred to as the forward-edge radius and back-edge radius of the footprint. The side-edge radius R_{SE} is less amenable to simple algorithm treatment but is also a key parameter in the sizing of the footprint (see later).

CHARACTERIZATION OF DEFENDED FOOTPRINTS WITH THREAT RANGE

A major consideration in designing the defense system relates to the radar detection range and whether or not that detection range is greater than or less than the threat range which the system is designed to intercept. This consideration results in two distinct regions with distinct differences in the size of the defended footprint (as measured by the forward-edge radius R_{FE}) as the threat range increases from values less than, to values greater than, the radar detection range R_D. The characteristic shape of how the defended footprint varies with the threat range for most defense systems is shown in Fig. 4.22. This shape is determined by the combination of the kinematics constraints of the interceptor, the threat range and speed, and the detection range of the defending radar. The exact threat range that determines when the footprint increases (left-hand part of the curve) to when the footprint decreases with increase in threat range (right-hand part of the curve) is not exactly at the point of the defense system detection range because of time delays and other factors in the kinematics of the intercept. Generally, the peak in the footprint curve occurs at a lesser value than the radar detection range R_D, as shown.

Unfortunately, it is not possible to derive closed-form solutions for the dimensions of the defended footprint. This is because of the many nonlinear features of the key parameters and the need to seek solutions of intersecting flight paths (usually in three dimensions and time) with a disparate set of constraints such as those just given. However, if the defended footprint area (see Fig. 4.20) is considered to be bounded by a forward-edge radius R_{FE}, a back-edge radius R_{BE}, and a side-edge radius R_{SE} it is possible, with some simplifying assumptions, to provide some reasonable estimates of the likely magnitude of the size and shape of the defended footprint. The forward-edge radius R_{FE} is the closest to being a closed-form solution based on the kinematics of the intercept. The back-edge radius R_{BE} and the side-edge radius R_{SE} can also be approximated to realistic values, but they require a combination of iterative techniques and graphical techniques in order to achieve the desired values.

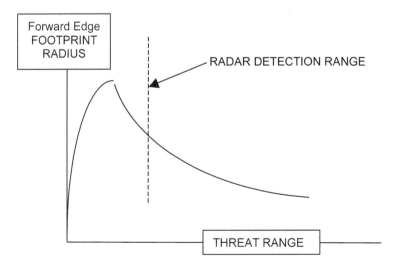

Fig. 4.22   Characteristic shape for forward footprint radius and threat range.

It is always possible, of course, to computerize all of the parameters and allow the computer to do the work for a consistent set of solutions, and this is usually done in any architectural analysis of systems. Although this is a desirable course, especially if time is of the essence, in such an approach it is not always obvious as to which parameter governed the final solution and dimension of the defended footprint area. It is more instructive from a decision-making viewpoint to see how or what part of the system is best to change by examining each dimension (R_{FE}, R_{BE}, and R_{SE}) in turn and accepting the inevitable approximate results. This would be useful, for example, if it were found that one component of the system was very expensive (see Chapter 5) and yet produced little improvement in performance. This might influence the choice of system. Each of these characteristics of the defended footprint will be examined in turn.

FORWARD-EDGE FOOTPRINT RADIUS

The two regions of interest are as shown in Fig. 4.23. In the top panel case the radar is of sufficient range that the missile flies directly into the radar search volume. In the bottom panel the radar detection range is shorter than the threat range, with a time delay before the missile flies into the radar search volume.

It is seen from Fig. 4.23 that the main characteristics that separate the two treatments of the defended footprint radius are the time delay before launch and the kinematics of the intercept. Even though in any actual system the changeover from one region to the other is not a hard or distinct point, they can be analyzed separately. Figure 4.24 shows the geometry of the intercept with the key determining parameters identified.

The particular arrangement shown in Fig. 4.24 is for the case of a collocated defense system (radar and interceptor). It is shown later how this basic arrangement can be modified for the case of remote launcher(s) and sensor(s), which can be used to advantage in certain scenarios. The key parameters are defined here:

R_{FE} = forward edge footprint radius
V_{ix} = average interceptor speed over the ground
V_{Tx} = speed of threat over the ground
t_F = flight time of threat (TOF)
t_d = time to detect threat
t_r = time to react, process, and launch interceptor
h_i = intercept altitude
θ_{re} = rentry angle of threat (at intercept altitude)
θ = (varying) flight-path angle along threat trajectory
γ = (varying) flight-path angle along interceptor trajectory

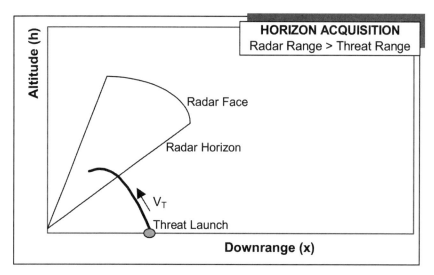

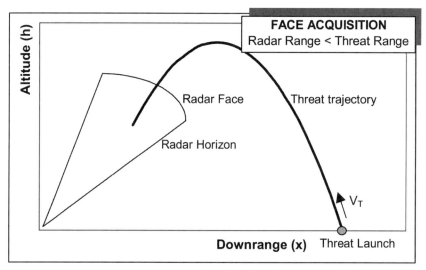

Fig. 4.23   Two regions characterizing forward-edge footprint radius.

R_D = radar (or sensor) detection range
ϕ_1 = radar minimum search angle

CASE 1: RADAR HORIZON ACQUISITION (DETECTION RANGE GREATER THAN THE
THREAT RANGE).   In this case it is easy to show that the footprint radius at the
forward edge R_{FE} of the footprint is given by

$$R_{FE} = V_{ix}(t_F - t_d - t_r) - h_i \cot \theta_{re} \tag{4.16}$$

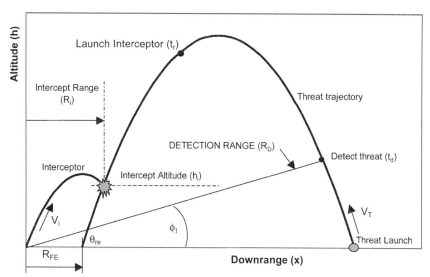

Fig. 4.24   Geometry of intercept (collocated defense).

with all else being constant, it is seen from Eq. (4.14) that, in this region $(R_D > R_T)$,

$$R_{FE} \sim \sqrt{R_T} \tag{4.17}$$

CASE 2: FACE ACQUISITION (DETECTION RANGE LESS THAN THREAT RANGE).   In this case it is easy to show that the footprint radius at the forward edge R_{FE} of the footprint is given by

$$R_{FE} = \frac{R_D \cos \phi - V_{Tx} t_r V_{ix}}{V_{ix} + V_{Tx}} - h_i \cot \theta_{re} \tag{4.18}$$

Although Eq. (4.18) appears as a closed-form solution, the parameters V_{Tx} and V_{ix} actually vary with time throughout the flight, and thus the solution R_{FE} normally requires a step-by-step integration process over time t from launch to intercept. Alternatively, an iterative technique can be used.

The specific parameters V_{Tx} and V_{ix} that define the threat and interceptor flight profiles need some explanation. They are the integrated values along the horizontal x axis of the combined speed and flight angles, such that

$$V_{ix} = \int_{launch}^{intercept} \frac{d(V_i \cos \gamma)}{dt} \, dt \approx K_1 V_{bo} \tag{4.19}$$

$$V_{Tx} = \int_{interceptor \, launch}^{intercept} \frac{d(V_T \cos \theta)}{dt} \, dt \approx K_2 V_T \tag{4.20}$$

As will be shown later in this chapter, the constants K_1 and K_2 can easily be estimated in most scenarios on inspection and with a knowledge of the characteristics of ballistic trajectories. See earlier treatment of trajectories and Appendix C. It will be shown later, especially for the other dimensions (R_{BE} and R_{SE}), that the use of graphical techniques combined with the iterations of the speed–time relationships allows for a good representation of the average ground speeds V_{ix} and V_{Tx}.

Although the relationships for case 2 are a little more complicated than for case 1, there is a trend, given the numerical values of most scenarios, that for this case ($R_D < R_T$)

$$R_{FE} \sim \frac{1}{\sqrt{R_T}} \tag{4.21}$$

It can be seen from the general trends shown by Eqs. (4.17) and (4.21), that the general shaping of the footprint radius plot is as shown in Fig. 4.22.

The crossover point of the two regions (cases 1 and 2) or peak of the curve can then be determined by satisfying the following condition:

$$R_D \cos \phi_1 = (t_F - t_d - t_r)(V_{ix} + V_{Tx}) + V_{Tx}t_r \tag{4.22}$$

This relationship shows the condition for the changeover from case 1 to case 2. No hard value exists because of the many variables that exist in any one particular defense system, but generally the changeover point occurs at that value of threat range R_T that is approximately less than one-half of the radar detection range R_D in most cases.

Although the preceding derivations have been done for the specific case of intercepting incoming BM, the analysis can be adapted easily to the intercept of airbreathing threats (CM and aircraft). Figure 4.25 shows the general nature of the approach to threats other than BM. It can be seen from Fig. 4.25, that the analysis will apply equally well to the intercepts of high-flying, deep-diving cruise missiles (CM) and also aircraft or uninhabited aerial vehicle (UAV)-launching ASM at the target to be defended.

As can be seen from the general relationships given by Eqs. (4.14) and (4.18), the numerical value of the forward-edge footprint radius R_{FE} can vary over a wide range depending on the characteristics of both the threat and the interceptor system (which is made up of the interceptor itself and the radar and any early-warning system). Some typical values, however, help to put the general sizing into perspective.

TYPICAL VALUES OF FORWARD-EDGE FOOTPRINT RADIUS.   Although the values of the forward-edge radius R_{FE} can vary over a wide range depending on the

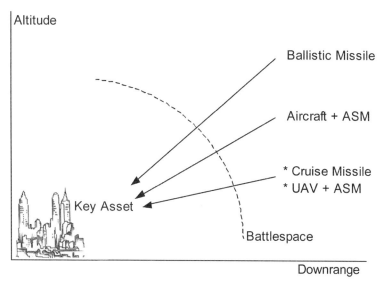

Fig. 4.25   Method applicability to intercepts of most threats.

specific design features of any particular design, it is helpful to see what typical values might produce in the way of footprints to gain an appreciation of both the magnitude and the trends in defended footprints as determined by likely values.

In the set of charts for this typical set, constant values of reentry angle θ_{re} of 45 deg, track time t_r of 20 s, and a radar lower limit search angle ϕ_1 of 7 deg have been assumed. These are not atypical. The intercept altitude h_i has been kept at the value of 15 km for the endo-system and endo-exo-system as an expected value for defense against WMD, and altitude is required to neutralize (as much as possible) the warhead fallout. For the exo-system the keep-out altitude has been raised to $h_i = 83$ km, that is, outside of the discernible atmosphere (see Appendix D) because the state of the art in cooling systems is not expected anytime soon to solve the potential distortion in the interceptor seeker window performance from aerodynamic heating.

Clearly, other values can be inserted in the general relationships just given, but these values shown here serve as typical values to illustrate the trends that are inherent in the intercept design process. Although it is possible to conduct a detailed step-by-step integration process of the respective flight paths of both the interceptor and the threat, it is possible to gain an appreciation of the relative sizing of the defense system by recognizing the general relationships that exist among the various parameters. The preceding analysis requires knowledge of the average interceptor speed V_{ix} that occurs during the flight up to intercept at the desired intercept altitude h_i plus an approximate treatment of the distance–time profiles of the interceptor and threat. A typical speed profile of the interceptor is as shown earlier in the collection of typical missile trajectories (see Figs. 4.6–4.10). Figure 4.26 shows a typical speed profile of the interceptor.

This particular example is for a 500-km class ballistic missile (with $V_{bo} = 2.22$ km/s) and has been taken from the set shown earlier in Fig. 4.7 and Table 4.2, except that a track time ($t_r = 20$ s) has been included to show the speed time profile after launch. The average speed of this interceptor V_{ix} can be determined by numerical integration over the time spent from launch to intercept, but typically it can be shown that to a good approximation $V_{ix} \approx 66\%$ V_{bo} for minimum energy trajectories of greater than 300 km range and for those cases in which the intercept is expected to be in the later stages of the interceptor flight. (This rule of thumb should be used sparingly, as in shown in a later case in which the interceptor is required to be launched in a lofted trajectory to maximize the back edge of the footprint, the ground speed of the interceptor can drop dramatically to values approaching 25% V_{bo} or less. For the shorter-range ballistic missiles the boost-phase time is a greater portion of the total flight time, and this approximation for V_{ix} breaks down.) This greatly simplifies the analysis for general sizing purposes, although it is to be emphasized that this is not a limiting condition of this analysis. One can conduct a more detailed flight profile analysis using the same equations if it were necessary to pursue different variations of flight

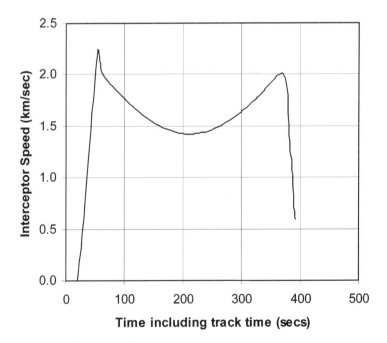

Fig. 4.26 Typical speed profile for interceptor.

profile of either the interceptor or the threat. With this additional assumption it now becomes possible to construct a set of footprint forward-edge radius R_{FE} charts as a function of radar detection range R_D and threat range R_T. Figure 4.27 shows the forward-edge footprint radius R_{FE} for three classes of interceptors, specifically $V_{bo} = 1.30$ km/s, 2.0 km/s, and 4.0 km/s. In each case in Fig. 4.26, the intercept performance is given for several radar detection ranges R_D from 100 to 1000 km. These plots are independent of the actual value of the radar cross section (RCS) of the incoming target. These charts apply once it has been determined that the given radar can indeed detect the target (at the particular value of RCS) at the radar detection range R_D selected. It is a separate design requirement to design the radar that can detect that target RCS at the center frequency selected for the radar. Also, these same charts can be used as a requirements set of charts to design a radar to meet the specified threat RCS at a specified detection range R_D.

The selection of $V_{bo} = 1.3$, 2.0, and 4.0 km/s interceptors was made to bracket the most likely set of interceptors either in operation or in development today. The lower-layer systems (called "endo" because they will spend the bulk of their flight time inside the troposphere) have values of V_{bo} that vary from about 1.0 km/s to less than 2.0 km/s, and a value of about 1.3 km/s represents a good average value for today's systems for the purposes of discussion. From the earlier discussion in this chapter, it is seen that these values represent those systems that operate almost entirely within the troposphere.

For the upper-layer systems (called "endo-exo" and "exo" because of the amount of flight time spent either transitioning or outside the troposphere), there are those systems that tend to operate in *both* the troposphere and the stratosphere (and have to contend with aerodynamic heating in their design) and those that tend to operate *solely* above the discernible atmosphere (to avoid such heating problems).* In Fig. 4.27, the last two examples of $V_{bo} = 2.0$ and 4.0 km/s provide good examples of both these types.

The combination of the three selected examples represents the typical interceptors for use in land and naval applications for both the lower-layer and upper-layer systems envisaged for an active-layered missile defense system (see Fig. 4.3 at the beginning of this chapter). Clearly, specific systems will have different characteristics than these typical values used here for illustrative purposes. In Fig. 4.27 both regions of cases 1 and 2 have been grouped so that one can see the likely regions for any particular case. As

---

*There are various ways of solving the aerodynamic heating problem, but the most common and obvious method is simply to ensure that the missile (or its kill vehicle) retains a protective seeker shroud of some kind that falls away outside the atmosphere and before "looking" at the threat. This feature requires that all intercepts be above the troposphere, which as shown in Appendix D would be above 70 km or more. In Fig. 4.27, the intercept altitude h_i for the exo system, that is with $V_{bo} = 4.0$ km/s, was selected to be at 83 km (to provide sufficient margin to ensure that its seeker would be unhindered during the intercept).

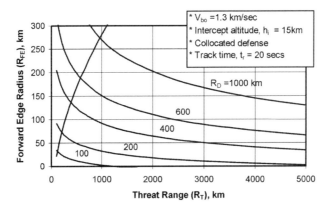

ENDO SYSTEM

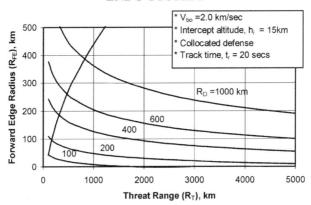

ENDO-EXO SYSTEM

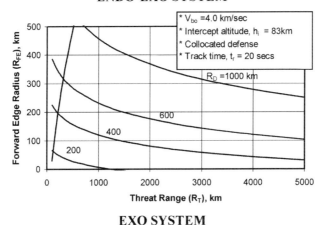

EXO SYSTEM

Fig. 4.27   Forward-edge footprint radius for three classes of interceptors.

noted earlier, it will be seen that the crossover between the case 1 region and case 2 region occurs at some range that is much less than the numerical value of the radar detection range for the reasons already given.

Some examples taken from Fig. 4.27 illustrate the general sizing of defended footprints for possible defense systems. Using illustrative threats taken from Appendix A, the possible point defense, area defense, and wide-area defense around some defended assets are as shown in Table 4.4.

PARALLEL AND POINT LAUNCH OF MISSILES. The use of the forward-edge footprint radius R_{FE} to characterize defended footprints has been in use since the early days of the SDI program in 1983. The complexities of generating defended footprints were recognized early on, and accordingly an important assumption was made at that time that has remained to this day in several parts of the literature. It is the assumption of "constant range." Guided by the knowledge that the strategic missiles were of very long range ($\sim 10,000$ km or more), it was decided at that time to conduct all analyses with this assumption.

Because the problem under evaluation at that time was the defense against long-range threats from one nation (Soviet Union) with a large geographical area toward another large nation (United States) again with a large geographical area that was thousands of kilometers away, the assumption of *constant-range* threats (and *parallel* trajectories) was not an issue. However, as the strategic problem gave way to a more tactical defense

Table 4.4   Defended footprint radius of illustrative (collocated) systems

	Defended footprint radius, R_{FE}	
Defense system	1000-km class BM threat (e.g., Al Abbas)	3000-km class BM threat (e.g., CSS-2)
Point defense (endo)		
$V_{bo} = 1.3$ km/s	30 km[a]	10 km
$R_D = 200$ km		
Area defense (endo-exo)		
$V_{bo} = 2.0$ km/s	360 km	230 km
$R_D = 1000$ km		
Wide-area defense (exo)		
$V_{bo} = 4.0$ km/s	525 km	380 km
$R_D = 1000$ km		

[a]This is a good approximation of the quoted performance of the lower-layer land-based interceptor systems PAC-2 GEM[+] (guidance enhanced missile)/PAC 3 for nominal values of detection range and threat range. See Ref. 35 for a comparison of the Patriot System performance in Desert Storm and the projected improvements expected in future developments.

problem in which the missile ranges were much shorter and the size of the nations launching the attack were much smaller, this assumption was no longer valid.

It made the analysis easier and allowed the analysts to provide solutions of a missile defense system against a specific (constant value) range threat. *It meant, of course, that the threat launch sites envelopes must equal the defended footprint area, as shown in Fig. 4.28.*

If one used the constant-range assumption in today's tactical scenarios, one would be immediately faced with the illogical analytical solution that the required matching of the defended footprint area with the envelope of the launch sites often produced launch site envelopes that were larger than the threat nation's geographical area!

Hence a new approach was required that led to the assumption of single-point launch as shown in Fig. 4.29. As can be seen from Fig. 4.29, this new approach was much more realistic for the tactical missiles; now the defense had to defend against missiles that emanated from a given threat launch site.

Although more realistic, the difficulty incurred by this approach was that it meant that the defense system was now not tied to a specific (constant) range threat but a *varying* range that varied from the shortest range, which determined the forward edge of the footprint R_{FE}, to the longest-range threat, which determined the back edge of the footprint R_{BE}. The forward-edge radius R_{FE} is the same, of course, for both the constant-range and the single-point launch assumptions. The results differ as the points are calculated around the outer edge of the footprint from the front to the sides to the back. It is interesting to compare these two approaches in a particular case. Figure 4.30 compares the results of a series of footprint calculations for two approaches (i.e., one set of calculations that assumed a constant range and one

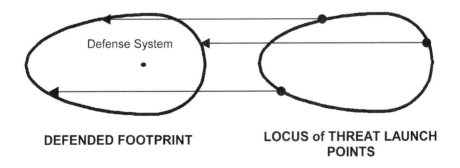

DEFENDED FOOTPRINT   **LOCUS of THREAT LAUNCH POINTS**

All threats of same range

Fig. 4.28   Defended footprint with constant-range threats.

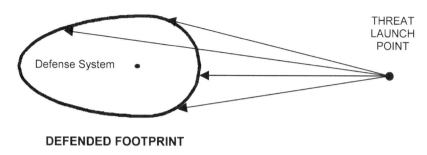

DEFENDED FOOTPRINT

Threats of various ranges

Fig. 4.29 Defended footprint against a single-point launch.

set that assumed the threat range varied but launched from a common point). This particular set of curves was for the case of interceptor $V_{bo} = 2$ km/s, a radar detection range of 500 km, and a response time t_r of 20 s.

To show the difference, it was necessary to calculate the actual area using another common artifice in footprint analysis: the *encapsulated radius* approach. This is simply a circle with a radius that fits inside the defended area to some reasonable degree of accuracy. The encapsulated radius approach will usually be larger than with just the forward-edge radius R_{FE}

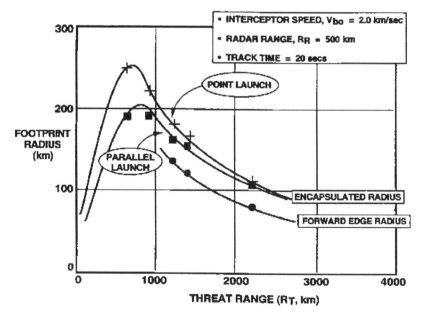

Fig. 4.30 Comparison of parallel and single-point launches.

and in many ways more representative of the total defended area, except that it is not amenable to direct calculation, which is the case of the R_{FE} calculation. [Each method of defended footprint calculation has its pros and cons. As will be shown later, the size of the footprint changes more drastically because of operational considerations (considerations of changes in threat axes; many-on-many raid sizes, etc.) than because of these basic assumptions that have been used by various authors over the last two decades.] The characteristic shape of the defended footprint radius curve just discussed in connection with the earlier Fig. 4.20 is seen to be retained for both the assumption of constant-range and for single-point launch.

As might be expected, the two methods approach a common solution as the range increases to strategic values. [See Chapter 2 for further discussion of ranges of strategic missiles and their implications in the now defunct Antiballistic missile (ABM) Treaty. This continuum illustrates the earlier arbitrariness of classifying missiles as either "strategic" or "tactical."] It has been found by many different iterations of this approach that the departure of the two results is the greatest for short-range threats (say, less than 1000 km), but the results start to converge as the range increases to 3000 km or more. This confirms the original assumption in the 1980s that as the threat reaches "strategic values" the assumption of constant range and parallel launches is a good representation of the defended footprint area—provided the attacking nation has sufficient land area to accommodate the envelope of launch sites! [In Chapter 2 it is noted that the original (now defunct) SALT/START agreements between the United States and the (former) Soviet Union had agreed that "strategic" meant missiles that could reach from the western boundary of the Soviet Union to the east coast of the United States, a distance of 5500 km.]

SEPARATING RADARS AND LAUNCHERS FOR IMPROVED FOOTPRINTS. It is often found that separating the radar from the interceptor launcher such that the defense system is no longer collocated can provide much larger defended footprints. [The problem with such an approach is that if the radars are placed too far forward, they become unprotected and targets in themselves. This could lead to the need for additional interceptors forward to protect the forward (remote) radar, and would have to be considered in any defense architecture.] This is a matter of both design and of operational needs. Most interceptors that are designed to date require that the interceptor, once launched, remain in the field of view (FOV) of the radar most of the time, especially in the end game at the intercept. Sometimes, the interceptor can be launched and then allowed to fly into the FOV of the radar, but invariably the system is designed such that during actual intercept the interceptor must already be in the FOV for guidance, end-game adjustments, and for kill assessment (to determine for example if a second shot is required). This can be accomplished in both possibilities of either forward-based radars or forward-based (remote) launchers.

FORWARD-BASED RADARS

If the radar is placed ahead of the interceptor launcher a distance of R_{FR}, then the forward-edge footprint radius is given by

$$R_{FE} = \frac{(R_D \cos\phi - V_{Tx}t_r)V_{ix}}{V_{ix} + V_{Tx}} - h_i \cot\theta_{re} + \left(\frac{V_{ix}}{V_{ix} + V_{Tx}}\right)R_{FR} \qquad (4.23)$$

The first two terms provide the value of the footprint given earlier for a collocated defense system. The third term accounts for the increase in defended footprint radius by placing the radar forward of the interceptor launcher by a distance R_{FR}.

FORWARD-BASED LAUNCHERS

If now the interceptor is placed forward and the radar remains at the defense site, then the forward-edge defended footprint radius is given by

$$R_{FE} = \frac{(R_D \cos\phi - V_{Tx}t_r)V_{ix}}{V_{ix} + V_{Tx}} - h_i \cot\theta_{re} + \left(\frac{V_{Tx}}{V_{ix} + V_{Tx}}\right)R_{RL} \qquad (4.24)$$

The equations for the increase in defended footprint radius whether by placing the radar forward a distance R_{FR}, or placing the interceptor launcher forward by a distance R_{RL} are very similar except that the influence of speed in each case is different by the ratio of the interceptor and threat speeds V_{ix}/V_{Tx}. Thus, which method provides the greater footprint depends on the speeds of both the interceptor and the incoming threat. The footprint radius R_{FE}, in this derivation, is always measured from the most rear component of the defense system (radar or interceptor).

DETECTION TIME AND REACTION TIME.   The time to detect the threat t_d and the time for the defense to react t_r are key parameters that are not directly calculable in any closed-form solution and depend on many defense system characteristics and the geometry of the defense system lay-down. There are some basic features that facilitate the analysis. The detection time t_d is the most complicated in that there are several considerations. In the derivation of the equations for the determination of the forward-edge footprint radius R_{FE} and in the later derivations, it is assumed implicitly in the equations that the detection is instantaneous in that it is implied if the radar is in the proper position, which it detects immediately (or at least as the threat arrives within range). In actuality, the radar might not be searching its volume at that precise moment in the expected time of detection t_d, and further time is required to identify and "classify" the threat before the defense system reacts (in time t_r). If the search beam of the radar were in the right position at the right time, the system would still have to wait until the threat burn time is complete and then also wait until the threat has been identified as heading in the direction of the defense system.

So the minimum detection time t_d is really something more like

$$t_d = t_{bo} + \text{time to classify threat direction} \qquad (4.25)$$

These times can be quite long depending on the geometry of the defense site in relation to the threat site (which must also take in the effects of radar horizon), the burn time of the threat t_{bo}, and the time taken to determine the direction of the threat. As an example, if the radar is cued by a satellite before it locks in its own detection range R_D, then the processing time of the satellite must be included in the detection time. Normally, three points on a curve are needed to ascertain the direction of the threat and if the (revolving) satellite takes 10 s say, to revisit the threat, then it should be expected that about 30 s would need to be included in the detection time. This time plus the time taken to complete the burn time t_{bo} of the threat is the final detection time. Frequently, this time is greater than that given by a simple geometric overlay of the radar. This will be used later in the example calculation.

BACK-EDGE FOOTPRINT RADIUS

The forward-edge radius R_{FE} already discussed is the easiest to represent in some form of readily analyzable approach, even though approximations have to be made on the speed profiles of interceptor and threat. The simplified approach for the determination of the forward-edge radius R_{FE} was caused in part by the constraints being determined almost solely by the kinematics of the intercept. This is not the case for the back edge of the footprint measured by the back-edge radius R_{BE}. A similar problem exists for the side-edge radius R_{SE}, which is discussed later.

As the incoming threat approaches the defended area and flies overhead, many different constraints come into play. These constraints include a minimum closing velocity between the threat and the interceptor. One can immediately visualize the problem of the interceptor being placed in a "tail chase" of the threat. Lethality constraints bring in the issue of the bearing or crossing angle between the threat and interceptor, which again becomes an issue as the threat moves toward the back edge of the defended area. The physical constraints of the radar also play a large part, in that if the radar has been designed to look forward to maximize the forward-edge radius R_{FE} then the radar is usually physically constrained from having the required 180 deg in elevation angle in order to detect the threat over such a wide area. Electronic scanning can minimize this problem. Many more constraints are brought about by both design and operational requirements.

Of all of the constraints, the normal requirement to keep the actual intercept in the field of view of the radar allows for an approach to determine the likely values of the back-edge radius R_{BE}. This requirement is sometimes tied to the design features of the interceptor, but it is also expected that for kill assessment and the need to determine whether a second shot is required or not that keeping the intercept in the radar FOV is a ready means of determining the back edge of the defended footprint.

The approach for determining the back-edge radius R_{BE} is a little different from the case for the forward-edge radius R_{FE} in that now the launch point of the threat is fixed at some distance D ahead of the defended area, and the problem is to determine how high the threat trajectory can be before it overflies the radar detection range R_D when set at the highest elevation angle ϕ_2 of the radar.

In what follows, the constraint of a keep-out altitude h_i is no longer necessary and does not appear in the equations. As will be seen in a following example, the intercept will occur at very high altitudes automatically as determined by the maximum range and maximum elevation angle of the radar.

Following a similar approach to that used before, it is easy to show that, in this case, the back-edge radius R_{BE} can be determined through ensuring three key conditions are satisfied, which are

$$V_{Tx} = \frac{V_{ix}(D - R_i)}{R_i + V_{ix}(t_r + t_d)} \tag{4.26}$$

$$R_i = R_{FR} - R_{D\,max} \cos(\pi - \phi_2) \tag{4.27}$$

$$R_{BE} = R_T - D \tag{4.28}$$

These general equations account for a forward-based radar R_{FR} and the defense site (of the interceptor launcher) being a distance D from the threat. The horizontal range of the interceptor, at intercept R_i, is given by Eq. (4.26). On inspection it can be seen from the preceding relationships that for a given set of defense site conditions (V_{ix}, D, R_{FR}, ϕ_2, and $R_{D\,max}$) there is a unique solution for the threat V_{Tx}. Note that the defense system reaction time t_r is the amount of time after detection time t_d. This should not be surprising because it is really a geometric constraint that the intercept should occur at the maximum detection range of the radar $R_{\text{D max}}$ set at the maximum elevation angle of the radar ϕ_2. It represents the last intercept that can be "seen" in the FOV of the radar.

The problem with the preceding set of equations is that they are only consistent provided the flyout times of both interceptor and threat can be matched to the actual trajectories. The equations are consistent when the (x, y, t) profiles of the interceptor match the (x, y, t) profile of the threat at the precise time and altitude required. If a computer simulation approach is not used, this can be accomplished easily by a graphical technique using the set of trajectories provided earlier in Figs. 4.6–4.10 and the equations in Appendix C. This will be illustrated by example later.

SIDE-EDGE FOOTPRINT RADIUS

As for the back-edge radius R_{BE}, it is necessary to include graphical techniques in the solution of the side-edge radius R_{SE} if computer simulations are not used. The preceding analysis has shown how the forward-edge footprint radius R_{FE} and the back-edge footprint radius R_{BE} can be used to indicate the general shape of defended footprints.

However, these simple measures might not suffice in all scenarios. It was shown earlier that the battlespace was affected by the form of the intercept on the sides of the footprint and that the crossing angle between interceptor and threat could have a strong influence on the dimensions of the protected area or defended footprint in this part of the battlespace. This crossing angle has been shown to have a role in determining the lethality of the interceptor especially if it is of the hit-to-kill type with no warhead or fragmentation capability. Figure 4.31 shows the concept of the crossing angle and the two related angles of look angle and strike angle.

The reader is referred to other texts for more detailed treatments on the lethality of warheads (both fragmentation and hit-to-kill) and the roles played by both the relative velocities and the crossing angle between interceptor and threat (for example, see Ref. 36). Here, attention will be given to how the end result affects the shape of defended footprints.

The two trajectories of the interceptor and threat cross at an angle Ω that will be called the crossing angle. As the interceptor approaches the threat it "looks" at the threat at an angle α to the trajectory and then "strikes" the threat at an angle β as shown. It is known by all boaters, pilots, and motorists alike that if the angles α and β remain constant as time proceeds then there will indeed be an intercept! The crossing angle Ω is the sum of the two angles (α and β). This concept can be used to advantage in structuring the intercept geometries on the sides of the defended footprint using the crossing angle Ω as the criterion.

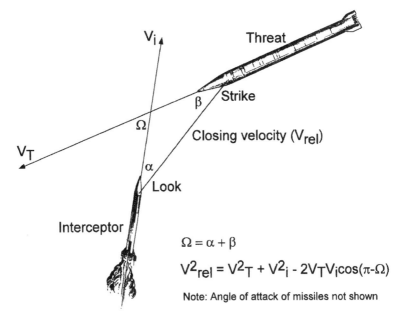

$$\Omega = \alpha + \beta$$

$$V^2_{rel} = V^2_T + V^2_i - 2V_T V_i \cos(\pi - \Omega)$$

Note: Angle of attack of missiles not shown

Fig. 4.31   Crossing angle in end game of intercept.

Typically, if the crossing angle Ω is too small or too large the inaccuracies in targeting can cause a missed intercept. The best value is a matter of individual system design but values in the range of $\Omega \approx 25–120$ deg can be regarded as typical, at least for the purposes of illustration. These limits are determined by different features such as the degree of ability to "see" forward by the interceptor seeker in its nose and the physics of the energy exchange between the interceptor and threat to ensure a high value of kill. This last feature will vary depending on whether the threat warhead contains conventional, nuclear or chemical or biological agents.

If the preceding concept of crossing angle from the end game is applied to the flight trajectories themselves, it is possible to gain an appreciation of the limits to the sides of the defended footprint. Using similar techniques to those used earlier to determine the front-edge footprint radius R_{FE}, it can be shown that by equating flight times of interceptor after launch to the flight time of the threat between interceptor launch and intercept, the projected interceptor range in the horizontal plane can be determined from

$$R_i = \frac{(R_{FR} + R_D \cos \phi_1 \cos \varepsilon - V_{Tx} \cos \beta t_r) V_{ix} \cos \alpha}{V_{ix} \cos \alpha + V_{Tx} \cos \beta} \tag{4.29}$$

This equation shows the key parameters of the defense system (R_D, V_{ix}, R_{FR} and t_r) together with the threat V_{Tx}. In Eq. (4.29), the interceptor flies out on a trajectory with an azimuth angle α in the horizontal plane to intercept the threat trajectory that is on a flight path at its azimuth angle β.

For the case $\alpha = \beta = \varepsilon = 0$, Eq. (4.29) reduces to the case for the intercept in the plane of the threat that results in the determination of forward-edge footprint radius R_{FE}. This equation is deceptively simple in that the parameters in the equation are not independent, as will be shown in what follows.

From the law of cosines, the projection of the intercept range R_i is a quadratic function as shown in the following equation:

$$\left(\frac{R_i}{\cos \alpha}\right)^2 = D^2 + \left(R_T - \frac{R_i}{\cos \beta}\right)^2 - 2D\left(R_T - \frac{R_i}{\cos \beta}\right)\cos \beta \qquad (4.30)$$

in which the left-hand side of the equation represents the square of the flyout range of the interceptor at intercept. This equation is valid provided α, $\beta \neq 0$. The geometry of the intercept is shown to be a function of the distance D between the threat and the interceptor launcher, the threat range R_T and the azimuths of the intersecting trajectories (α, β). This shows that the intercept range R_i is a quadratic equation in the intercept geometry parameters. The two roots of this equation provide the required relationship between α and β, but the algebra is quite cumbersome. If access to a computer that can provide the necessary high-speed iterations is not available, it is possible to do it manually even if somewhat laboriously! Here a graphical technique is used to the advantage (in absence of the complete computer simulation) to ensure that the trajectory parameters are treated consistently.

Additionally, the side-edge footprint radius R_{SE} that results from this flyout intercept range $(R_i/\cos \alpha)$ is given by the geometry of the intercept:

$$R_{SE} = \frac{R_i}{\cos \alpha}(\sin \alpha + \cos \alpha \tan \beta) \qquad (4.31)$$

The radar detection range R_D must also satisfy the following relationship.

$$R_D \cos \phi_1 \cos \varepsilon = (D - R_{FR}) - V_{Tx} \cos \beta t_d \qquad (4.32)$$

In this derivation the placement of the radar at a distance R_{FR} ahead of the interceptor launcher has been assumed. The radar detects the threat at a radar elevation ϕ_1 and azimuth ε. The azimuth ε must be within the azimuth search limits of the radar (see later). Other variations are possible but omitted here for clarity and with the intent to show only the type of setup that can be developed rather than all possible combinations.

The new parameters introduced by the preceding equations are the distance D of the defense site from the threat, the azimuth of the threat trajectory β, and the interceptor launch azimuth α. At intercept, the relationship $\Omega = \alpha + \beta$ holds true, and it is assumed that these angles hold throughout the latter parts of the end game, that is, no midcourse guidance is required. It is possible to combine all of these equations algebraically, but the result is cumbersome and also tends to lose the insight that can be extracted from examining each in turn.

In any actual case, Eqs. (4.29–4.32) require step-by-step integration procedures that are standard in any computerized analysis. When possible, it is always best to generate the solutions on a computer where the needed iterations can proceed without the frustration of manual solutions. Alternatively, one can use iterative procedures in a spreadsheet until all parameters come within stated ranges set by the designer. This approach was taken in the following example by using the trajectory parameters developed earlier for both the interceptor and the threat trajectories.

EXAMPLE PROBLEM SHOWING FORWARD, BACK, AND SIDE EDGES OF A DEFENDED FOOTPRINT

As stated earlier, it is not possible to generate closed-form solutions of the various dimensions of the defended footprint because of the time-varying aspects of the intercept and the many constraints that apply. However, the preceding equations do provide a means of making rapid estimates of the forward-edge radius R_{FE}, back-edge radius R_{BE}, and side-edge radius R_{SE} provided some approximations are made on the characteristics of the trajectories, which can be made easily using the derivations provided earlier in this chapter and in Appendix C.

EXAMPLE PROBLEM 4.1—DEFENSE OF EUROPE

In this example assume that it is required to defend against a set of ballistic missiles of varying ranges R_T that are expected to be launched from a site that is 1900 km away from the defense site. The defense system is made up of a launcher at the defense site equipped with interceptors that have a V_{bo} performance capability of 4.0 km/s. Because of the high-speed features of this interceptor, it is restricted to making intercepts above the atmosphere to avoid aerodynamic heating from obscuring its seeker window.

An intercept altitude h_i of 83 km has been selected. A surveillance and fire control radar with a maximum detection range ($R_{D\ max}$) of 1000 km is placed forward approximately 900 km from the defense site. (To keep the calculations simple, it is taken that it is known that this radar can indeed detect the incoming missiles with their inherent RCS values at the stated detection range of 1000 km.)

The pertinent characteristics of the radar are as follows:

Low elevation angle (for search), $\phi_1 = 7$ deg;
Maximum elevation angle, $\phi_2 = 129$ deg (assumes electronic scanning beyond the boresight);
Maximum detection range, $R_{D\ max} = 1000$ km; and
Initial detection time, $t_d = 78$ s. *

The pertinent characteristics of the interceptor are as follows:

$V_{bo} = 4.0$ km/s,
Launcher located at the defense site, and
Launch angle γ determined from time to intercept at $h_i = 83$-km altitude.

The pertinent characteristic of the battle management system is reaction time, after detection, to launch interceptor, $t_r = 20$ s. The pertinent characteristics of the threat are minimum energy trajectories, with speeds from 2.90 to 5.80 km/s.

FORWARD EDGE OF DEFENDED FOOTPRINT
From Eqs. (4.16) and (4.18), the value of R_{FE} can be calculated. The results are shown in Fig. 4.32 for several ranges R_T of the threat. The particular case in this example (threat speed, $V_T = 2.90$ km/s) results in a value of $R_{FE} = 1006$ km. As can be seen from Fig. 4.32, this value of R_{FE} is sensitive to the warning and reaction time t_r. Also, Fig. 4.32 shows the effect of this time t_r varying from the example of 20 s to 120 s to indicate the impact of long track times for the defense system to react.

In this example calculation of the forward-edge footprint radius R_{FE}, it can be seen that changes in the reaction or track time t_r can impact the expected radius. This example was constructed to show a defended footprint in which the threat range R_T was close to the crossover point of 1 being already in the FOV of the radar to 2 having to wait until the threat missile entered the defense system radar. In this particular example, if the track time t_r would be increased from 20 s to 120 s, the forward-edge footprint radius R_{FE} would drop to about 859 km (about 15% drop in protected radius).

Once the initial estimate of the forward-edge footprint radius R_{FE} has been made, it is possible to determine what type of interceptor trajectory must have been used in order to intercept at the desired intercept altitude h_i.

---

*As noted earlier, this detection time is much longer than the geometry would indicate (i.e., $t_d = 4.2$ s) because of the burn time ($t_{bo} = 45$ s) and the need to have three hits by a cueing satellite (for an additional 30 s) to ensure that the threat was indeed headed for the defense site and thus required an interceptor launch. Also, whether done graphically or by simulation, there was the additional constraint to ensure that the intercept was at a high altitude ($h_i = 83$ km) to ensure that the interceptor seeker window did not overheat through aerodynamic heating.

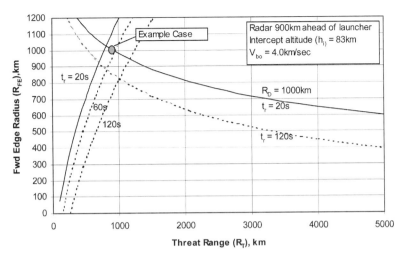

Fig. 4.32   Example forward-edge footprint radius.

This can be done by using the trajectory characteristics developed earlier using the speed V_{bo} and flight time information. In this particular case, it is found that the initial trajectory angle at burnout θ_{bo} would have to be about $\gamma_{bo} = 19$ deg, which indicates that the interceptor was on a depressed trajectory in order to intercept the threat ($V_T = 2.70$ km/s) in time at the desired intercept altitude h_i of 83 km. Such calculations done after the basic footprint analysis provide insights on the required features of the defense system.

BACK EDGE OF DEFENDED FOOTPRINT

On inspection of the relative geometries of the defense site location in relation to the threat and the high elevation angle ϕ_2 of the defense radar, it should be expected that any intercept of such long-range threats would require the interceptor to be lofted at a high angle. This proves to be the case. Using Eqs. (4.25–4.27) and the graphical matching of the trajectories, it is found that the back-edge footprint radius R_{BE} is approximately 2700 km against a long-range threat with $V_T = 5.80$ km/s. Figure 4.33 shows the geometry of this example set of calculations with both the trajectories that determine the forward-edge footprint radius R_{FE} and the back-edge footprint radius R_{BE}. For clarity, the location of the forward-based radar (at distance of $R_{FR} = 900$ km) ahead of the interceptor launcher and defense site) has been omitted from Fig. 4.33.

Compiling both the forward-edge and the back-edge footprint radius values into Fig. 4.33 dramatically shows how the defended footprint is constructed and how the interceptor geometries are quite different. In this constructed example the interceptor had to be launched on a depressed

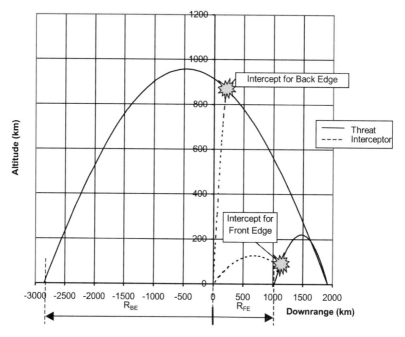

Fig. 4.33   Example showing both forward- and back-edge footprint radius.

trajectory ($\gamma \cong 19$ deg) to intercept the shorter-range threat and then on a lofted trajectory ($\gamma \cong 79$ deg) in order to intercept the longer-range threat.

It should be cautioned that this method for estimating the back-edge footprint radius is less reliable than that for the forward-edge radius, because of the sensitivity of the geometry in attempts to match the intercept at the precise point of $R_{D\,max}$ at the high elevation angles shown and to match all trajectory parameters in a graphical technique. Small changes in the key parameters (such as V_{ix} and V_{Tx}) can cause large changes in the endpoint on the ground R_{BE}, but the trends are representative for preliminary indications of the probable size of defended footprints.

SIDE EDGE OF DEFENDED FOOTPRINT

The parameter β is used as the independent variable to guide the calculations. That is, the threat axis is rotated away from the line connecting the defense site with the threat launch site at varying amounts of azimuth β with corresponding increase in threat ranges R_T to satisfy the constraints of the problem. The width of the defended footprint R_{SE}, measured at the defense site and perpendicular to the $\beta = 0$ threat line, is then determined from the preceding equations.

There are several solutions to the preceding equations but only one that will satisfy matching the interceptor $V_{bo} = 4.0\,km/s$ and the fixed *parameters of radar detection range R_D and distance from the threat site D. There are two solutions, one of which will maximize the fly-out range of the interceptor for the combination of α and β that gives a reasonable value of the crossing angle Ω. The iterative procedure on Eqs. 4.29–4.32) rapidly converges to the following solution. The defense system includes the following elements:*

> *R_{SE} of 2071 km,*
> *t_r of 20 s,*
> *Ω of 72 deg,*
> *β of 42 deg,*
> *α of 30 deg,*
> *R_D of 1000 km, and*
> *D of 900 km.*

The threat range varies from $R_T = 895$ km at $\beta = 0$ deg (and matches that for R_{FE}) to $R_T = 3000$ km at $\beta = 42$ deg to give value of $R_{SE} = 2071$ km.

If this result is now combined with the earlier results for the forward-edge footprint radius R_{FE} and the back-edge footprint radius R_{BE}, the defended

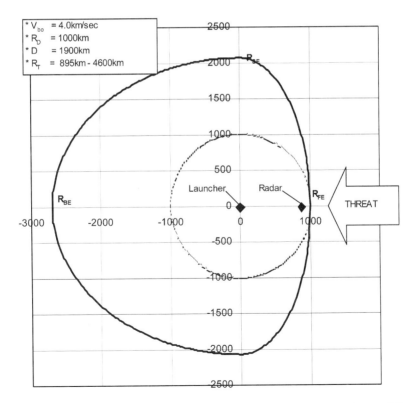

Fig. 4.34   Example defended footprint with the three bounds R_{FE}, R_{BE} and R_{SE}.

footprint for this example is then defined by these three footprint dimensions:
$R_{FE} = 1006$ km, $R_{BE} = 2700$ km, and $R_{SE} = 2071$ km. Figure 4.34 shows
this new representation, in which the additional criterion of crossing angle Ω
has been taken into account.

Assuming that these three bounds of the defended footprint are connected
by ellipses, the defended footprint area A_{FP} is given by

$$A_{FP} = \frac{\pi}{2} R_{SE}(R_{FE} + R_{BE}) \qquad (4.33)$$

With the computed values of R_{FE}, R_{BE} and R_{SE}, the defended footprint area
A_{FP} becomes approximately 12.1 million km². Superimposed on Fig. 4.34 is a
circle with radius R_{FE} for comparison and later discussion.

Note that the defended footprint is now quoted as the defended area
against a range of threats emanating from one threat site (Sabha in this
example) that vary from the more realistic range of threats ($R_T = 895$ to
4600 km), rather than a constant value as would be given by the parallel
footprint approach.

FOOTPRINT SUPERIMPOSED ON EUROPE

The preceding example of a simplified defended-area footprint shows
that it is possible to prepare a rapid construction of the performance
capabilities of a given defense system by using rapid spreadsheet techniques
and knowledge of ballistic trajectories in a variety of situations. There are
many shortcomings from such methods in that they do not take into account
all factors, constraints, or many of the operational scenario architectures, but
they do provide a rapid means for first evaluation. It is possible to introduce
the various modifications necessary to the preceding approach by taking the
preceding defended-area footprint and superimposing it on a map of Europe.
(Because the footprint radius calculations were done in a two-dimensional
plane, there will be some loss of precision when transferring the results to the
spherical surface of the Earth. In this example, the results were superimposed
on a modified polyconic projection map of Europe to minimize the
distortion.) This has been done in Fig. 4.35.

In the constructed example it was taken that the defense site was 1900 km
from the threat site and that the radar was 1000 km from the threat site (or
900 km from the defense system launcher). If the threat site is taken to be
somewhere in the center of Libya (near Sabha for example), the 1000-km
detection range radar to be on the southern part of Sicily (say near Ragusa)
looking south across the Mediterranean, and the interceptor onboard a ship
cruising in the Ligurian Sea (north of Corsica), this would match the
geometry of the example scenario reasonably well. Such a superposition

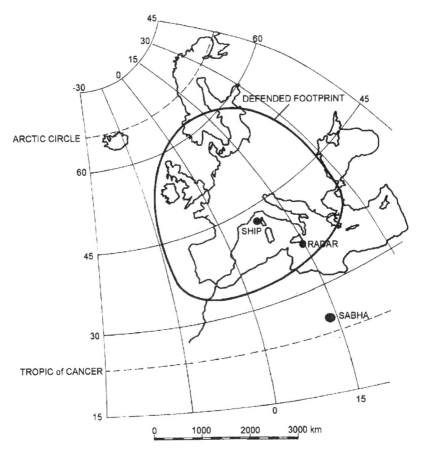

Fig. 4.35   Example of a defended footprint superimposed on Europe.

allows for an appreciation of the scale of the defended footprint. In the constructed example all sites (threat and interceptor) were very nearly in the same plane, which is rarely the case in actual scenarios. Although not done here, in a simple extension one can modify the formulation to bring in the effects of flight out-of-plane and to include the effects of different azimuths of each component. As discussed later, this apparent shortcoming is not as limiting as might be expected when one considers the more likely scenario of many-on-many and varying threat axes in any real scenario.

Another feature that can be seen in Fig. 4.35 is the potential for a flexible defense system in which the launcher is onboard a ship that can cruise to different locations depending on the location of the threat axis, but can still use the forward-based land radar. The use of the rapid techniques shown here and in the constructed example allows for quick appreciation of the likely defense architecture before more detailed analyses are embarked upon.

Two conditions that were used in the preceding example highlight some important aspects of designing systems to provide some stated or required defended footprint area. The first is the requirement to keep the intercept altitude h_i at 83 km. This was not included in the derivation of R_{SE} because it would have restricted all intercepts to a very high value of α, and this was considered not useful in this context. Fortunately, to maximize the value of R_{SE} the intercepts would automatically be at much higher altitudes and thus not constrained by the intercept altitude limit. The intercept altitude h_i increases from the value of 83 km for the forward-edge radius R_{FE} to over 850 km for the back-edge radius R_{BE} (see Fig. 4.33). A check on the trajectories used for the determination of the side-edge radius R_{SE} will show that the intercept altitude would be above 500 km. If the intercept altitude would be restricted to $h_i = 83$ km for R_{SE} determination, then the interceptor would have to be on a flight path with values of α in excess of 85 deg. This would also mean that the intercept would be far outside the FOV of the radar.

The second condition that was assumed was to drop the requirement to keep the intercept in the FOV of the radar. If that condition were retained, the footprint radius R_{SE} would be smaller. This can be seen most readily in Fig. 4.36, where the geometry of the example intercept is displayed.

The intercept geometry of the example problem is shown displayed on a map of Europe to illustrate a sense of scale. The interceptor launcher is shown onboard a ship in the Ligurian Sea north of Corsica. The (land-based) radar in Ragusa, Sicily, is also shown, as is the assumed threat of varying ranges coming from Sabha, Libya. The forward based radar (900 km from the ship) is shown with a typical ± 60 deg azimuth search angle.

It is immediately seen that the threat launch has been observed by the radar and that the interceptor has been launched (at time $t_r + t_d = 98$ s) within the FOV of the radar at Ragusa. The threat (of range R_T) has traveled approximately 300 km by that time. The intercept, however, has occurred just outside the FOV of the defense radar (in Ragusa), as shown. The side-edge footprint radius R_{SE} is shown in the approximate location as indicated in Fig. 4.35. In this illustration, only an approximate adjustment of the two-dimensional representation onto the polyconic projection map of Europe has been made for simplicity.

The result that the intercept has occurred outside of the radar FOV does not necessarily invalidate the solution because in an actual architecture it would be possible to have a second radar (shown dotted) located (in this example) in Greece to the Ragusa radar such that the intercept would occur inside the second radar FOV for any kill assessment, where a decision for a second shot could be made. An appropriate netted BM/C^3 system would provide the needed assessment of the intercept that could then be relayed

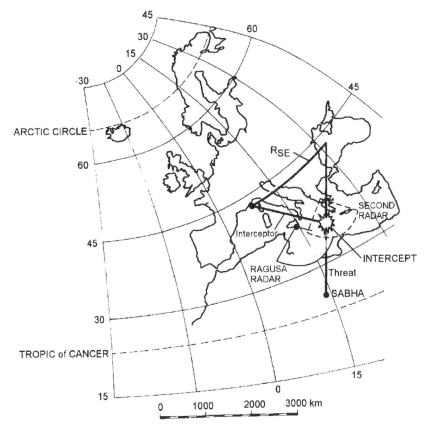

Fig. 4.36   Geometry of intercept for example problem.

back to the interceptor ship in the Ligurian Sea to determine if a second shot were required. This shows the advantage of cooperative engagement architectures between system components (interceptors, radars) and between nations seeking a cooperative defense.

Example problem 4.1 illustrates how the varying constraints can be used to advantage in the construction of the defended footprints and in the specification or requirement for the governing dimensions R_{FE}, R_{BE}, and R_{SE}. Such techniques help to illustrate the key driving parameters in the defense system architectural components before major computer simulations are embarked on. An additional advantage of the various formulations for the individual dimensions of the defended footprint (R_{FE}, R_{BE}, R_{SE}) is that they can be used, once a complete simulation is available, to indicate the possible improvements by varying the final simulation results with the modifications that can be examined with the simplified equations.

SOME OTHER CONSTRAINTS LIMITING SIZE OF
DEFENDED FOOTPRINT AREAS

There are many other constraints on the size of the defended footprint area, most of which tend to reduce the size from that predicted from the simple method shown earlier. Although not an exhaustive list, some of the more prevalent constraints placed on defended footprints are listed here.

1) *Threat not in FOV of radar (or long enough to establish track),*
2) projected intercept was above maximum allowable altitude,
3) *projected intercept was below minimum allowable altitude,*
4) *projected intercept was not in interceptor's fly-out envelope,*
5) *interceptor time of flight is less than allowable,**
6) interceptor time of flight is more than allowable,
7) *crossing angle between interceptor and threat too large or too small,*
8) closing velocity (between interceptor and threat) is too low,†
9) strike angle is less or more than that required for desired, P_k,
10) look angles exceed gimbal limits of interceptor seeker,
11) projected intercept is below the Earth's exclusion angle (see later for effect of radar horizon),
12) insufficient time for a second (or more) shots,
13) long flight times through the atmosphere would cause overheating, and
14) projected intercept at too high an altitude for aerodynamic controls to function, etc.

The italicized constraints are those that have been used in the construction of the simple method shown earlier with the examples that use R_{FE}, R_{BE}, and R_{SE} as the governing constraints.

The other constraints are equally as important but are quite design-specific (e.g., gimbal limits, aerodynamic heating, and shroud design limits), and use must be made of more complete simulations by using the complete equations and trajectories in step-by-step integration techniques. Generally, these constraints tend to reduce the size of the defended footprint from those shown here.

Two other "constraints" related to the radar also can have a significant impact on the size of the defended footprint. Figure 4.37 shows these effects. The top panel of Fig. 4.37 shows the effect when the radar pattern is the normal tear-drop shape rather than the idealized pie shape used in some simulations and illustrated in Fig. 4.36. In a detailed simulation that took

---

*Some interceptor designs require sufficient time for their protective shroud to fall away (to avoid seeker window heating as the interceptor flies through the atmosphere).
†Some interceptors (especially those using hit-to-kill techniques) require certain strike angles and closing velocities to ensure a high value of kill.

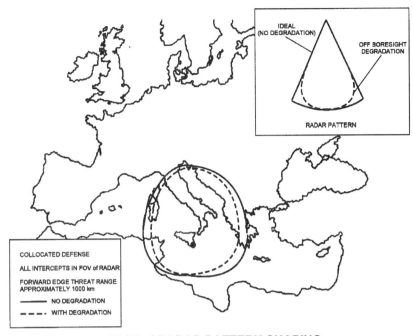

EFFECT of RADAR PATTERN SHAPING

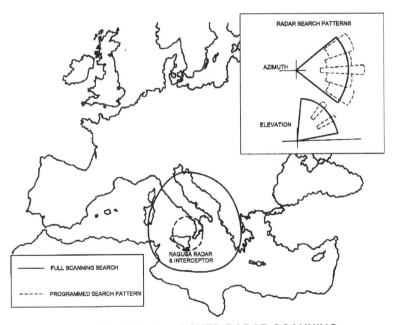

EFFECT of FULL and CUED RADAR SCANNING

Fig. 4.37 Two characteristics of radar search that impact footprint size.

proper account of the losses in radar propagation at the edges of its FOV, it would be expected that as the intercepts moved toward the side edges of the footprint the system could not intercept because the radar would not "see" the threat.

The lower panels of Fig. 4.37 shows another characteristic that must be considered in actual simulations of specific systems — the time loss of the radar during its search pattern. It is normal to program any radar, especially one with a large FOV, to either randomly or programmatically search the expected area of the incoming threat with high priority on those parts of the sky where the threat is expected to enter the battlespace. This is illustrated in Fig. 4.37 by bold lines that indicate the full search envelope in both elevation and azimuth, and dotted lines that show the more typical programmed search volumes. The two dotted segments in elevation shown would be for expected threats coming in at high θ_{re} (say, for BM) and low θ_{re} (say, for CM and aircraft). This latter adjustment in the radar search pattern could also be caused by external sensor (say, a satellite), which cues the radar as to where to concentrate its search and eventual acquisition of the threat. As shown in this hypothetical example, such cueing can have a marked effect on the size of the defended footprint.

These aspects of the radar will be returned to later in this chapter when considerations of early warning and the battle management/command, communication, and control (BM/C^3) aspects are integrated into the defense system features. Although most of the constraints and characteristics discussed all tend to reduce the size of the defended footprint, the impact of incorporating early-warning features into the defense system can (in many cases) significantly increase the size of defended footprints.

The preceding analyses have been done for a single threat launched at a single defense site. It is very unlikely that in any future conflict involving missile attack that the threats will come from *only* one site. There are two aspects to a missile attack: 1) there are likely to be multiple sites, and 2) more than one missile at a time will be launched toward the defense site(s). Figure 4.38 is a simple illustration of the effect on the defended footprint if there are other threats from different threat axes being launched from different locations toward the same defense site. For the purposes of illustration, the same footprint generated by the example shown earlier will be used to demonstrate the likely effect.

The effect is self-evident in that the shaded area in Fig. 4.38 is the net defended area from the single defense system against threats from different locations. (Note that this netted defended footprint area would be even smaller if it were not for the addition of a second radar observing the actual intercept.) The net defended area is considerably reduced from the single threat calculations used earlier. This very typical effect suggests

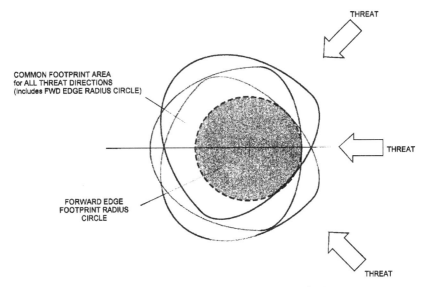

Fig. 4.38   Reducing footprint area as threat axis changes.

that the simple method shown earlier by using just the forward-edge footprint radius R_{FE} actually provides a reasonable idea of the likely defended footprint once the effects of multiple (or moving) threats and other limitations are included.

Note that the general effect, although not universal, tends to reduce dimensions of the defended area but less so on the forward-edge footprint radius R_{FE}. Hence, the increase in the defended footprint area A_{FP} as a result of including the crossing angle Ω effect shown earlier in the example has been considerably reduced again if the (likely) effect of changing threat axes is included.

Because of this effect, the forward-edge radius R_{FE} is often used as the main figure of merit of any defense system. In Fig. 4.38, the dotted circle of radius R_{FE} has been retained in the net defended area even as the threat axis shifts.

The preceding treatment of defended footprints has only touched the surface of all of the possibilities with concentration on the basic principles that can be applied to a variety of scenarios. Other topics that influence the final outcome of defended footprints include

    1) defense against depressed and lofted ballistic missile trajectories;

    2) cooperative defense between land, sea, and airborne defense systems;

3) netted architectures between many defense systems, where intercept solutions are shared between interceptors and radars;

4) cueing of radars by other systems (e.g., satellites, airborne early warning);

5) use of multistage interceptors with different burn profiles;

6) effect of boost-phase (and ascent-phase) intercepts; and

7) ship-launched and submarine-launched interceptors.

Each of these topics introduce new features that can radically change the final form of defended footprints, but they all use essentially the same principles outlined here in the simple derivations of defended footprint dimensions. The emphasis has been on first establishing the basic principles. Some of these other topics are referred to and treated later in this chapter.

SYSTEM PERFORMANCE IMPROVEMENT WITH COMPONENT CHANGES

Part of the decision maker's set of choices is the effect on the system performance by changing the performance of each of the main system components. One example would be the decision on the choice of interceptor speed V_{bo} and the choice of radar detection range R_D. An obvious question would be, *If it is desired to increase the footprint area from some baseline value, is it better to increase the interceptor speed or the radar detection range?* This would require a detailed analysis of the various options, but it is possible to gain an appreciation of the possible improvements by using the equations developed for the forward-edge footprint radius R_{FE}.

INTERCEPTOR SPEED AND RADAR DETECTION RANGE

Consider a nominal case of an interceptor system with a $V_{bo} = 1.3$ km/s, a reaction time $t_r = 20$ s, and a design threat range of 1500 km. The forward-edge footprint radius R_{FE} was calculated and presented in the earlier derivations in Fig. 4.27. If now the particular subset for the threat range of 1500 km is chosen for a set of V_{bo} and radar detection ranges R_D, it is possible to generate a set of curves as shown in Fig. 4.39.

The nominal case shown is for an interceptor speed $V_{bo} = 1.3$ km/s and radar detection range $R_D = 200$ km and a reaction time of $t_r = 20$ s. From Eq. (4.18) it can be shown that the forward-edge footprint radius R_{FE} against a 1500-km range threat is approximately 16 km (as displayed on Fig. 4.39). Suppose the decision maker would like to *double* this defended footprint radius to 32 km. Is it better to increase the interceptor speed or to increase the radar detection range, all else being equal? As seen in Fig. 4.39, to accomplish this the interceptor speed would have to be increased to greater than 2.0 km/s, an approximate 55% increase in speed. If instead of increasing the interceptor speed the detection range were increased, Fig. 4.39

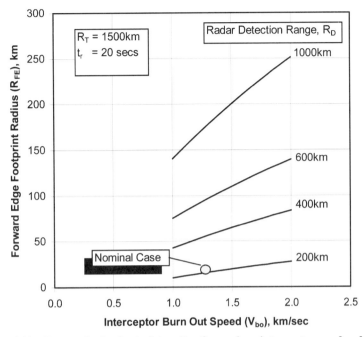

Fig. 4.39   Forward-edge footprint radius for various interceptors and radars.

shows that a radar detection range $R_D = 280$ km would be required. This is an approximate 40% increase in radar detection range over the nominal value of 200 km. This provides the input to a trade study to determine if the cost of developing the faster interceptor to give $V_{bo} = 2$ km/s is more or less than that of developing a longer detection range radar from $R_D = 200$ to 280 km.

SOME DESIGN TRADES

In the selection of the main components of the missile defense system there are several design trade possibilities to consider. The choice between developing a higher speed interceptor or a longer detection range radar (or IR sensor) is one key choice that needs to be made. Using the same equations that have been developed earlier it is possible to combine the charts to show the effect on the defended footprint in a direct manner. Figure 4.41 shows a "carpet plot" of interceptor burn-out speed (V_{bo}); detection range (R_D) and Forward Edge Footprint Radius (R_{FE}). The "carpet plot" is a convenient way of showing three dimensions on an interpolative two-dimensional chart.

This particular chart has been constructed for the specific case of defending against a threat ballistic missile with a nominal minimum energy range of 1500 km. By the use of this plotting technique, the relative effects of

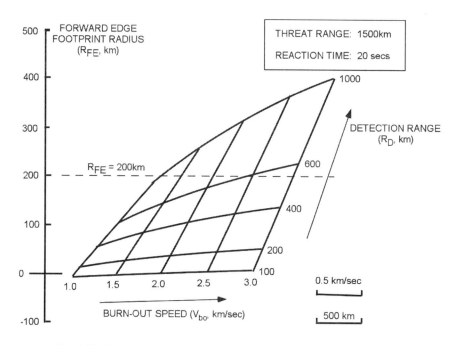

Fig. 4.40   **Design trade between interceptor speed and detection range.**

both V_{bo} and R_D on R_{FE} are immediately apparent. If it is desired to provide a defended footprint radius of, say 200 km, it is seen from Fig. 4.41 that there are several possible combinations of interceptor performance and detection range to achieve this value. In this example, the value of $R_{FE} = 200$ km could be achieved with an interceptor with $V_{bo} = 3$ km/sec and a radar detection range of $R_D = 540$ km. Conversely, the same defended footprint could be achieved (as shown in Fig. 4.41) with a slower interceptor with $V_{bo} = 2$ km/sec if the radar could be improved to provide a detection range of $R_D = 660$ km. The design trade is then one of determining the relative merits of achieving one set of system parameters over the other. Would the cost and time required to develop a higher speed interceptor be more or less than the corresponding development for a longer range radar? In this particular example, it is seen that for short detection ranges (R_D less than 200 km) even significant increases in interceptor burn-out speed (V_{bo}) cannot compensate for the high speed of the incoming TBM and only very small defended footprints are possible. For longer detection range systems ($600 < R_D < 1000$ km) then it is possible to achieve sufficiently large defended footprints with different combinations of V_{bo} and R_D. Other threats and conditions would generate other results but the technique is the same.

A second key design trade would be that of deciding how to compensate for the various time delays that occur throughout the entire BM/C3 network as the defense system goes into effect. Time delays can occur through the processing of the signals, the time taken for discrimination of the warhead from countermeasures (see later) and delays caused by operational functions such as simply ensuring the system is "in position". Charts such as shown in Fig. 4.42 help in this regard. Again, the example of defending against a 1500 km range TBM is used.

In this example, it is seen that while it would be possible to achieve a reasonable footprint (approximately 50 km radius) with a radar of 200 km detection range and a system reaction time of 5 s; this rapidly degrades into no footprint if the system time delays approach 60 s. In this example, it is seen that the example desired footprint radius of $R_{FE} = 200$ km could be achieved with a radar detection range, $R_D = 650$ km if the BM/C^3 network could process and transmit all data within 5 s. It is unlikely that such a processing time is possible in practical situations and it is of interest to see that if the radar detection range could be increased to $R_D = 780$ km, then time delays up to 60 s could be tolerated. This would be a significant relief on the BM/C^3 network but would require a more capable radar. Such charts allow for rapid appreciation of the possible design trades that could be pursued in the selection of the required characteristics of all the major components in the missile defense system.

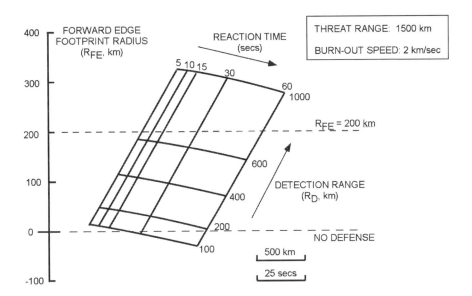

Fig. 4.41   Design trade between BM/C3 time and detection range.

LAUNCHING THE INTERCEPTOR EARLY OR LATE

The advantage of the use of remote launchers to increase the footprint area A_{FP} has already been discussed [see Eqs. (4.23) and (4.24)]. If instead of, or as well as, displacing the launchers or radars the additional possibility of launching early (i.e., a displacement in time instead of distance) would also improve the size of the defended area. This would be possible, for example, if the system had advanced signal-processing capability that would allow early launch with in-flight corrections. Using the same nominal case, Fig. 4.42 shows the effect of varying the reaction time t_r from the nominal value of 20 s.

If the same case of a threat range of 1500 km is selected, the value of the forward-edge radius R_{FE} of 16 km can be seen in Fig. 4.42 for the nominal case of $t_r = 20$ s. If now the defense system is equipped with an advanced capability of signal processing such that the interceptor can be launched early (say 10 s after the threat launch), the forward-edge footprint radius R_{FE} is seen to be increased to 21 km. This is a 30% improvement in defended footprint radius. Conversely, if the system is delayed by 10 s, the footprint radius drops to 11 km, which is a 30% less radius with the related decrease in area A_{FP}.

These and other variations are possible with use of the earlier developed equations for a rapid indication of where improvements might best be sought in the overall system makeup.

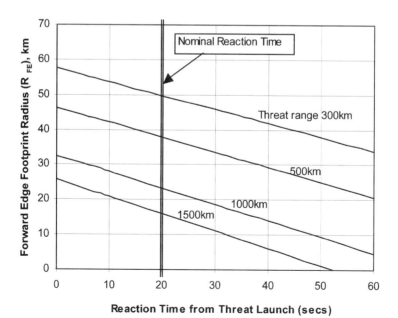

Fig. 4.42 Effect of varying reaction time.

PROBABILITY OF SURVIVAL IN DEFENDED FOOTPRINT AREA

The probability of an asset surviving a missile attack within the defended footprint just derived will depend on many factors associated with both the interceptor itself and its supporting components of radar, BM/C^3 system, and the characteristics of the incoming missile. Additionally, if a particular survival value is required over the defended asset this in turn influences the size of the footprint as will be shown. The two parameters *survivability* and *defended area* are not independent if reasonable values of survivability are to be expected within any foreseeable state-of-the-art of missile design.

The survivability of the defended asset can take many forms. It can mean a completely *passive defense* form, where various types of protective cover are assumed, which can vary from air-raid shelters to protective clothing to a completely *active defense* form or some combination of both forms of defense. Chapter 3 discusses the various forms of defense and survivability.

In the case of active defense, there are also several possibilities. These include the following.

1) Knock the incoming missile off course.
2) Destroy the incoming missile by a fragmentation or similar interceptor warhead.
3) Destroy the incoming missile by a hit-to-kill or kinetic energy means.
4) Destroy the incoming missile's propulsion booster (usually early in its flight).
5) Destroy the incoming missile's electronic systems (guidance, control, etc.).

Some of these means of defense fall into the category of *mission kill*, which means that the asset is defended. This method could incur collateral damage and place another asset or even neighboring nation at risk. Of the *direct kill* methods, the defense means used in the preceding list also vary from an interceptor that is also a missile in itself to other techniques such as directed energy, which can be one of several technological designs (lasers, high-power microwaves, etc.). As can be seen from this list of means of defense, there will be variations of performance capability from both operational and technical aspects. There will be a wide variation in the values of survivability achieved from one means to another. For the immediate discussion attention is given to the ability to achieve a given level of survivability to be achieved by a simple or direct intercept.

Chapter 3 showed the values of survivability that could be expected in a multilayer defense system. It is shown in Chapter 2 that the probability of survivability of the defended asset P_{sa} could be expressed as

$$P_{sa} = 1 - (1 - SSPK_1)^{n_1} * (1 - SSPK_2)^{n_2} * \cdots\cdots (1 - SSPK_i)^{n_i} \quad (4.34)$$

which is a string of multiple probabilities of a number i of layers, a salvo n_i in each layer, and in which each interceptor has a single-shot probability of "killing" the threat with a value of $SSPK_i$ in each layer. The word "killing" really means "intercept" in this formulation. Whether a "hit" is a "kill" depends on the dynamics of the intercept, the mass distribution of the two missiles, and the content of the incoming ballistic missile warhead. In the case of biological or chemical warheads, the level of lethality that might be achieved given a hit, even if $SSPK = 1$, might be far from sufficient to ensure the survivability of the defended asset. This is discussed later in this chapter.

For the case of a two-layer defense system (say, a lower-layer system for endo threats and an upper-layer system for the endo-exo or exo threats), Eq. (4.34) can be reduced to

$$P_{sa} = 1 - (1 - SSPK_1)^{n_1} * (1 - SSPK_2)^{n_2} \quad (4.35)$$

For example, if each interceptor in each of the two layers is assumed to have a performance capability of $SSPK = 0.70$ and a salvo of $n_1 = n_2 = 2$ is used in each layer, then $P_{sa} = 0.9919$. That is, if 100 ballistic missiles are launched at the defended asset, which is protected by a two-layer defense system, then after four interceptors (two upper and two lower) are launched to intercept, this formulation states that 99 ballistic missiles would be hit or killed, but one ballistic missile would probably get through to the defended asset.

An alternative firing doctrine, which is considered common practice, is to use a shoot-look-shoot or salvo-look-salvo or S-L-S-firing sequence. In this firing doctrine it is desired to achieve the same probability of survival as for the salvo-firing doctrine just given but to use less interceptors. This technique requires a kill assessment of the first intercept *before* expending another interceptor. If the first shot with its probability of $SSPK_1 = 0.70$ is successful (which by definition will be 70% of the time), then there will be no need to fire the second shot (or salvo). In this case, as shown in Chapter 2, the number of interceptors required (NR) to achieve the same value of P_{sa} is given by

$$NR = k\left(1 + \sum_{i=2}^{n}(1 - SSPK)^{(i-1)k}\right) \quad (4.36)$$

in which $k =$ number of rounds in a salvo, $n =$ number of salvos, and $i \leq n$.

In this formulation, for simplicity, it has been taken that the interceptor performance capability SSPK is the same for each layer. In Chapter 2 a set of different values was provided to indicate the effect of using the S-L-S-firing doctrine instead of a salvo-firing doctrine. If a S-L-S-firing doctrine was used instead of a salvo of two interceptors firing doctrine, the number of interceptors required would be considerably less in most cases. For example, if in a single-layer system, a salvo of three interceptors were used to achieve a probability of survival of $P_{sa} = 0.9927$, then only 1.39 interceptors would be required using the S-L-S-L-S-firing doctrine. Because of the high cost of interceptors (see Chapter 5), this could be an attractive approach.

Unfortunately, not obvious with this calculation is the fact that looking before firing the second shot means that the defended footprint area reduces, sometimes by a significant amount. This, in turn, means that although the immediate defended asset might enjoy the probability of survival P_{sa} with a lesser number of interceptors used then large areas around that defended asset will now be undefended. This undefended area will now need an additional set of interceptor sites, which would increase the number of interceptors required! This circular argument might well mean that it would be cheaper to simply salvo rather than shoot-look-shoot! This can only be assessed in any particular scenario by analyzing the defense both ways and adding up the results to determine which technique is best.

To illustrate how this defended area would reduce in order to accommodate a shoot-look-shoot firing doctrine, use will be made of the earlier treatment of defended footprints.

EXAMPLE PROBLEM 4.2—DEFENSE OF THE UNITED KINGDOM AND EUROPE

Using example problem 4.1 (ballistic missiles launched from Libya into Europe) as a baseline, show the effect on the defended footprints if a shoot-look-shoot firing doctrine is used instead of the earlier assumption of shoot. Show the effect of a defense with a high-speed interceptor system ($V_{bo} = 4.0 \, km/s$) located close to the threat and of a slower speed interceptor ($V_{bo} = 2.5 \, km/s$) placed some distance from the threat. Example problem 4.1 used a defending ship cruising in the Ligurian Sea at a distance of about 1900 km from the threat site as a means of defending Europe. In this example problem the use of a land-based defense system located in the United Kingdom (north of London) some 2900 km—an additional 1000-km separation from the same threat site—will be used. This will illustrate several features of the defense and the impact on the defended footprint with either a S- or a S-L-S-firing doctrine.

The same equations derived before will be used in this case except that the defense system reaction time t_r will now be increased for the second shot footprint by the amount of time required to 1) first witness the intercept or nonintercept of the first shot and 2) the time required to react to the second shot.

For the first case (see preceding example problem 4.1) of defending Europe with the high-speed interceptor (V_{bo} = 4.0 km/s), the reaction time t_r was assumed to be 20 s after detection time (t_d = 78 s). From matching the threat and interceptor trajectories, it is found that the time of intercept occurred at t_i = 403 s. Because the defense system now does not need to be cued or various components brought on line for the second intercept, the additional time to react can be reduced considerably. A conservative amount of time of 5 s will be assumed, in which case the second shot can now be launched at a time of 330 s after the first detection or 408 s after the threat launch. By substituting these values of reaction time into the preceding equations, it is possible to show the S-L-S defended footprint, with the approximate dimensions of R_{FE} = 495 km, R_{BE} = 2035 km, and R_{SE} = 1485 km. This is displayed in the same polyconic projection in Fig. 4.43, together with the original S defended footprint from example problem 4.1.

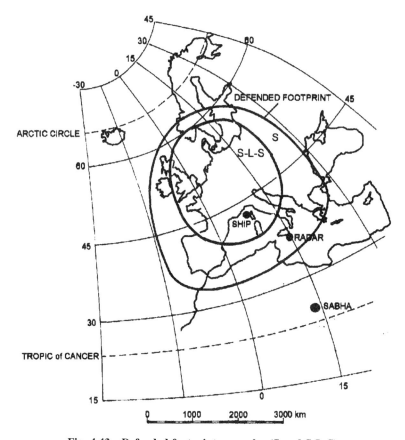

Fig. 4.43   Defended footprint examples (S and S-L-S).

For the second case assume that the defense interceptor system ($V_{bo} = 2.5 \, km/s$) is placed just north of London and that the radar is located on the coast near Hastings some 55 km south of the defense site. In this case it is found that the detection of the threat is delayed a considerable amount of time by the radar ($R_D = 1000 \, km$) even if cued because of the large distance to be traveled by the threat. The forward-edge radius (R_{FE}) for the S-firing doctrine case is then as shown in Fig. 4.43.

Unlike preceding example problem 4.1, in which the defense site (in the Ligurian Sea) was very close to the threat, the defended footprint around London is here very clearly a face-acquisition solution for the radar. Numerically, the forward-edge radius is found to be approximately $R_{FE} = 247 \, km$. Using similar techniques to the preceding example, the values for the back-edge radius R_{BE} and side-edge radius R_{SE} are computed to be $R_{BE} = 593 \, km$ and $R_{SE} = 347 \, km$.

Figure 4.45 shows the complete results for the case of a single defense system with an interceptor ($V_{bo} = 2.50 \, km/s$) located near London and with a radar ($R_D = 1000 \, km$) near Hastings. The top panel of Fig. 4.45 shows the defended footprint as calculated, whereas the bottom panel of Fig. 4.45 shows how this footprint would look when superimposed on a polyconic projection map of the United Kingdom. Additionally, the results of calculations for the S-L-S firing doctrine defended footprint, using the same techniques as before, have been superimposed on the bottom half of Fig. 4.45.

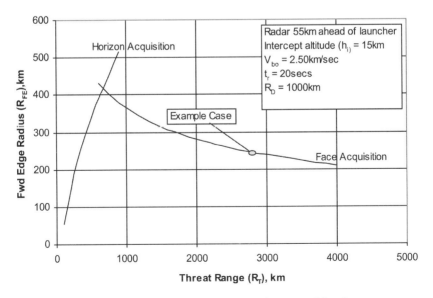

Fig. 4.44   Forward-edge footprint radius around London.

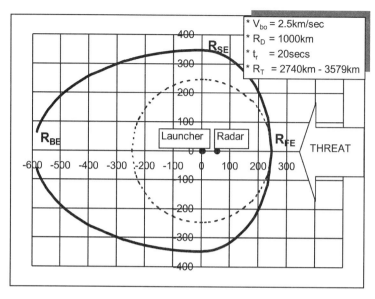

Defended Footprint ("S" Firing Doctrine)

Defended Footprints ("S" and "S-L-S") in UK

Fig. 4.45   Example defended footprints (S and S-L-S) around London.

As can be seen for the case of a defense system in the United Kingdom, in this example the defended footprint using a S-L-S-firing doctrine gives a much reduced defended area. This is because the defense system performance is such that the interceptor cannot intercept the threat soon enough before the threat has entered the battlespace to an advanced degree. The probability of survival P_{sa} for the S-L-S-firing doctrine has been retained to the same level as for the S-firing doctrine, but the defended area has been greatly reduced because of the time delay incurred by waiting until the intercept has been observed from the initial shot.

From example problems 4.1 and 4.2, the following conclusions can be drawn. In the map showing the defended footprints over Europe (Fig. 4.43), it can be seen that although the inventory of interceptors (NR) can be reduced with the use of a S-L-S-firing doctrine for the same level of survivability many parts of Europe would become undefended. In this example Greece and half of Italy and Spain and part of Scandinavia are now undefended with the S-L-S-firing doctrine as compared to that with a S-firing doctrine. A second conclusion can be drawn from the defended footprints over the United Kingdom with a lower performing system and greater distance from the threat: the difference between the S and the S-L-S footprints is much greater than the preceding case. An additional observation from these examples is that the forward-edge radius R_{FE} is a greater part of the defended-area geometry the further away the defended site is from the threat.

These example problems show that if the probability of survival P_{sa} is to be increased by the use of a salvo-look-salvo-firing doctrine and the inventory of interceptors is to be kept to a minimum, then a significant price is to be paid in reduced defended area. Even though the cost of interceptors might be high, it would appear to be better to use more interceptors in a salvo-firing doctrine rather than a salvo-look-salvo-firing doctrine in many scenarios because of the much reduced defended area that results. As will be shown later in this chapter, this effect can be alleviated to some degree by the use of a defense system architecture that employs what is called a *netted BM/C³ architecture*. In this case the various interceptor sites and radar sites are "netted" such that any combination of interceptors and radars can be used to intercept the threat. This operational feature was illustrated earlier in Fig. 4.36, where an additional radar was integrated into the defense system to determine if a second shot would be required.

Important observations from even the just simplified calculations are as follows:

1) A significant number of interceptors are required, whether in a salvo-firing doctrine or in a salvo-look-salvo-firing doctrine to achieve a survivability level above 99.99%.

2) The achievement of a high probability of survival using a salvo-look-salvo-firing doctrine leads to a significant loss of defended area. In Chapter 3 it was noted that in the case of the potential use of WMD the defense will require very high values of P_{sa}. Recognition of this likelihood has led to the development of boost-phase-intercept (BPI) systems.

BOOST-PHASE AND ASCENT-PHASE INTERCEPT

There are several advantages to intercepting ballistic missiles during the initial phases of its flight after launch. First, the signature [both RCS and in particular infrared (IR)] is larger during this phase and therefore less stressful on the defender, and, second, if the threat is destined to release a reentry vehicle (RV) or submunitions it is easier to intercept the *single* booster during the early flight phases of the ballistic missile than to intercept the *multiple* RV (plus any decoys) during its terminal phase. If the submunitions are likely to contain chemical or biological agents or other devices of WMD, there is the added benefit that an intercept during the boost phase, or possibly during the ascent phase, could cause the offending submunitions to fall back on the intending attacker with unfortunate results for him. Perhaps, if the attacker knew that the defender had such a [BPI or ascent-phase intercept (API)] capability this might be a sufficient *deterrent* to attack in itself.

Again, although the preceding benefits of BPI/API in the defense against WMD are clearly worthwhile there is the additional benefit of providing an additional defense-in-depth feature to augment the midcourse and terminal defense systems (discussed in the preceding sections of this chapter). With the uncertainty, even with today's high technology, in the ability to guarantee high probability of intercept with today's defense systems, the addition of such a capability could be worth the investment. Of course, as will be shown later, in order to intercept during the boost or ascent phases, it will require the defender to be quite close to the attacker's territory and frequently, because of world geography, the defender would have to be *inside* enemy territory (airspace) in order to accomplish such an intercept.

BASIC GEOMETRIES FOR BOOST-PHASE (ASCENT-PHASE) INTERCEPT

Figure 4.46 shows the essential elements of the intercept. The example in Fig 4.46 of an aircraft or ship conducting this mission is chosen merely to emphasize that because of the geographical features of the world such defending platforms are more likely candidates for such a mission than land-based systems. In certain limited scenarios it is possible for land systems to intercept during boost or ascent phase, in which case the calculations for ship-based defense systems would be analogous to the land-

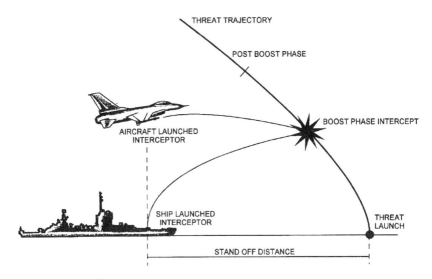

Fig. 4.46   Boost- and post-boost-phase intercept.

based system. The advantage of the aircraft is that if *air superiority* could be achieved early enough, that is, *before* any ballistic missile attack, the aircraft could get closer to the threat launch site by flying over enemy territory. This becomes an operational issue as to whether or not sufficient warning is provided that a threat missile will be launched *after* air superiority can be achieved.

As depicted in Fig. 4.46, the flight paths of both the threat and the defending interceptor will be generally curved (ballistic or otherwise) shapes. It will be shown here that a simplified treatment, where the interceptor travels in a straight line, provides an appreciation of the intercept requirements that lends itself to decision making on the key parameters. It is possible to conduct the analysis with the more complex trajectories of air-to-air missiles, but the essential features can be illustrated with the simplified analysis without loss of important conclusions. Figure 4.47 shows this slight variation on the actual paths of intercept. [The aircraft is used here (instead of ship or land systems) because analytically the difference is minimal with aircraft altitudes of approximately 10 km compared to the high altitudes of ballistic missiles (see earlier in this chapter and Appendix C).] Because the actual interceptor path will always be longer than the straight (idealized) path, the actual interceptor speeds will always be *greater* than predicted by such an idealized treatment. During the boost phase, the differences between actual and idealized interceptor speeds will be negligible, but as the threat trajectory continues into the ascent and midcourse phases the actual

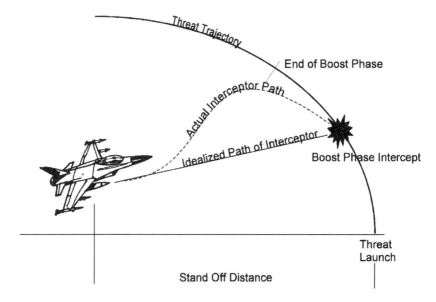

Fig. 4.47 Idealized boost-phase intercept.

interceptor speeds can grow to 40–50% more than that predicted by the idealized method. Because the interest here is in the boost and ascent phase, such discrepancies are not important.

In an analysis of boost-phase intercept, the case of an aircraft as the launch platform of the interceptor will be used. In some scenarios a land-based system or a ship-based system might be able to intercept threat missiles in the boost phase, but this would not be the most likely scenario when the Earth's geography and likely threat areas are considered. Also, because the effect of altitude on the solutions to boost phase (and ascent phase) intercepts is quite small, the selection of the aircraft for the example analysis reflects the general result. The actual flight path of the interceptor once launched from the aircraft will normally follow some form of curved path as the guidance system ensures that the interceptor closes in on its target for the intercept. For the purposes of illustration here, the interceptor will be assumed to fly in a straight line directly toward the threat missile, as shown in Fig. 4.47. Clearly, any determination of the required interceptor speed using this assumption will always result in a speed that is *lower* than in any real case. Further, in this simplified treatment it is assumed that the aircraft is flying at some typical altitude (say, 10 km) and is already in position in the correct combat air patrol (CAP) circuit and has already been vectored by some external early-warning system into the correct position (of pointing directly toward the threat launch site), all of which are time-consuming functions that will require *higher* speeds than predicted here. Additionally, it is assumed that the threat missile has been launched vertically and only begins its turnover, again towards and in the plane of the aircraft, at the point of burnout (see later discussion on launch angles).

An example case is taken of an aircraft flying at an altitude of 10 km that is required to intercept a 2000-km ballistic missile launched vertically. The illustrative 2000-km trajectory shown in Fig. 4.9 will be assumed. The end of boost phase for that missile was shown to be at a downrange distance of $x_{bo} = 106$ km and at an altitude of $h_{bo} = 70$ km. The burnout time for that example missile was $t_{bo} = 65$ s. In Appendix C it is shown that depending on the technological maturity of the threat missile the burn time could be expected to vary from about 40 s to greater than 120 s (in very old designs). Values of 65–85 s would be considered to be the more likely case in any foreseeable future.

Figure 4.48 shows the idealized geometry of an aircraft already in position that is launching an interceptor in a straight line toward a threat missile that has been launched toward and over the aircraft. In this idealized treatment both the aircraft interceptor climb path and the threat trajectory are in the same plane. In this idealized treatment it is easy to show that the (average)

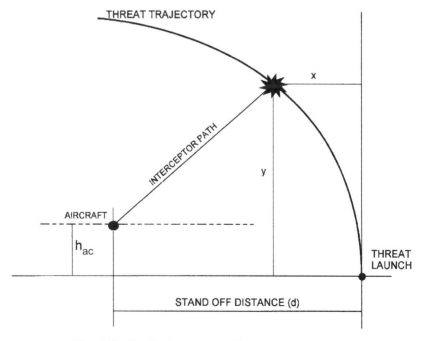

Fig. 4.48   Idealized geometry of boost-phase intercept.

speed of the interceptor along its flight path is given by

$$V_i = \frac{1}{(t - t_r)} \sqrt{(d - x)^2 + (y - h_{ac})^2} \qquad (4.37)$$

where

V_i = required interceptor (average) speed
t = time from threat launch
t_r = detection (and processing) time for interceptor system
d = standoff distance interceptor platform is from threat launch site
x = horizontal distance traveled by threat toward interceptor
y = altitude reached by threat
h_{ac} = altitude of aircraft-launching interceptor

The speed of the aircraft will help in the operation, but even at Mach 3, which is approximately 1 km/s at this altitude, the main requirement will still be for a high-speed interceptor (see later) if any intercept is to be expected during the short burn time and for any reasonable standoff distance by the aircraft. In this treatment it is assumed that the threat is launched at time $t = 0$ and that the interceptor is launched at time $t + t_r$, where t_r is the time consumed by the defense system to first detect and process the threat coordinate data and then to initiate the interceptor launch.

In this simplified treatment it is also assumed that the interceptor is launched at a constant (average) speed V_i in the plane of the threat.

The results of this simplified treatment of an aircraft launching an interceptor on a straight flight path toward the threat is shown in Fig. 4.49. First, it is immediately seen that if the defending aircraft is to standoff at reasonable distances from the threat territory and to not fly over threat territory, say, 200–400 km back from the threat launch site, then very high interceptor speeds are required (greater than 6 km/s) if an intercept is to be made during the boost phase. It is possible to intercept after boost phase and engage in an API, but two difficulties immediately occur. First, after boost the IR signature of the threat drops dramatically, making targeting more difficult. Second, if the threat missile is going to release any submunitions containing various forms of chemical or biological agents this can occur at any time after the boost phase. A third, and indirect issue, would be that if the intercept is delayed too long there is the increased likelihood of collateral damage and debris falling on neighboring friendly nations. Hence, all emphasis should be on BPI.

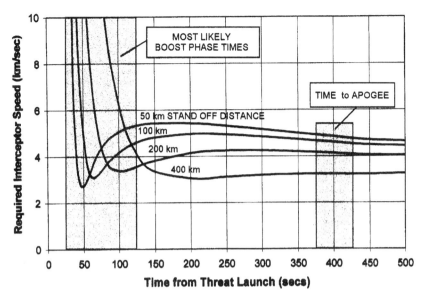

Fig. 4.49 Interceptor speeds required for BPI.

As the threat missile flight continues past apogee, the interceptor speeds required level out (see Fig. 4.49) to values achievable with today's missile systems. But this means that the intercept operation is no longer in boost (or ascent) phase, and the solutions would be those already discussed for the terminal phase. For this example case of intercepting a 2000-km range threat ballistic missile launched vertically, the required interceptor speeds rapidly increase from about 1.5 km/s if the standoff distance is 50 km to greater than 6 km/s if the standoff distance is about 400 km. For standoff distances greater than 400 km, the required interceptor speeds increase exponentially to extremely high speeds (10 km/s and higher).

For this example the acquisition time or delay time t_r assumed, as before, is 20 s. This is a critical parameter in the success or failure of any BPI. The actual calculations to produce the interacting fly-out curves of interceptor and threat missiles require step-by-step calculations of many parameters (and available with any high-speed computer), and, while necessary in any particular analysis for a given system, they are not required to provide the basic insight to make decisions on key features. By making a few simplifying assumptions, as shown here, without sacrificing the essential elements, it is possible to analyze the main drivers in the choice for BPI or API.

In this simplified analysis no allowance has been made for the time for the interceptor to accelerate to its design speed. Each of the assumptions just identified contributes, in this idealized treatment, to *underestimate* the interceptor speed required by some amount (up to 40–50% in postapogee intercepts). As mentioned earlier, any differences between these idealized values and those of an actual case are minimal during the boost phase and only become significant for large standoff distances and for later phases of the threat trajectory. Because the interest here is in the BPI/API phases and for reasonable standoff distances (say, of the order of 400 km or less), these differences between the idealized solutions and those in any specific case are minimal and can be safely ignored.

It was assumed in the preceding analysis that the threat missile was launched vertically from the launch site ($\theta = 90$ deg) and only turned toward the aircraft at burnout at time $t = t_{bo}$. It is possible that the threat missile could be launched immediately at time $t = 0$, at its maximum range angle with angles approaching $\theta = 45$ deg. This would be favorable to the defending aircraft. However, it is also possible, as an evasive maneuver and as a means of enticing the defending aircraft closer in to enemy territory, to launch *away* from the aircraft. These three possibilities are shown in Fig. 4.50, labeled as A, B, and C trajectories.

Also, as shown in Chapter 2, Fig. 2.2, the energy management system similar to that used by the defense interceptor THAAD to burn off excess fuel could also be used as a BPI evasion maneuver by a threat missile. Such a scheme would be similar to that shown as possibility C in Fig. 4.50.

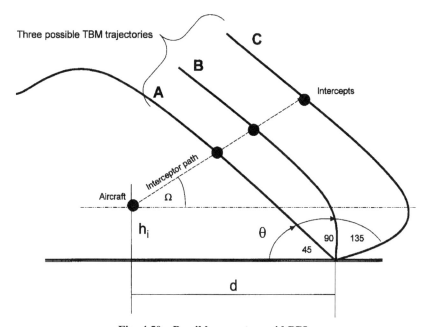

Fig. 4.50   Possible ways to avoid BPI.

DESIGNING A MISSILE FOR BOOST PHASE INTERCEPT

There are some challenges to the design of a missile that can be launched from an aircraft and accelerate to the high speeds required ($V_{bo} \approx 6$ km/s or Mach ≈ 20) as indicated in Fig. 4.49. The majority of air-to-air missiles that exist in the world's inventory to date have design speeds less than 2 km/s, which is much less than that required for BPI. The problem is alleviated somewhat if intercept during the ascent phase is acceptable, but this places the mission at risk if the intent is to intercept before any submunitions that might contain chemical or biological agents are released. The issue of proximity to the threat launch also increases the risk to the aircraft if it has to fly over the enemy's air defenses in order to be within range of the threat ballistic missile's boost phase. The example shown in Fig. 4.49 was for a 2000-km range threat. Shorter-range BM would have even shorter burn times, and longer BM would have higher speed. Both factors insist that the defending air-to-air missile must have a high-speed capability of the values shown here. Chapter 2, Fig. 2.14 shows that to obtain high speed that range has to be sacrificed, which impacts the required standoff distance d to a significant degree.

It is informative to collect the various pieces of data on missile speed and weight from the large database on CM, which is germane to the design problem facing the designer of BPI interceptors. Figure 4.51 shows the pertinent data.

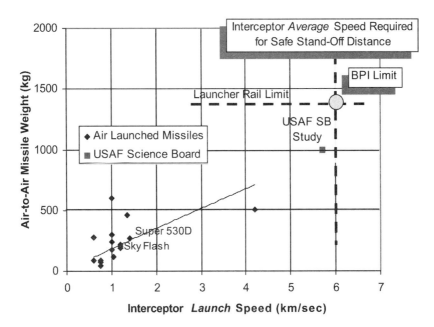

Fig. 4.51   Air-to-air missile weight and speed.

Several important factors have been included in Fig. 4.51. Most of the existing missiles are shown in the left hand part of the chart and only limited design studies have ventured into the realm of about 6 km/s. The U.S. Air Force Scientific Board[37] conducted a detailed study into the requirements for BPI, and their work supports the results shown here in Fig. 4.49 and in the surrounding analysis. The data point from the U.S. Air Force Board is included in Fig. 4.51. An important note to make on Fig. 4.51 is that for both the U.S. Air Force Scientific Board and in the analysis conducted here the 6 km/s speed shown is the *average* speed over the standoff distance traveled, whereas all of the other data points are the *initial* speed or design V_{bo} speed. The ratio of the initial speed to the average speed can be quite high, by a factor of $1 \rightarrow 1\frac{1}{2}$, depending on the distance traveled and the shape of the flight path. With this important caveat in mind, the chart clearly shows that it will be research and design (R&D) challenge to obtain these speeds and at the same time retain the required light weight such that the interceptor can be carried by fighter aircraft. A design goal or limit is shown on Fig. 4.51, which recognizes both the speed requirement and the weight limit posed by the aircraft launcher rail limitation. Fleeman[38] discusses this launcher rail limit in some detail, showing that the maximum weight that can be tolerated on the aircraft launcher rails varies from 250 to 1500 kg depending on the aircraft, the position on the wing (inboard or outboard), the nature of the mission and other factors. The maximum (best-case) value of 1500 kg has been used in Fig. 4.51. At the low limit of 250 kg, it can be seen from Fig. 4.51 that this would severely limit the ability to achieve the required missile design using current technology. This is not to say that with an aggressive R&D program that such a missile could not be achieved. The U.S. Air Force Scientific Board advocated that a multistage missile (at least three stages) and a high specific impulse I_{SP} propellant would be required to accomplish the high speed required (of the order of 6 km/s) and the range (say, 400 km) and stay within the aircraft launcher rail weight limits. The related problem of such a high-speed missile being unable to maneuver at such high speeds adds a further complication into the design process.

Typically, the aircraft could accommodate between two and four such missiles per load out. If the same firing doctrine is used for the aircraft and interceptor combination as for the land- and sea-based systems, then it is expected that a salvo of at least two interceptors per ballistic missile would be required. This results in *one* aircraft for *one* or at most *two* incoming threat missile(s) before the aircraft would have to return to base for reload. From time to time, designs have been put forward to produce a fighter or possibly a fighter-bomber in which the air-to-air missiles are housed in an internal bay on the aircraft. This would give the aircraft more capability than does the current design of *one* aircraft for *one* ballistic missile.

Such a result leaves the decision maker with a dilemma. In Chapter 3 it is pointed out that for certain types of WMD extremely high values of SSPK are required to ensure that most submunitions containing the deadly agents have been destroyed. Values higher than 0.9997 might be required. Even with the most optimistic value expected from today's interceptors, this might be difficult to achieve. (It is recognized that with proper warning and with proper protective clothing and shelters, such addition of *passive defense* measures alleviate the need for such high values of *SSPK*. However, it should also be recognized that such passive measures might not always be in place at the right time for either the military forces or for the civilian population.) The alternative approach is to intercept the threat ballistic missile while it is still accelerating in its boost phase and before it can release its deadly submunitions. The most likely defense system to provide this capability is the aircraft equipped with the high-speed interceptors as just described, *but*, as shown, such a capability requires intensive R&D that introduces risk (see Chapter 7) and prolonged development schedules (see Chapter 6).

These and other problems of designing such a missile clearly indicate that significant R&D should be embarked on if the risk can be reduced to manageable levels in any foreseeable future.

BM/C^3 AND EARLY WARNING

Chapter 3 states that a missile defense system is made up of three main components, which are 1) early warning, 2) interceptor systems, and 3) BM/C^3.

Much of the work in this book has concentrated on the various features of the possible interceptor systems (missiles and directed energy). It has also been clearly shown that such systems could not function, or at best could not function well, if it were not for the critical functions of early warning and the connecting BM/C^3 networks. Chapter 5 on cost provides the basic elements of the costs of early-warning systems that include space-based systems (i.e., satellites in various orbits), radars (land-based and shipborne), and airborne IR sensor systems.

Although the performance capabilities of the early-warning systems and BM/C^3 systems are touched on briefly in Chapter 5 on cost, the subject matter is too broad and scenario dependent to be treated adequately in the space available for this book. There are some overarching features of both early warning and BM/C^3 that can be summarized, and they provide a set of boundary conditions for most of the interceptor systems considered.

In this context both early warning and BM/C^3 will be discussed in the same section. In some communities the sensors are treated separately from the BM/C^3 networks, and in others the sensors are seen as an integrated part of the network that must function seamlessly if the interceptor systems are to do what is asked of them. The key fact that binds all three elements of the missile defense together, of course, is that no matter what the detail characteristics of any one of the three components are they must all accomplish their function within the timeline of the flight of the incoming missile. *This is an absolute; otherwise, the defense system has no purpose.*

There are several possibilities for dividing up the timeline from *threat launch* to *interceptor successful intercept* but the following 10 steps cover the basic functions, which are described in Table 4.5 for the needed steps up to the first intercept and in Table 4.6, which completes the intercept process from the kill assessment of the first intercept and the need for any successive shot to complete the intercept process.

As will be seen in both Tables 4.5 and 4.6, there is an entwined relationship between all three components of the interceptor system: the early-warning sensor, the interceptor system itself, and the connecting BM/C^3 (or BM/C^3I) network. Whether certain sensors are considered a part of the BM/C^3 network or part of the early-warning system is not important in the analysis of the timelines. That is a simple programmatic issue and possibly a cooperative issue if the total defense system reaches across national boundaries within the Alliance and shares system components.

Figure 4.52 shows these 10 main functions that the defense system must perform for a successful intercept once the threat missile has been launched toward the defender. In more detailed analyses there are other intermediate time functions incorporated in the firing procedures of interceptor systems, but the essential elements are captured with the preceding 10 steps and shown in Fig. 4.52. Again, the total time allotted is fixed by the characteristics of the threat missile and the reaction times of the defense system.

Each of the 10 steps for a successful intercept requires a finite amount of time, but the total time required for all 10 steps from detection of *threat launch* to *final kill assessment* must fit within the TOF of the incoming threat missile. From the earlier analyses of ballistic missile trajectories provided earlier in this chapter (and Appendix C), these times are known to a reasonable accuracy. The most likely time of flight of BM is given in both Chapter 2 and in Appendix C but are repeated here for convenience in Table 4.7. Note that these are for minimum energy trajectories.

In Appendix C it is shown that if the trajectories are *lofted*, then the flight times can be up to 78% longer than the minimum energy trajectory times; and if the trajectories are *depressed*, then the flight times can be 40% shorter than the minimum energy trajectory times shown in Table 4.7.

Table 4.5   Seven steps on the way to intercept

Steps for intercept	Function
1) *Sensor detects*	The early-warning sensor detects the launch of the threat missile. This can be after *intelligence* information has been received. This sensor can be a satellite, radar, or airborne IR system. This is logged in as time $t = t_d$ (see earlier analyses on defended footprint).[a]
2) *Sensor cues*	Time proceeds as the early-warning sensor takes one to three sensor hits on the threat missile to determine the threat missile characteristics and direction.
3) *Defense interceptor assigned*	Through an agreed-upon command structure and battlefield protocol before the battle, a system is set up for deciding which interceptor system will be assigned to go after the threat missile. This takes a small but finite amount of time.
4) *Handover to interceptor system*	Once the threat trajectory has been processed by the early-warning system and the preferred interceptor system selected, the threat data are transferred to the preferred interceptor system and its fire control radar.
5) *Interceptor launched*	During this function, the selected interceptor system has been alerted, and after the threat missile enters the FOVs of its fire control radar the interceptor is launched. This is called time to react t_r in the earlier defended footprint analyses. The elapsed time at this point is $t_r + t_d$.[b,c]
6) *Interceptor midcourse guidance*	In some instances it might be possible to give midcourse guidance to the launched interceptor.
7) *Intercept*	The interceptor intercepts the threat missile at time t_i and at an altitude of h_i in the terminology of the earlier defended footprint analyses.

[a]Sometimes the BM/C^3 system includes intelligence functions (and equipment) and is then called by the acronym BM/C^3I. As pointed out in Chapter 5, such differences need to be understood when costing the system.

[b]Under certain circumstances it is possible to launch the interceptor early and rely on later midcourse guidance to correct the interceptor trajectory as a time-saving device. This assumes that there is sufficient energy management system onboard the interceptor to accomplish this without sacrificing the divert propulsion needs in the end game as the interceptor warhead or hit-to-kill vehicle maneuvers itself into the proper position for intercept.

[c]Sometimes it is decided NOT to launch an interceptor if, for example, it is determined that the threat trajectory is not headed for the defended asset, and it is allowed to make a harmless splash in the ocean or similar fate for the threat missile. This very real battlefield decision must be properly accounted for when comparing the number of intercepts vs the number of threat missiles launched.

From short-range missiles to 3000-km range missiles, the time available for the 10 steps toward a successful intercept and for early-warning systems interceptor systems and the BM/C^3 systems to function *must be accomplished within 2 to 15 min.*

Table 4.6   Final three steps for a successful intercept

Steps for intercept	Function
8) *Kill assessment*	A finite amount of time is required for the interceptor radar system to determine if an intercept has indeed occurred and whether or not a second shot is required.
9) *Second, ..., nth shot if needed*	At this time a second interceptor is launched under a shoot-look-shoot-firing doctrine if it is determined that a second shot is possible or needed. After this the process from Step 7 can be repeated for as many shots as required.
10) *Final kill assessment*	At this final stage of the defense, a kill assessment is made to ensure that the desired P_{surv} of the defended asset has been achieved.

EXAMPLE PROBLEM 4.3—MISSILE DEFENSE RESPONSE TIMES

Assume that a 1500-km range ballistic missile is launched toward the defense. What are some of the key times available for the defense to act?

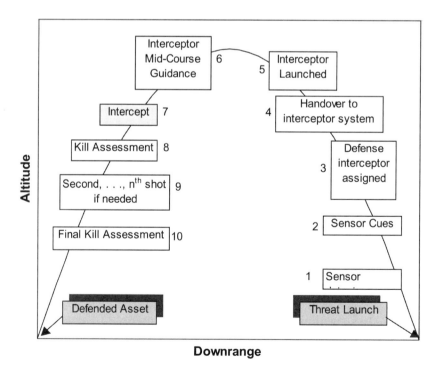

Fig. 4.52   Typical events along threat missile trajectory.

Table 4.7   Flight times of minimum energy ballistic trajectories

Maximum range, km	Ballistic flight time, s	Ballistic flight time, min	Boost-phase time, mins (average $n = 6 g$)	Approximate total TOF, min
120	153	2.6	—	—
300	243	4.0	—	—
500	313	5.2	0.50	5.70
1000	443	7.4	0.83	8.23
1500	542	9.0	1.00	10.00
2000	626	10.4	1.08	11.48
2500	700	11.7	1.17	12.87
3000	767	12.8	1.25	14.05

From the earlier trajectory equations it can be shown that the burn time for a 1500-km range ballistic missile is approximately 60 s, and the height at end of burn-out is approximately 58 km. From Appendix F this gives the radar horizon as 995 km at the moment of burnout. In a practical sense any radar would have a lesser value of range because of the need to have some minimum elevation angle to avoid clutter, and hence a much longer-range radar would be required. Alternatively, a satellite would "see" the launch even if in low Earth orbit at 1000-km altitude with a radar horizon of greater than 3500 km. If it assumed that the satellite is the early-warning sensor then it would require approximately three hits to determine trajectory. If the satellite were the DSP, then 30 s would be a reasonable assumption for type, range, and direction determination. (Later satellite systems such as SBIRS will greatly reduce this time requirement.) If there were no clouds, then it is reasonable to assume that 90 s would be consumed in providing this data to the interceptor system.

The TOF for a 1500-km-range ballistic missile is approximately 542 s (Table 4.7); hence, 452 s remain for the interceptor to acquire the target from the satellite, launch an interceptor, intercept the ballistic missile, and leave time for kill assessment and launch any second interceptor.

If the interceptor is a 500-km-range missile (with a V_{bo} of 2.0 km/s), it would have a flight time of approximately 313 s, which is less than the time available but leaves little room for error.

Such calculations indicate the nature of the problem of missile defense and the need for much improved early-warning systems and a seamless BM/C^3 network to transfer the data throughout the defense system.

BM/C³ SYSTEM FOR MISSILE DEFENSE AND TERRORIST DEFENSE

It is difficult to generalize on the type of BM/C³ systems (and all its acronym variations) that are required for all of the defense systems likely to be considered in defense against both BM and CM in the future. Many networks have been employed in the past in each of the nations of the Alliance. Some of these systems have been relatively seamless in transferring data throughout their subsystems, and some have been self-contained. Up until the present time, most of the available networks have concentrated on being air defense systems capable of handling airbreathing threats only. The advent of ballistic missile defense is a new phenomenon that has only gained momentum since the 1991 Gulf War where ballistic missile threats became a sudden reality. The high speeds of ballistic missiles, which are an order of magnitude higher than airbreathing vehicles (see Chapter 2), and the high altitudes reached, which are more than an order of magnitude greater than airbreathing threats, has brought in a new dimension to BM/C³ networks that has not as yet been put into practice. Figure 4.53 shows a pictorial of what a semiglobal BM/C³ network might look like to handle the evolving threats of both BM and airbreathing threats (ABT).

In Fig. 4.53, the use of the term extended air defense (EAD), which means conventional air defense against any airbreathing threat (such as aircraft and cruise missiles) and extending that defense against the evolving threat of BM. There are several key features in Fig. 4.53, some are underway as a result of the normal development of defense systems, and some are

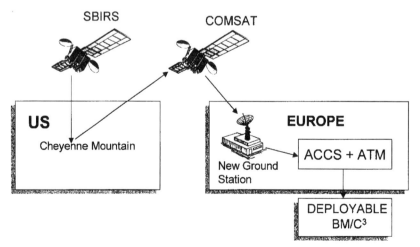

Fig. 4.53   A possible broad based BM/C³ network for extended air defense.

postulated here as possible new avenues to pursue in the light of the current world situation involving missile threats.

The NATO Air Command and Control System [(ACCS) see Chapter 5], is the current program that is underway and is expected to become operational within this decade. It was primarily designed to replace all of the earlier BM/C^3 systems for conventional air defense. This system will be the central core of BM/C^3 network and sensors for Allied defense. Plans are being discussed to incorporate the new ballistic missile threat. There are also plans to incorporate an out-of-area capability when NATO is called upon for such scenarios. Currently, it is a military system. After the 11 September 2001 attack in the United States with the use of civilian aircraft as "missiles" (see Chapter 2 for more discussion of this threat), the newly created U.S. Homeland Security Office proposed that all military BM/C^3 systems now be compatible with the civilian air-traffic-management (ATM) systems that operate civilian airports. Figure 4.53 shows a possible addition of a similar ATM capability with ACCS in the NATO scenario.

The final feature shown in Fig. 4.53 is the possibility of linking the U.S. early-warning satellite system (shown as SBIRS in the diagram) with this NATO network for a semiglobal BM/C^3 network. The particular idea shown is of a satellite link through some form of ground station into this network. There are other possible ways of making such a connection. The main point to be made here is that any such arrangement will require a timeline from the early-warning satellite through any ground station (or other device) into an Allied network that then is in sequence and must pass the data ultimately to the fire control units on the battlefield (or at sea or in the air). These timelines are not reproduced here in any generic sense because of the obvious dependence on the scenario and choice of systems. The overriding requirement in all cases, however, is that all processing times, relay times, etc., must be within the times already summarized and discussed earlier for the final integration of early warning, interceptor, and BM/C^3.

DISCRIMINATION

In much of the analysis in this book, the performance of the missile defense system is calculated on the assumption that the defense system radar (or other sensor) has indeed detected the incoming missile at a specific detection range (R_D) for the given signature level (RCS or IR signature). Or, in some cases, the defense system radar has received a "hand over" set of target information from some form of Early Warning system such as a satellite. In either case, the performance of the defense system then proceeds from this essential input. While these are correct assumptions, an essential part of the acquisition process not discussed up to this point is that the signal received, by either the Early Warning system or the defense system fire control radar, is indeed that of a threat missile and not that of a decoy or other form of countermeasure.

In the chain of target acquisition, there are five basic elements that must be satisfied before the performance analyses discussed thus far can go into effect, they are:

Awareness
Detection
Classification
Discrimination
Tracking

Each of these five elements have a significant effect on; (a) the starting time (t_d) used in the performance analysis, (b) the reaction time (t_r) for the defense system to respond, and (c) the probability of kill (P_k) of the defense system.

AWARENESS

This element is that associated with the Early Warning system. Because of many other events that are probably ongoing at the same time, the Early Warning system must determine whether the "target" is really something that the defense system must be aware of, as opposed to say, a satellite launch or some other benign event.

DETECTION

Here, the radar or Early Warning system has now applied certain criteria (such as speed and signature value) that determines to some level of acceptance that the "target" *detected* is or is not a missile launch.

CLASSIFICATION

To determine if the suspected missile launch just detected is a missile that the defense system should be worried about, the Early Warning system or the fire control radar must *classify* the threat. This is normally done by determination of speed and trajectory. The question that this system must answer is: is this a ballistic missile of a certain speed and range, and is it headed in the direction of the defense system? In the earlier discussion on the subject of BM/C^3 and Early Warning (see Table 4.5), these questions are to be answered by the first two steps in target acquisition. It is commonly referred to as "cueing". Depending on the level of information achieved by what time, the names "coarse cueing" and "precise cueing" are frequently used in the literature. The time taken for these functions depends on the characteristics of the sensor system doing the cueing. If, "three points on a curve" are needed to classify the launched threat in terms of speed and direction, and the sensor system needs, say 10 s (using the aging DSP satellite as an example) to establish each point in terms of space (x,y,z) and time (t), then some 30 s are required to *classify* what the threat is (e.g. Taepo Dong I headed NW toward London) for this part of the target acquisition process, as discussed earlier.

DISCRIMINATION

Unfortunately, that is not the end of the story, in the detection of the threat, and thus the determination of the time "to detect" (t_d). Because of use of countermeasures (such as decoys, chaff, jammers, etc), the object just detected may not be the actual missile or warhead. The defense system must now contend with the much more difficult subject of *discrimination*, which is the ability to discriminate the actual threat warhead from the accompanying objects meant to confuse and defeat the defense.

Various techniques are used in modern day technology to determine which is the threat and which is the object designed to throw the defense "off course". Most of these technologies take time during the acquisition process and thus degrade either the time allocated to detect (t_d) or immediately after during the reaction time (t_r). One example, is the use of the relative drag values (or ballistic coefficient β) of the heavy warhead and the lighter weight decoy or other countermeasure. As the decoy slows down and pulls away from the warhead, then the determination of the actual warhead becomes clearer, but time has continued to degrade the reaction time (t_r). Further, if the radar determines that, say, Object A is not the warhead and the defense interceptor must now divert (through its onboard divert propulsion motors) to intercepting Object B, which has now been clearly defined as the warhead, the issue is whether or not there is sufficient onboard power and time to cover the intervening distance that has occurred between Object A and Object B. This will change either or both of the times (t_d and t_r) and the probability of kill (or hit) of the incoming missile.

TRACKING

Once the various functions of *awareness*, *detection*, *classification* and *discrimination* have been properly accounted for in the acquisition process, then the performance analyses that are based on a given set of (R_D, t_d, t_r and σ) can then be safely used. Because of the above real world factors, one should be careful about assuming a "perfect defense" with all parameters taken as ideal. To this end, the various sensitivity explorations in all the performance analyses have been included to determine the likely outcome in wartime or less favorable conditions.

The various elements in the important subject of *discrimination* are quite complex and contain many classified features that preclude a detailed treatment. However, there are some basic characteristics that can be summarized to indicate the scope of the problem and how it affects the performance analyses included in this book.

Figure 4.54 shows a collection of possible objects that could be released from the threat missile. Such objects could be released at any time after the end of burn-out or boost phase of the missile, which normally would occur above the atmosphere (see Appendix C, Fig. C.4 for typical burn-out altitudes for ballistic missiles). The objects drawn in Fig. 4.54 are indicative only and demonstrate that the "target cluster" is usually made up of closely spaced objects (CSO) during the initial parts of the ballistic missile trajectory. Such a cluster could contain:

Warhead

Separate decoys

Decoys shrouded in gas filled, metal-coated balloons

Radar reflecting Chaff strips

Jammers emitting signals that interfere with defense radars

Other electronic countermeasures (ECM) to confuse the defense

Not all threat missiles would necessarily contain all such devices but all are possible in the expected range of threat missiles in the future. In addition to the possible contents of the threat launched object, there is the additional

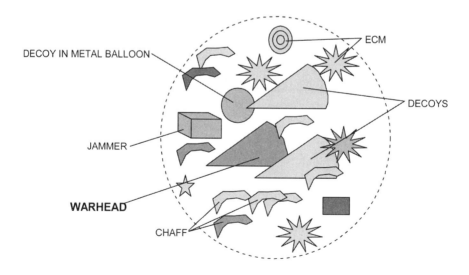

Fig. 4.54   Possible target cluster released by threat missile.

complication of the likelihood of nearby debris from earlier launches and space object remains that could confuse the defense sensors.

In the diagrammatic sketch of a possible cluster of closely spaced objects, the actual warhead is surrounded by the various countermeasures. The problem facing the defense system is to discriminate early in the trajectory between the warhead and the other objects. This essential function of discrimination must be done in sufficient time to launch the interceptor and further to ensure that the interceptor once launched intercepts the warhead and not the other objects. In the particular case of nuclear warheads, there is the possibility that the threat missile contains a salvage fuse, which could detonate if the defense interceptor intercepts the missile. This could cause complications in the defense if the chosen altitude of intercept is not selected properly. Such a detonation could produce the electromagnetic pulse (EMP) and transient radiation effects on electronics (TREE) (see Chapter 2 for more discussion on this real threat from nuclear explosions). Such an effect could be exactly what the attacker intends, in that the EMP and TREE effects incapacitate the defense communication systems.

Because the target cluster is released above the atmosphere, all objects will tend to continue on the same ballistic path until they re-enter the atmosphere near the defended asset. As shown in the earlier trajectory calculations (see Figs. 4.6–4.10) this occurs at and below about 30 km altitude, which is in the terminal phase of the trajectory. This is shown in Fig. 4.55.

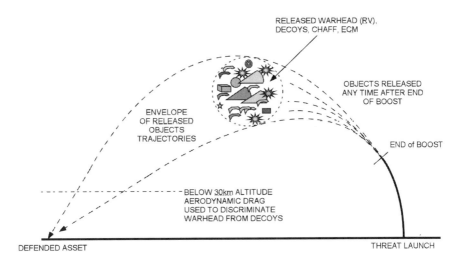

Fig. 4.55   Trajectories of released target cluster.

Clearly, waiting until the aerodynamic resistance of the higher drag, lighter weight decoys and ECM devices becomes the discriminator is not an attractive defense because of the proximity to the defended asset. Such time delays in the defense are indeed possible but assumptions of long times for detection and reaction (t_d, t_r) would severely degrade the size of the defended footprints as shown earlier in this Chapter. Many other techniques have been well tested but are classified and thus not possible to be discussed in this book. An indication of the type of an interceptor defense system trade that can be accomplished can be seen by the treatment of a defense radar that uses radar discrimination techniques other than the aerodynamic techniques mentioned.

DISCRIMINATION USING RADAR TECHNIQUES

Skolnick (Ref 51) provides the detection range (R_D) of a radar designed to search the sky, in order to detect an object, in the form:

$$R_D = \left(\frac{P_{av} A_e t_s \sigma}{4\pi F_n k_B T_o L_s (S/N) \Omega_s} \right)^{1/4} \tag{4.38}$$

where
    R_D = detection range (m)
    P_{av} = average power of the radar (kW)
    A_e = effective area of radar antenna (m^2)
    t_s = time to search radar volume
    σ = radar cross section of object (RCS) (m^2)
    F_n = radar system noise figure
    k_B = Boltzmann's constant (1.38×10^{-23} J/deg)
    T_o = reference temperature (290°K)
    L_s = system losses
    S/N = signal-to-noise ratio required at receiver output (single hit)
    Ω_s = scan volume (steradians)
This shows that the radar detection range is dependent upon the average power (P_{av}); the effective antenna area (A_e) and the volume of the "sky" that the radar must search (Ω_s). In this form, the radar equation is not dependent explicitly on the choice of (center) radar wavelength (λ) but on the RCS, which is a direct function of the wavelength (λ). Now, for *discrimination* purposes, it is known that the resolution or angular width of a radar beam or beamwidth (B) is a key parameter in determining the size of the object and is given by:

$$B = \frac{\lambda}{D} \tag{4.39}$$

where D is the diameter of the receiving antenna $\left(\approx \sqrt{4A_e/\pi}\right)$. The beamwidth (B) is also determined by the angle at which the beam power falls to one half (3dB) its maximum value at the center of the beam. For this reason, this is known as the 3dB beamwidth. It is sometimes more informative to express this as the separation distance (ΔX) between the two objects, such that:

$$\Delta X = \frac{\lambda R_D}{D} \tag{4.40}$$

Consider now the case of two objects of the target cluster shown in Fig. 4.54; one being the actual warhead, the other a decoy that is in close proximity to the warhead. The RCS signatures might look like that shown in Fig. 4.56. In the left hand sketch of Fig. 4.56, the decoy is immediately in front of the warhead masking it from the surveillance radar. A single beamwidth (B) is shown, which is the smallest width that the radar can use for resolution of the target. In the right hand sketch of Fig. 4.56, the decoy and warhead are separated by one full beamwidth (B) and now the radar can discern two distinct shapes.

As the two objects (in this example) become more separated but remain in the radar field of view (FOV) as they continue along their trajectory, the radar can determine more characteristics of the two shapes and from previously determined algorithms the defense can determine which is the warhead and which is the decoy, and take appropriate action. To lend some

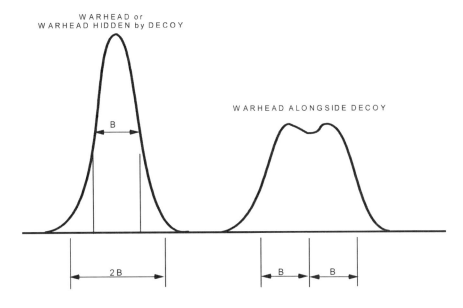

Fig. 4.56   Two closely spaced objects in radar FOV.

practicality to the discussion, consider an X-band radar being used in the search of the objects. The wavelength (λ) in X-band is about 3 cm. If the effective antenna diameter of the radar is, say, 12 m then the beamwidth (B) is approximately 0.0025 radians or 0.14°. If this example X-band radar has a detection range (R_D) of 1000 km, then the beamwidth will be $\Delta X = 2.5$ km ($= 0.0025 \times 1000$). That is, the two objects (warhead and decoy) would have to be separated by 2.5 km in order for the example radar to differentiate between them. There are other factors that contribute to the discrimination. These include measurements of range resolution, pulse compression, Doppler measurements, etc that can be found in the standard texts on radar systems. Each of these techniques provides different features that improve the discrimination between warhead and decoy (and other countermeasures) to a fine resolution. The actual size, length and signature can all be determined and with the proper use of discrimination algorithms determine which of the objects is the warhead. A key element in such techniques is the time taken to make the final determination while the objects are still moving along their trajectories and separating. The final consideration in the "end game" analysis is the determination that the interceptor can divert, if need be, from locking on to a decoy and then switching to the warhead once the discrimination process is complete. Figure 4.57 shows diagrammatically what could be the situation.

In Fig. 4.57, the interceptor is shown diverting from the decoy which its seeker had originally "locked on" and was now diverting to the warhead once the on board discrimination process was complete. The time taken for this action and the effect on the probability of kill now depends on many characteristics of the interceptor itself (seeker capability; capability of its divert system in reaction time, amount of fuel remaining) and also on the radar detection features (wavelength, antenna size, signature processing capability, etc). Such features of the system are system dependent precluding general results but a normal part of the design trades that must be made for any missile defense system given a threat set that it must be capable of defeating. The end result is to minimize the key performance parameters of detection time (t_d) and the total reaction time including the time for discrimination and divert, (t_r). Depending upon the ability of the defense system to maneuver the interceptor into the correct position to hit the warhead at the "sweet spot" will determine the value of the probability of kill (both SSPK of the interceptor and P_k of the system).

DISCRIMINATION USING OTHER TECHNIQUES

There are other techniques of discrimination other than that described for radar systems. The use of aerodynamic discrimination has already been mentioned for the latter stages of the missile trajectory with its obvious

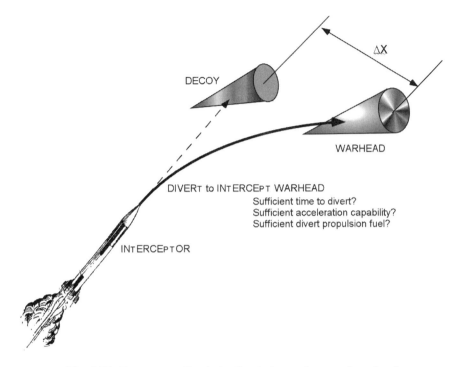

Fig. 4.57   Interceptor discriminating between decoy and warhead.

limitations. If IR sensors are used, then very similar techniques as for the radar systems can be used, except that now the radiation emittance of the various objects in the target cluster serve as the discriminators. The infrared (IR) signatures at wavelengths just to the right of "red" in the visible part of the electromagnetic spectrum provide a unique capability of identifying various bodies. Like the recognizable RCS from body shapes, the IR radiant emittance (measured in watts/cm^2) also show recognizable patterns for all the objects in the target set shown in Fig. 4.54. Typically, IR signatures are grouped into four categories, as shown in Table 4.8.

Table 4.8   IR signatures

Category	Wavelength	Typical Objects emittance
Short Wave IR (SWIR)	1-3 microns	Missile exhaust plumes
Medium Wave IR (MWIR)	3-8 microns	Missile bodies immediately after burn-out
Long Wave IR (LWIR)	8-14 microns	Missiles during mid-course trajectories
Very Long Wave IR (VLWIR)	14-30 microns	Long range missiles late in the trajectory

The function of IR decoys is to duplicate as much as possible the emittance characteristics of the missile and warhead in the various IR wavelength categories. The actual temperatures in the above categories of IR signatures range from $> 1000°K$ for exhaust plumes to $< 250°K$ for cold bodies deep in space. Large objects cool at slower rates than small objects, so a large warhead would cool more slowly than, say, a small decoy inside a metal-coated balloon. Because of this characteristic, IR sensors take measurements in different wavelengths and times, so as to determine the "signature" of the object and thus use this in an algorithm to identify it and discriminate it from the warhead. The use of using two or more wavelengths to form a signature (or as sometimes called, the "fingerprint") of the object is a useful discrimination technique. The use of different wavelengths is referred to as "multi-colored" sensing and is essential to determine the temperature differences and thus the identification of the object (warhead or decoy).

All such discrimination techniques (radar, IR, aerodynamic) are all used in different systems and sensors to rapidly sort out which of the objects is the offending warhead. Data fusion of all sensor data is also used. The object is to minimize the time taken (t_d, t_r) to identify the warhead such that defensive measures can be taken.

SPECIAL TOPICS OF INTEREST

History is a constant reminder that weapon systems and defense systems, once introduced into the world's military arsenals, will frequently be used differently than the designer originally intended. When, for example, the aircraft was first introduced into warfare during WWI, it was seen simply as an *observation platform* for the "real battle" that was ongoing on the ground. It was several years before the aircraft was used as an invaluable part of the inventory of weapons. It is a large step from *observation platform* to a *supersonic, stealthy fighter equipped with smart weapons*. It should be expected that as missile defense "matures" and is used in various modern-day conflicts that the nature of missile defense systems will also change. The bulk of this book deals with missile defense that assumes the interceptor is launched to intercept some missile launched by the enemy. The equations are straightforward and amenable to computer simulation and solution. One should not get comfortable with this idea too readily as the technology matures.

During the writing of this book, for example, the concept of unmanned air vehicles carrying interceptors (the concept of a *missile carrying a missile*) has already been introduced into warfare. There are other innovations underway that will change the very nature of missile warfare and the means of missile defense. The issue of not being misled by missile range has already been mentioned in Chapter 2. Some possibilities for both offense and defense are provided in the text that follows to stimulate thought.

DEBRIS CONTROL

The kinetic energy and momentum of the products of an intercept are such that the debris can still continue toward the original target with some degree of likely damage. If the defense includes a set of defense systems with trajectory capability almost orthogonal to the threat trajectory, it becomes possible to "herd" the debris into safe zones. In the example shown in Fig. 4.35, in which the threat is launched from Sabha, Libya, into Europe, there is the danger of collateral damage to European nations if intercepted in a generally head-on manner. Alternatively, if defense systems are located, say, in the Black Sea, much of the debris can be directed into the Mediterranean, or if a high enough V_{bo} of the interceptor is used, into the Atlantic Ocean.

SLINGSHOT AIRCRAFT

It has been shown that the aircraft is a very flexible platform for use in missile defense, in that it can be rapidly moved to new strategic and tactical locations as the threat axis moves, and is the most likely solution to BPI, etc. It has the limitation however of limited payload capability (say two or four interceptors only). This problem could be bypassed if the aircraft had the capability to fly to the threat site and then, on the pilot's command, launch his interceptors—except that the interceptors are now in launchers sitting on the ground or onboard ship. The aircraft can "slingshot" his interceptors over his shoulder and be limited only by the aircraft endurance and not the launcher rail limit. The aircraft's payload of two to four interceptors can be effectively increased to tens or even hundreds of interceptors.

TWO-LAYER DEFENSE SUBMARINES

A problem in many scenarios is the ability to place the two-layer defense system in place, especially if the conflict is out of area. Getting land-based systems and surface ship systems into place has been a problem in the past. The delay time of getting the land systems in place in Israel and Saudi Arabia during the 1991 Gulf War is a good example of this problem. There is a program underway at this time to convert strategic submarines to tactical missile use. If this concept is extrapolated further, one can foresee a submarine equipped with both upper-layer interceptor systems and lower-layer interceptor systems (in addition to the current plan to install cruise missiles) and produce a stealthy, submerged two-layer defense system that can travel to many trouble spots on the globe and provide an invulnerable defense system. The improvements in satellite early warning, introduction of global positioning system (GPS), and improved midcourse guidance techniques make this a feasible option.

NETTED ARCHITECTURES

Not shown in the analyses in this book because of lack of space is the leverage that can be achieved on the interceptor systems, with greatly improved defended-area footprints, if all systems are netted in a seamless manner. This allows one interceptor battery for example to use the radar of another battery and other variations. Such a scheme allows for faster response and supports a more graceful degradation of performance if one system or the other is incapacitated. Such schemes will require transnational cooperation and interoperable systems, and hopefully, development has started in this direction.

FIGHTER AIRCRAFT AND DIRECTED-ENERGY WEAPONS

During the 1960s, the technology of directed energy weapons (specifically lasers) could not support the weight penalties of such systems onboard ships. In the 1990s (one generation later) the technology is placing such systems onboard *transport* aircraft. Embryonic research is underway to reduce this weight problem even further, such that possibly within the next generation it will be feasible to produce *fighter* aircraft with similar capability. This will greatly change the time constants and defended footprint values discussed in this book.

The preceding thoughts are meant to stimulate discussion and also to alert the analyst not to get too comfortable with his computer programs on missile defense during these early years of missile defense evolution.

If specific program information on the performance of the various interceptor systems is not available to the reader, then the performance values given in this chapter can serve as the default values for the evaluation method in Chapter 8.

EXAMPLE PROBLEMS

4.1   In the defense of a nation with 500-km radius circumscribed circle, how many upper-layer systems are required if the interceptor $V_{bo} = 4$ km/s and the radar detection range R_D is 1000 km?

4.2   In example problem 4.1, how many upper-layer systems required if the radar detection range R_D drops to 600 km?

4.3   Using the generalized expression for the average speed of an air-launched interceptor V_i [as given by Eq. (4.37)] for boost-phase intercept, derive the specific relationship assuming that the (x, y) coordinates of the ballistic missile trajectory are a parabola with the launch point at $(0, 0)$. Assume the apogee is h_{apogee} and the range is R.

4.4   For the boost-phase intercept analysis shown in Fig. 4.47, it is shown that for a burn time t_{bo} of 85 s and a standoff distance of 400 km an average interceptor speed V_i of 5.5 km/s is required if the system reaction time t_r is 20 s. What is the required speed if the reaction time increases to 30 s.

Chapter 5

Cost

All good things are cheap; all bad are very dear.
Henry David Thoreau, 3 March 1841

Almost invariably, "cost" is a significant consideration in the decision on any defense system. The problem, however, is that cost means different things to different people, and it is important to understand how cost is to be treated before any computerized analysis is done to compare systems. Depending on the decision maker and where he is in the decision hierarchy, cost can be viewed in either absolute or relative terms. Further, where in the various phases of the acquisition cycle does the cost of the defense system come under consideration? In everyday personal life cost is often discussed relative to the question "*how much down, how much a month?*" Such a question, applied frequently to the personal purchase of homes, automobiles, sailboats, and other personal investments/consumables, applies equally well to the cost of defense systems at the government level. A related question concerns not only the cost in absolute terms, but how much the system costs in relation to a recognized budget?

Some questions pertinent to the examination of the defense system at the highest level in decision-makers minds are as follows:

1) How much is a nation prepared to spend on its defense? or collective defense?
2) How much of the defense budget should be spent on missile defense?
3) How should cost of missile defense be related to the cost (in lives, property, national honor) of having no (or inadequate) defense?
4) How should the cost of missile defense be tied to the threat? or its effects?
5) Is it better to wait until a missile defense capable of defending against the "worst" threat is available, at (possibly) a higher cost, or proceed with a lower-cost system with "a most likely threat capability" fielded as soon as possible?
6) Should the operations and support cost (over say a 20–30-year operating lifetime) receive a greater emphasis than the initial procurement cost? (This is the national defense department's equivalent of the personal, "how much down, how much a month" dilemma.)

Such questions, and many more, must be included in any cost analysis of missile defense systems. Such questions must be analyzed and debated in parallel to the normal detailed analyses of the cost of individual systems being developed or procured. In popular defense parlance such considerations are described as "top-down, bottom-up" analyses.

COST RELATED TO BUDGETS

Because budgets are related to income, it is natural to consider the national income of the nations that wish to establish some measure of determining the level of defense required or desired. A nation's economic well-being is often expressed by its gross national product (GNP), which is the measure of total output of goods and services. There are five main elements to GNP: 1) consumer spending; 2) business outlays on investment; 3) consumer investment on housing, etc.; 4) government spending on goods and services, *including national defense*; and 5) net exports. If the measure is restricted to just the national output produced domestically, then the gross domestic product (GDP) is used. Frequently, the cost of defense is compared to the GDP. (The switch from using GNP to GDP occurred in the mid-1990s.) How much a nation is willing to spend on defense as a fraction of its total GDP varies depending on how the nation views its situation in the global market environment. Nations will typically spend a higher percentage of their income in time of war than during periods of peace.

U.S. BUDGET TRENDS

The most complete historical data readily available are those of the United States. Figure 5.1 shows the monies allocated to U.S. defense expressed as a percentage of the GDP for the period 1940–2000 with projections to 2005.[39] It covers the periods of WWII, the Korean War, Vietnam War, Post Cold War, the Gulf War, and the periods of (relative) peace in between.

In Fig. 5.1 it can be seen that during WWII (1941–1945 for the United States), the national defense budget reached values of *more than 35% of GDP*. Over the succeeding period from the end of WWII to today (even through the Korean, Vietnam, and other wars), this expenditure has been steadily declining from about 12% of GDP to today's value of *approximately 3% of GDP*. It becomes a matter of national acceptance and political will as to what is the correct value for a sustained and adequate defense for the future. At the highest decision-making level such a value must be set for each nation for its defense. Again, for the United States, during this same period (1940–2000) the GDP has increased exponentially from about $100 billion in 1940 to more than $10,000 billion ($10 trillion) in 2000.

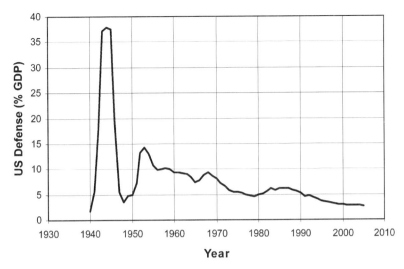

Fig. 5.1   U.S. defense budget as share of GDP.

Such percentages for defense can be also expressed in absolute terms as shown in Fig. 5.2, which shows that the U.S. defense budget has hovered around $300 billion (in constant 1996 dollars*), plus or minus $50 billion or so, over the last 40 years.

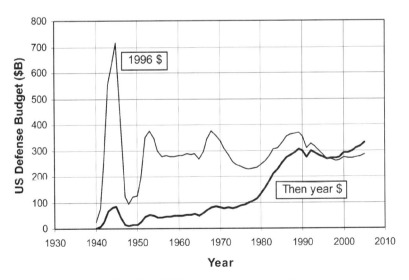

Fig. 5.2   U.S. defense budget trend.

*Because of the time needed to collect data and express it in consistent terms, it is not always possible to put all data in a common set of data banks in a timely manner. In this instance the U.S. Budget Reports for 2001 provide data only up to constant dollar charts for 1996. It is felt that such variations do not impede the general understanding or decision making on budgets as treated in this book.

There were peaks in the defense budget to approximately $350 billion during the Korean and Vietnam wars, after which the budget was decreased to approximately $250 billion during the succeeding periods of peace. The defense budget was again increased during the 1980s when missile defense gained prominence with President Reagan's Strategic Defense Initiative (SDI), which contributed to the end of the Cold War, the breakup of the Former Soviet Union (FSU), and the removal of the Berlin Wall (9–10 November 1989). The value of $300 billion per annum represents the average value through this period to today. (The current situation in the world is calling for greater defense budgets. In the United States for 2004, the request is for approximately $379 billion for defense, which equals about 17% of the proposed total budget of $2.23 trillion.)

It is a little difficult to appreciate the cost of defense when expressed in terms of GDP because GDP, like life-cycle cost (LCC), is not a dollar amount that can be saved, spent, or otherwise recognized by the general public. So another way to view the defense budget is to see what *share of the nation's annual budget* is to be allocated for national defense. Figure 5.3 shows the annual *national defense* budget for the United States over the years from 1940 to the present day when compared to the U.S. *total federal outlays.*

It is seen from Fig. 5.3 that the actual numerical value of the available funds for national defense has not grown as rapidly as the total budget. This can be seen more clearly in Fig. 5.4, where the national defense budgets have been expressed as a percentage of the total budget.

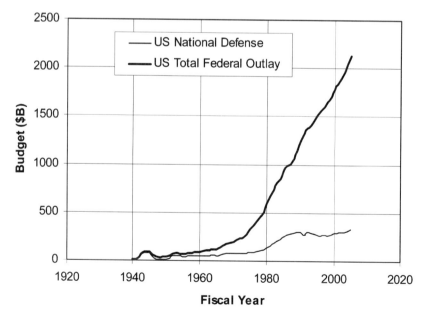

Fig. 5.3   U.S. Total federal outlay and U.S. national defense.

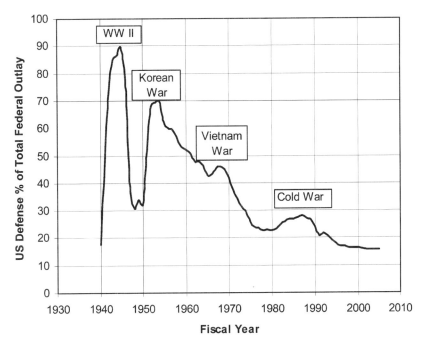

Fig. 5.4   U.S. National defense as percentage of total federal outlays.

Figure 5.4 shows a different perspective on how much a nation is willing to spend on its defense. During WWII (1939–1945) the United States increased its *annual budget for defense* to a peak value of 90% of the total federal outlay. (The United States entered the war in 1941.) After WWII this share rapidly fell back to prewar values of approximately 30% of the total budget until the Korean War, or military action as it was officially called (1950–1953), when the funds allocated to national defense rose to about 70% of the total federal budget. The share of the total federal budget for defense has slowly declined from this value (70%) over the intervening years to today with small increases during the Vietnam War (1955–1975) with a peak in 1968 to approximately 46% of the total annual budget, and the Cold War (1983–1989) when the budget rose again to a local peak of 28% in 1987. Since then, the budget for national defense has continued to decline to today's value of about 16–17% of the total annual budget. These numbers have been taken from the "*Budget of the U.S. Government, Fiscal Year 2001.*"[39]

NATO BUDGET TRENDS

Statistics such as those just displayed for the United States vary among the other nations of NATO (as well as in other non-NATO nations). The *average* share of GDP expended on defense in the European nations of NATO is less than that of the United States when the nations are taken together, although individual nations expend a higher share of GDP on defense. Figure 5.5 compares the U.S. share of GDP and the share of GDP expended by the combined total of the 17 European nations of NATO over the period 1980–2000. (Note that Iceland does not maintain a defense force. The number of nations joining NATO is increasing. Currently 19 nations are members with several more ready to join.)

In Fig. 5.5 the U.S. data have been compiled from the *"Budget of the U.S. Government Fiscal Year 2001,"*[39] which includes all data for each year since 1940. For the other NATO nations the data are only readily available in average values over certain time periods, hence the plateau values shown in Fig. 5.5, but the trends are clear.[40] From Fig. 5.5 it is seen that the United States, today, spends approximately 3% of GDP annually on defense and that the European nations collectively budget approximately 2% of their GDP annually for defense.

Figure 5.6 compares the specific values of share of GDP for each of the NATO nations for the year 2000. The United States is highlighted, with 3% GDP allotted to defense. Two nations have budgeted a higher share than the United States (>3%) of their GDP, and 15 nations have budgeted a lower share (<3%), with the average share for all nations of the Alliance being about 2.2% GDP for national defense.

The defense expenditures in absolute terms for each of the NATO nations are provided in Table 5.1 (Ref. 40) .The values are shown in Table 5.1, in which each currency unit is expressed in billions and based on current prices and exchange rates. It is important to recognize that in the collection of the data, there are likely to be differences (sometimes significant) in numerical values because of the different accounting methods used in NATO and those

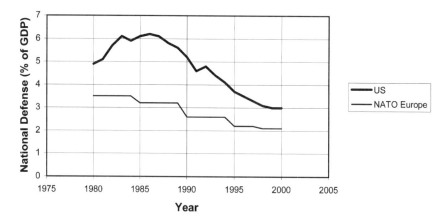

Fig. 5.5   Defense expenditures of NATO.

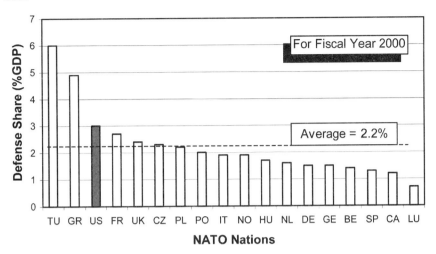

Fig. 5.6 Defense share of GDP for NATO nations.

used in the individual nations. At the time of publication of the *NATO Handbook* (1998), the nations of Czech Republic (CZ), Hungary (HU), and Poland (PO) had not yet joined the Alliance.

For example, for those nations providing military assistance abroad the relevant amounts are included in the quoted values. For those nations receiving military assistance, the amounts are not included. Also, the fiscal years vary among nations, again leading to variations in the quoted values from national values. Because France is not a participant in NATO collective force planning, the NATO values quoted here (from Ref. 40) are indicative only. Such differences can sometimes lead to confusion, and so this highlights the care that must be taken when comparing costs from one source or another. The purpose here is to indicate levels and trends and not precision in the data. In Table 5.1 the defense expenditures have been quoted in the nations' individual currencies, because the data had been provided in those units. As of 1 January 2001, the transition to the *euro*, the common European currency, was complete (for business transactions), and it is expected that future budget charts will use this new common currency. For the general populace a transition period in the use of the old currencies (non-euro) was allowed until March 2002.

The defense budgets of all nations have been declining relative to the GDP as just discussed. Today, there appears to be no sign of any significant increase in the defense budgets within NATO in the near future. The dollar value of most European defense budgets is down 7% in the 1999 budgets because of the fall of the *euro*, and follows a 22% decline in real terms since 1992.[41] Such statistics provide the needed background for any decision making on cost.

Table 5.1   Defense expenditures of NATO nations (1980–1997)

Nation	1980	1985	1990	1993	1994	1995	1996	1997
Belgium (BEF)	115.8	144.2	155.2	129.6	132.0	131.2	131.3	134.8
Denmark (DK)	9.1	13.3	16.4	17.3	17.3	17.5	17.9	18.6
France (FF)	110.5	186.7	231.9	241.2	246.5	238.4	237.4	242.5
Germany (DM)	48.5	58.7	68.4	61.5	59.0	59.0	58.7	57.9
Greece (DRA)	97.0	322.0	612.3	932.0	1052.8	1171.4	1343.3	1510.7
Italy (LIT)	7.6	17.8	28.0	32.4	32.8	31.6	36.2	37.2
Luxembourg (LFR)	1.5	2.3	3.2	3.7	4.2	4.2	4.4	4.6
Netherlands (DFL)	10.5	12.9	13.5	13.1	13.0	12.9	13.2	13.4
Norway (NKR)	8.2	15.4	21.3	22.5	24.0	21.4	23.7	23.6
Portugal (ESP)	43.4	111.4	267.3	352.5	360.8	403.5	401.1	448.5
Spain (PTA)	350.4	674.9	922.8	1054.9	994.7	1078.8	1091.4	1099.2
Turkey (1000 TL)	203.0	1.2	13.9	77.7	156.7	302.9	611.5	1101.7
United Kingdom, £ sterling	11.6	18.3	22.3	22.7	22.5	21.4	22.1	21.8
NATO Europe, $US	**112.0**	**92.2**	**186.2**	**172.8**	**172.1**	**184.2**	**186.6**	**184.8**
Canada, CA$	5.8	10.3	13.5	13.3	13.0	12.4	11.5	10.7
United States, $US	138.2	258.2	306.2	297.6	288.1	278.9	271.4	273.0
North America, $US	**143.1**	**265.7**	**317.7**	**308.0**	**297.6**	**288.0**	**280.0**	**280.8**
Total NATO, $US	255.1	358.0	504.0	480.8	469.7	472.2	466.5	465.6

NATIONAL DEFENSE PER CAPITA

It should be cautioned that there is no single measure that should be used to determine the "correct" amount that should be spent on defense. The *defense share of GDP* and *defense share of annual federal outlay* are just two measures. Another perspective on the defense expenditure that provides important insight would be the *defense per capita*. Such a measure indicates the relative amount of the available national income that should be spent (*per person*) compared to that which should be spent on other national or government "services." Using the data from the U.S. budget and from the *NATO Handbook*, it becomes possible to construct such a perspective on defense expenditure. Figure 5.7 shows this comparison.

It can be seen from Fig. 5.7 that there is a wide disparity between the nations regarding the amount deemed appropriate for defense. Using data for the year 1999, it is seen that the average value of *defense share of GDP* is 2.2% and that the average value of *defense per capita* is approximately $19,700. If there is any trend to the data, it could be argued that those nations that expend the largest percentage of GDP on defense tend to expend the smallest amount of defense per capita and that those nations which enjoy the highest standards of living tend to expend below-average shares of their GDP. Such statistics have prompted the United States, in its assessment of

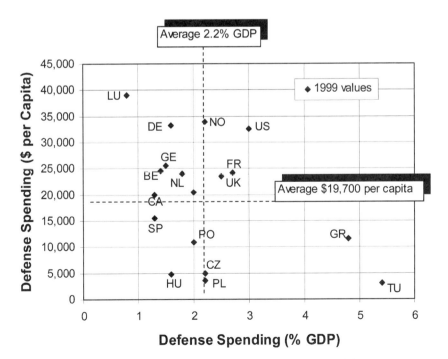

Fig. 5.7   Defense expenditure per capita and share of GDP.

allied defense, to suggest that those nations with the strongest economies and wealthiest populations should carry a proportionately larger share of the burden of providing for a *common defense* of the alliance.[42]

It is not the purpose of this book to place judgment on what is the correct amount that should be spent on defense by each nation, but it is the purpose to show the various considerations that the decision maker must take into account when assessing the cost of defense at the macro level.

COLLECTIVE DEFENSE

As discussed later in this chapter and again in Chapter 8, "collective defense" is an important consideration. Because attacks by long-range missiles will exceed national boundaries, it is necessary to con-sider the cost of defense of the Alliance in a different light than for straight-forward (individual) national defense. Decisions will have to be made as to how the cost of missile defense must be shared among all nations at risk. Using the published data for all 19 nations of the Alliance (less Iceland) for the year 2000 and using current exchange rates, it becomes possible to show the relative amounts of actual budgeted defense expenditures for a combination of NATO nations. This is shown in Fig. 5.8 for a particular combination of nations selected for the purposes of comparison.

It will be noticed from Fig. 5.8 that the larger nations in Europe have defense budgets which individually are less than 10% of the total NATO

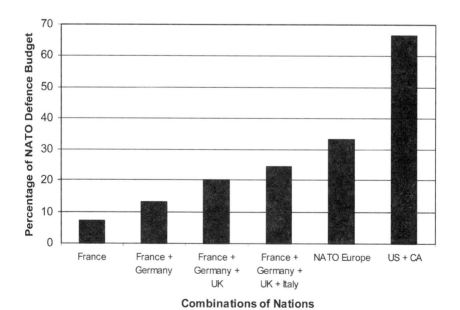

Fig. 5.8   Combined defense budgets of NATO nations.

defense budget, that if all European nations' defense budgets were combined the total would be about one-third of the total, and that North America (United States and Canada) amount to two-thirds of the NATO total. That is, North America contributes to the Allied defense twice the combined contribution of Europe.

PROCUREMENT BUDGET

A further consideration in the understanding of the cost of defense systems is the makeup of the defense budget as it is applied to the cost of military personnel, the cost of maintenance, and other important needs that are in addition to the actual cost of the hardware embodied in the defense systems under consideration. These total costs must be allocated to the U.S. Army, Navy, Air Force, and to other departments of the Department of Defense. Using the U.S. data as the most complete data available, Fig. 5.9 shows the same total defense budget as shown in Fig. 5.2, but the share allocated to just the procurement account of all hardware (systems) is also shown.

In the last decade or so, the procurement account amounts to less than 20% of the total defense budget or approximately $50 billion per annum for the United States. Again, this statistic varies among other nations. If this same percentage is applied to the NATO defense budget (see Fig. 5.5), it could be expected that approximately $100 billion per annum might be applied to the procurement of all defense systems in the Alliance. Of course, this amount must be used for all defense missions, of which missile defense is just one part. The amount that could conceivably be allocated to extended air defense (EAD) is a matter of conjecture, but if it were taken that approximately 20%

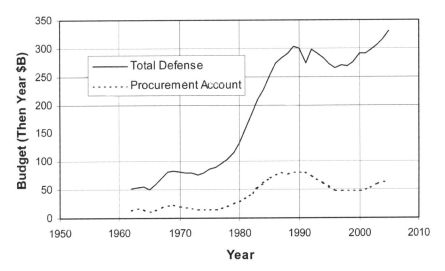

Fig. 5.9   U.S. total defense and procurement accounts.

of the procurement account within NATO was set aside for EAD this would provide of the order of $20 billion per annum for EAD. [EAD is defined by NATO as conventional air defense (AD) against airbreathing targets plus defense against ballistic missiles (BMD).] Such a crude approximation is only provided for "scoping" of the problem and should not be taken as a prescriptive analysis of the final cost of missile defense or final budget requirements. To add another benchmark on how much is actually being spent on missile defense, the budget for FY 2003 for the U.S. Missile Defense Agency (MDA) is approximately $7.8 billion, which is 2% of the federal budget. Of course, such a number does not include sunk costs, costs that are now borne by the services (once MDA has transferred responsibility of the efforts), and costs of ships and aircraft that serve as platforms for sea- and air-based missile defense systems.

LIFE-CYCLE COST

If one now changes the viewpoint from the macrobudgeting level within the nations to examining what individual missile defense systems are likely to cost, it is convenient to use the standard form of LCC, which can be expressed as

$$LCC = (R \& D \ Cost) + (Procurement \ Cost)$$
$$+ (Operating \ and \ Support \ Cost) - (Scrap \ Value)$$

In most cost analyses it is normal to drop the scrap value term (see later) and to consider just the three main components: 1) research and design (R&D) cost, 2) procurement or production cost, and 3) operating and support cost. (In some nations the full term RDT&E is used for the more complete research, development, test, and evaluation. Hopefully, this is a name change only, but one should check the individual nation's accounting system.) The sum of the first two components 1) and 2) is often referred to as the *acquisition cost* of the system. (This is an approximation for ease of discussion. See Chapter 6, where a more detailed and correct combination of RDT&E and procurement is provided to give acquisition cost.) Because most nations keep R&D and procurement in different accounts within the nation's defense budget, the same logic will be used here. Also, because the expenditure of defense R&D funds is purely to reduce risk in the acquisition of a defense system, the R&D cost is discussed more fully in Chapter 7.

The general phasing of this life cycle for all defense systems is as shown in Fig. 5.10. The degree of overlap of the three main phases of the life cycle of any defense system and the timing of the start of each phase (marked by the milestones B and C) vary depending on the specific program. This is discussed in more detail in Chapter 6. Each bloc or phase has been shown as

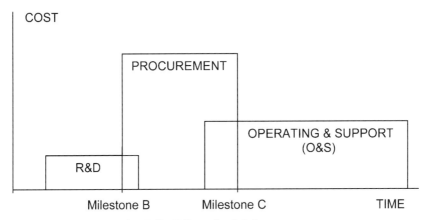

Fig. 5.10   Life cycle of defense system.

rectangles in Fig. 5.10 for simplicity, whereas in reality there is usually a ramp-up in costs at the beginning and tapering off of costs at the end of each phase. The amount of monies for each phase also varies among the different types of programs. Typically, R&D amounts to about 10% of the total LCC. The procurement costs and the operating and support (O&S) costs tend to be of about the same magnitude during a reasonable operating life (say, 20–30 years) at amounts of approximately 40–50% of LCC for each phase. These are general guideline numbers provided here for discussion only, and any particular system can change significantly from these values.

Although there is no universally accepted definition of the makeup of each of these three main components to LCC, it can be generally taken that the main ingredients are as listed here:

1) The first component is the R&D cost, which involves research costs, technology costs, and development costs, and typically is approximately 10% of the LCC.

2) The second component is the procurement cost, which involves contractor costs of nonrecurring production, recurring production, system test and evaluation, initial spares, software, etc., and government costs of GFE. The procurement cost is typically approximately 40–50% of the LCC.

3) The third component is the operating and support cost, which involves mission personnel, energy (fuel, POL), intermediate maintenance, depot maintenance, contractor support, sustaining support, indirect support, disposal of waste material, etc., and typically is approximately 40–50% of the LCC.

It must be emphasized that the typical values shown in the preceding list vary widely among systems and nations. For example, *development costs* can include prototypes and demonstrators if needed. The U.S. Air Force normally includes the cost of demonstrator aircraft as part of the R&D costs, but the U.S. Navy would not include the "first-of-the-class" ship in R&D. For missile defense programs it is normal to include prototype missiles in the test program of the R&D account. In some cases in missile *procurement cost* it can be that for some systems the contractor supplies the main body of the missile, but the warhead is added later as part of government-furnished equipment (GFE). Hence, in such a case it would be necessary to add both the contractor cost and the government cost to ascertain the actual procurement cost of the missile. For the *O&S cost* there are even greater variations possible. The O&S cost must be computed for some agreed on lifetime on the (missile) system, which can vary from 10–30 years, with a commensurate significant difference in the stated cost. If the missile defense system under analysis requires satellite early warning, then the costs of such satellites should be included, especially if sustaining launching of satellites is required to keep the system operational. (Satellites typically have short lifetimes, on the order of five years, and the lifetime of a missile defense system might be 15–20 years or more. In such a case it might be prudent to include the scrap value of the satellites to offset the LCC of the total defense system.) Some nations include mission personnel in the cost of the system, whereas other nations (including many in Europe) prefer to collect all personnel costs in a separate accounting element from the procurement account and not attributable to individual systems. Because manpower or personnel costs can be a significant contributor to the O&S costs of any particular system, this cost component requires careful attention in any cost comparisons or estimations. In the U.S. Department of Defense accounting system, these O&S costs are paid out of the *operating and maintenance (O&M)* account. Chapter 6 provides further description of the makeup of RDT&E, procurement, and operating, and support functions. The issue of *multimission systems* also tends to complicate the compilation of system costs. For example, it is normal to build ships that have a multimission capability (e.g., antiair warfare and antisubmarine warfare), as well as being a platform to carry missile defense systems. It requires some clever accounting to ensure that the cost of missile defense is not burdened with the cost of other missions. Hence, it is important when compiling costs of a given missile system to take account of the difference between *dedicated* systems and *multipurpose* systems. If a missile defense system can be used for the defense against ballistic missiles (BM) and also against the airbreathing threat (ABT), say, the cruise missile (CM), then attention must be made as to the cost of just BMD or the cost of the broader mission of EAD.

Despite such difficulties in consistent cost accounting, it is possible to discern trends in the data. These are outlined to understand cost of a missile defense system.

COMPONENT COSTS

Within the general accounting category of *procurement cost*, the defense system can be thought of as being made up of three major components: 1) early-warning system; 2) interceptor system; and 3) indigenous battle management, command, control, communication, and intelligence (BM/C^3I) system. This categorization was introduced in Chapter 3 and applies mainly to *active defense*, although there are overlaps with other mission areas that will be introduced as appropriate. In any particular instance some components might already exist, and in other cases the entire defense system needs to be developed or produced. For example, if the defense system under consideration functions under the requirement that it will utilize an existing satellite early-warning system then all that is needed, perhaps, are the downlinks and ground station. In any case it is necessary to compile the costs of all of the components such that a database is available for a variety of systems and combinations.

For all of the components to function as a cohesive whole, there is the need for a system (within the defense system) set of communication features, fire control mechanisms, etc. These elements tend to be grouped under the acronym BM/C^3I, which over the years has become the label for such a system whether it does or does not contain all of the functions and components embodied in "*battle management, command, communication, control, intelligence.*" (In some communities even more "Cs" are added to include "consultation" and "computers," which from a cost-accounting viewpoint requires attention if a comparison of common system elements is sought in any BM/C^3I system. One needs to check if the cost of computers is included or not. Also, some systems include the cost of sensors, whereas others prefer such components to be allocated to early-warning costs.) For the purposes of this book, the BM/C^3I system will mainly refer to this indigenous set of components, including the link to a broader network that connects dispersed systems, other systems, and the all-important link to the command centers. Usually, this broader BM/C^3I system is treated as a separate entity. Some examples of these broader networks in Europe would be ADGE, NADGE, ACCS, SCCOA, etc.

Each of the three main components of a defense system will be treated here as separate entities. Their combination into systems will be treated later in this chapter and further in Chapter 8.

CAUTIONARY NOTE ON COST MODELING

Cost modeling or cost estimating is an important part of defense planning and requires a detailed treatment of all of the component makeup of each of the systems. For example, for the interceptor missile it is made up of components such as airframe, propulsion booster, guidance and control system, seeker, dome, warhead (if applicable), divert propulsion (if applicable), and many other parts. Generally, the seeker section and guidance and control contribute the largest "share" of the cost, say, 20–30–40%, with the airframe, stabilizers, etc., contributing around 10%. Such details vary significantly from one interceptor missile type to another. Those missiles with fragmentation warheads would have a different cost makeup to those relying on kinetic energy (HTK) warheads. Once sufficient databases have been compiled, then cost-estimating relationships (CER) can be statistically derived to give functional forms for the CER, such as

$$\$_{\text{interceptor}} = f(W^n, V^m, \ldots) \tag{5.1}$$

in which the functional form f depends on such parameters as the weight raised to some power W^n, the speed raised to some other power V^m, and other key parameters.

Two historical examples illustrate the types of CER that are generally compiled *once a sufficiently large database has become available*. The first example is the cost of aircraft (in 1975 dollars),[43] which could be expressed as

$$\$_{\text{a/c}} = \text{constant } (W^{0.41} S^{0.20} F^{0.55}) \tag{5.2}$$

in which $\$_{\text{a/c}}$ is the cumulative average procurement cost (in dollars) for a quantity of 100 aircraft. The key parameters were the aeronautical manufacturers planning report weight (AMPR) in pounds, called W in this formulation; the maximum speed in miles per hour S of the aircraft at its design altitude; and the maximum thrust in pounds F at sea level. Another example[44] compiled for the cost of general vehicles (ships, aircraft, etc.) once a large enough database had become available is expressed as

$$\$_V = 1000W_e + Q^{-0.33}(1200P_I) \tag{5.3}$$

in which the cost of the "platform" or vehicle $\$_V$ before any payload was expressed as a function of its empty weight in tons W_e, its installed power in horsepower P_I, and the effect of the production learning curve through the number Q of vehicles built.

Similarly, in the case of sensors and radars the cost depends on many parameters such as choice of center frequency, number of transmit/receive (T/R) modules, and rotating or nonrotating antenna. The cost of satellite early-warning systems depends on whether or not the satellite is designed for geosynchronous (GEO), medium-Earth (MEO), or low-Earth (LEO) orbits, whether or not it is powered by batteries or solar power, and many other important components and design features.

Unfortunately, such CERS as those just shown for aircraft and other vehicles cannot be reliably prepared at this early stage of missile defense system preparation. This is because there have been significant technology improvements in the various systems over the last two decades coupled with an insufficiently large database of common types to establish any reliable trends.

There have been a sufficient number of missiles, radars, satellites, and other important defense systems components built however so that at least a *general trend* can be established. Simplified CERS will be developed here suitable for use in *feasibility cost estimation* and can be used as default values pending the establishment of more accurate, reliable, and repeatable data. Each of the main ingredients (early warning, interceptors, etc.) is treated in what follows, and the data can be used at least for feasibility-type investigations. Once a large enough database has been established over time, then reliable forms of CER of the types just shown can be substituted in the Evaluation (Chapter 8) to advantage.

EARLY-WARNING COMPONENT COSTS

Early-warning systems are made up of radar and IR systems. They can be on satellites, in aircraft, on the ground, or onboard ships. Some available data from unclassified sources have been collected here to indicate the trends and level of the costs of such components.

SATELLITES. It is taken by most that the space age began formally on 4 October 1957 when the former Soviet Union successfully launched *Sputnik I*. This was the first man-made satellite to go into orbit around the Earth. A short 12 years later man stepped on the moon on 20 July 1969.[45] Since that time, many types of space vehicles and satellites have been successfully launched and used for a variety of purposes, such as meteorological and weather forecasting, Earth images and high-resolution photography for assessment of visible Earth resources, radio communication, and military purposes. These military purposes include both communications and remote sensing that could theoretically include radar, infrared, and visual. In practice, designs for radar satellites have proven too heavy, and the main emphasis is on IR and optical means.

The choice of orbit for the many satellites orbiting the Earth varies between four main types of Keplerian trajectories. Other types of orbits that are used for special purposes. An example is the heliosynchronous orbit that maximizes the light conditions for the satellite, but concentration here will be on the four main types: highly elliptical orbit (HEO), GEO, MEO, and LEO.

Which orbit is selected depends on the mission of the satellite and international agreements.

The *LEO satellite* often used for commercial communications tends to be small and launched to operate in orbits that are nearly circular at altitudes from less than 1000 km to altitudes of approximately 2000 km above the Earth's surface for communication satellites. Two examples are the *Iridium* 66+ satellites), each weighing approximately 700 kg orbiting at approximately 760-km altitude in six orbital planes, and *Globalstar* with a weight of 500 kg (48+ satellites) at approximately 1400-km altitude in eight orbital planes. From a military viewpoint such relatively low altitudes allow for better communications and sensing of CM and other aerodynamic objects. Also, the period is short (usually 1 and one-half–2 h).

The *MEO satellite* operates at a higher altitude (about 10,000-km altitude) and has a longer period (6–7 hours) and provides a longer time than the few minutes for the LEO satellite, of a few hours above the local horizon for communication purposes with a user on the ground. Fewer satellites are used for this type, and this is an issue in hand-over errors between satellites. An example of a MEO satellite is the *Inmarsat* (a U.S. maritime satellite communications system) with 10 satellites orbiting at about 10,000+ km altitude inclined in two planes. *Inmarsat 2* launch mass on Arianne is 1310 kg.

The *HEO satellite* operates with an extreme variation between perigee (approximately 500-km altitude) and apogee (approximately 50,000-km altitude). Such orbits are used to provide sensing and communications mostly in the polar regions. An example of this type is the Russian *Molniya* satellite system with three satellites in 12-hour orbits around the Earth. The *Molniya* system has a perigee of 1000 km and an apogee of about 39,000 km. A sister satellite MEO system is the Russian *Tundra* system, which uses three satellites in 24-hour orbits around the Earth. The *Tundra* system has a perigee of almost 18,000 km and an apogee of over 53,000 km. The European Space Agency satellite *Exosat* (launch mass 500 kg) and an apogee of 200,000 km is another example of HEO satellites.

The *GEO satellite* has the unique feature that its period is one sidereal day (23 h, 56 min, and 4.09 s). A special case of the geosynchronous orbit is the geostationary orbit with zero inclination and zero eccentricity. Such a satellite appears stationary above a fixed point on the surface of the Earth. The geostationary satellite operates in an orbit of 35,786-km altitude. From this altitude a single GEO satellite commands a view of about one-third of the Earth's surface (approximately $-75°$ latitude to $+75°$ latitude). Three satellites thus provide an almost complete Earth coverage. There are many examples of the GEO satellite, and Agrawal[46] provides a good summary of the many geosynchronous satellites that are either in orbit or planned to be in the near future and their specific characteristics (their history, specific design features, and use).

For use in missile defense, there are several satellite programs of interest. These include the Defense Support Program (DSP), a geostationary satellite that is already in service providing much needed sensing data. It is an old design and rotates once every 6–10 s, providing a slow but accurate measurement (three "hits") suitable for threat missile characterization and trajectory determination. (The low value of 6 s is a rounded-up value of 5.7 s as quoted for DSP in Ref. 47.)

The United States has several planned satellite programs with different features that are expected to enter service this decade. Specifically, the U.S. Air Force is planning on an improved satellite system, the Space-Based Infrared System (SBIRS), which will provide a faster "revisit time" than the DSP and will be in service at various dates in "blocks" and "increments" up to 2011 (data for the SBIRS systems extracted from Ref. 48). SBIRS increment 3 is the deployment of a *LEO* infrared-sensing satellite constellation and supporting ground-processing system. It will provide threat missile booster detection, midcourse tracking, and discrimination data for the defense systems. Because of the LEO orbit selection, it is called *SBIRS low*, and the first satellite launch is expected in 2006. The *SBIRS high* satellite will be a similar system but at a *GEO* orbit. The SBIRS high system will be integrated with the DSP (until the DSP program phases out later in this decade); the SBIRS low and some additional SBIRS satellites will be on an *HEO* orbit. The total SBIRS system will eventually be made up of 30 satellites, and the full constellation is planned to be in service by 2011.

Figure 5.11 shows line drawings of the general features of the DSP and SBIRS satellites. The 92-cm aperture telescope of the DSP can be seen in the sketch. The DSP can "see" in two IR wavelengths with a sensor array of 6000 pixels, suitable for detecting both the plume of the threat missile during launch and after boost phase after burnout. Each pixel can scan a 3-km-wide spot on the Earth. The DSP was used successfully, detecting all of the 166 SCUD missile launches in the 1991 Gulf War and in the earlier Iran–Iraq War in 1988. The sketches for the SBIRS high and SBIRS low satellites with

  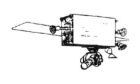

DSP Satellite   **SBIRS High Satellite**   **SBIRS Low Satellite**

Fig. 5.11   DSP and SBIRS early warning satellites.

their improved (classified) performance are also shown in Fig. 5.11. Because of the geostationary feature of the DSP and the main SBIRS high satellites, the polar regions of the Earth (above $\pm 75°$ latitude) cannot be seen; hence, two of the SBIRS high satellites will be in HEO orbits.

For many of the defense systems discussed in Chapter 4 to react to the threat in a timely manner, precise cueing of the type planned for the SBIRS Early-Warning Satellite system is required. Such a system is designed to provide the necessary precise cueing to provide enhanced warning of missile attack in several scenarios including defense of the U.S. homeland, the Allies, and deployed forces in out-of-area situations. The SBIRS low will provide the needed information for CM defense, and SBIRS high will provide the needed information for BM defense.

Although still in the early preliminary design phases (started in 2000), there is also a cooperative effort underway between the United States and Russia to develop an IR satellite program designed to address both ballistic missile defense and national security. The Russian–American Observation Satellite (RAMOS) Project seeks to develop an early-warning system that will include both aircraft and space elements. The RAMOS will operate in the short wavelength and in the mid- to long-wavelength infrared bands. This will aid both boost phase and midcourse signatures identification. The satellites are planned for launch in 2006 for a nominal two-year life expectancy in orbit.

Because of the short life expectancy of satellites, two cost elements are of interest: 1) the *procurement cost* and then 2) the operating and support costs, of which *launch costs* are a large part. Figure 5.12 shows a collection of satellite procurement costs that are for a variety of LEO, GEO, and MEO satellites. Figure 5.13 gives the available data on satellite launch costs expressed in the form of cost per payload (satellite weight) launch. Although these data are not in the form that can be sorted in terms of cost drivers such as type of payload, power supplies, antenna design, choice of transmission frequency, and many other important factors, they do provide a general costing for preliminary use in cost estimating.

The data in Figs. 5.12 and 5.13 have been compiled from a wide range of published sources including aerospace magazine articles and interpreted published contract values (not always the most reliable source, because frequently they can contain hidden factors such as built-in escalation cost factors, block estimates for quantities produced, and other factors that are difficult to unravel). The published data in most cases are not definitive, and the values must be assumed to have a wide variation in accuracy. As just one example, the *Titan 4* launch vehicle is quoted as delivering a payload of between 18,000 and 21,000 kg into LEO orbit or a 5800-kg payload into GEO orbit for a common launch cost of \$250–400 million.[49]

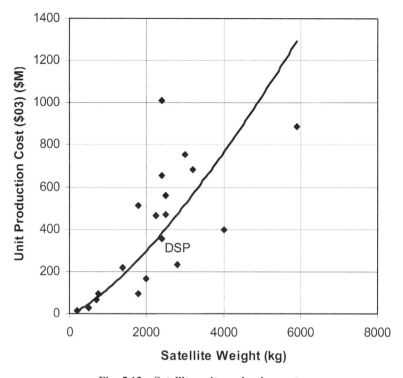

Fig. 5.12   Satellite unit production costs.

Hence, given such a wide range of values the published data can only be used as a guide to satellite costs, but the data are collected here for general use. The lifetimes of satellites also exhibit a wide range, from 1–15 years depending on the type and mission.[49]

Even though the data are somewhat sketchy, trend lines have been added for satellite unit production costs in Fig. 5.12 and the launch costs in Fig. 5.13, which have the following forms.

Satellite unit production cost:

$$\$_{satellite} = 0.01 \cdot W^{1.36} \tag{5.4}$$

Launch cost into LEO orbit:

$$\$_{launch} = 0.53 \cdot W^{0.74} \tag{5.5}$$

Launch cost into GEO orbit:

$$\$_{launch} = 1.25 \cdot W^{0.52} \tag{5.6}$$

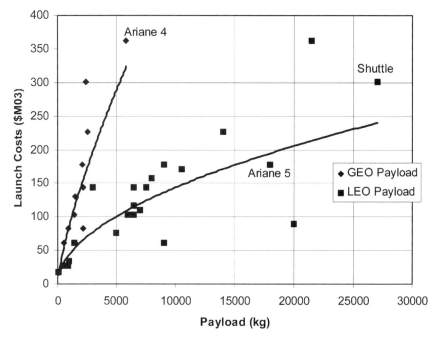

Fig. 5.13   Launch costs of LEO and GEO satellites.

in which $\$_{satellite}$ is the unit production cost of the satellite in M\$03, $\$_{launch}$ is the launch cost in M\$03, and W is the gross weight of the satellite in kilograms. No clear distinction was found between LEO, MEO, and GEO satellites in terms of production costs, although GEO sizes tend to be smaller. For example, the DSP satellite (model numbers −18, −20, and −21 launched in the period 1997–2001) just referred to has a weight of approximately 2363 kg and a design life of 5–9 years.

The launch costs have changed significantly over the years. In 1994 the shuttle launch costs were quoted as approximately \$225 million, which would be approximately \$300 million in today's dollars with an assumed 4% annual inflation rate. However, recently published data[49] quotes \$550 million for a shuttle launch. Hence, the *trendlines* provided here should only be used as guidelines pending more detailed information being made available.

Although there is no definitive trend in launch costs, a set of approximate values such as \$28,000 per kilogram of payload for LEO satellites and \$88,000 per kilogram of payload for GEO satellites as *average* values of the preceding data can also be used for preliminary sizing purposes.

EARLY-WARNING RADARS.   The use of *r*adio *d*etection *a*nd *r*anging (radar) has seen phenomenal growth and use in all civil and military applications since the original Hülsmeyer patent in 1904.[50] This original work was greatly expanded upon by such researchers as Marconi, Taylor, and Young in the years leading to WWII, in which radar was used to such advantage in many military situations. The history of radar development can be ably studied in Skolnick.[51]

Although weight is obviously a key parameter that drives the cost of satellites, radar costs are driven by very different parameters. This is best seen through the use of the familiar radar equation, which in itself can take many forms. The most useful form for the purposes to be used here is

$$R_{\max} = \left(\frac{P_{\text{avg}} A_e^2 \sigma}{4\pi\lambda^2 S_{\min}}\right)^{1/4} \tag{5.7}$$

This form of the radar equation shows the maximum detection range $R_{\max}$ of the radar in terms of the following key parameters:

P_{avg}, average transmitter power;

A_e, effective antenna area;

σ, radar cross section (RCS) of target (threat);

λ, wavelength of transmission; and

$S_{\min}$, minimum detectable signal.

Even in this form, the radar equation hides several interdependencies of each of the parameters. The antenna *gain* G is a measure of the efficiency of the radar and expresses the ratio of the power radiated in the direction of the target (threat) to that power which would have been radiated from an isotropic antenna, that is, that antenna which radiates uniformly in all directions. This antenna *gain* is related to the other key parameters in Eq. (5.7) as expressed by the following equation:

$$G = 4\pi A_e / \lambda^2 \tag{5.8}$$

The next key parameter is the radar cross section σ of the threat. This is a measure of the reflected energy back to the antenna from the target and has the units of area. It is a complicated characteristic that is not always related to the physical size of the target. The reader is referred to the various textbooks on RCS for discussion of how the RCS or σ relates to physical size, shape, frequency, distance from receiving antenna, etc. It is sufficient to say here that σ is a function of radar frequency or wavelength λ. These interrelationships between all of the parameters make it difficult to express simple or direct relationships that can be stated in any global form.

It is seen, however, from the preceding equations that radars can be designed with many different purposes in mind that directly relate to the end value of cost. If the radar is to be designed to detect a certain RCS (σ) at a certain distance $R_{\max}$, then there are different combinations of aperture size

A_e, choice of wavelength λ, and average power transmission levels P_{avg} that can accomplish this purpose. These various combinations lead to different costs. Radar system manufacturers do not always provide all of the different characteristics in sufficient detail to make the necessary corrections to the functional form of the cost. Generally, one would expect the radar cost to have the following form:

$$\$_{radar} = f(R_D, \lambda, \sigma, A_e, \ldots) \qquad (5.9)$$

in which the radar detection range R_D has been used in Eq. (5.9) to reflect the form of the performance analysis given in Chapter 4. Usually, in the design of the system the maximum range R_{max} is sought and equated to R_D. Here another complication comes into play in that the radar detection range depends also on the volume to be searched and the integration time to search that volume. In another form of the radar equation, these relationships can be determined, but it can be said that the range depends on what is called the "power-aperture product" or $P_{avg}A_e$. That is why, when radar detection ranges are quoted, it needs to be said that the range R_D is the range which can be achieved by concentrating the beam to a narrow azimuth and elevation angle, and thus concentrate the power P_{avg} for a longer range, or is that range quoted for the case when the power is being used to search a much larger volume. This feature was shown in Chapter 4 (see Fig. 4.37) when the subject of cued range by some external sensor (such as the satellite just discussed) was discussed. Clearly, the range quoted for any particular radar must identify this characteristic and to what size target σ to which this is being applied.

Most radar manufacturers have their own formula for the cost of the radar expressed in each of these key parameters. Unfortunately, many of the design requirements are classified (especially when revealing the RCS being used in the design), which limits the ability to provide general formulas for the radar costs. Fortunately, there are some basic characteristics that dominate the designs, and these provide sufficient guidance in interpreting cost data from the various manufacturers reports, brochures, etc. Two key parameters are the final detection range R_D and the center frequency of the radar (or as is usually used, the wavelength λ). [The wavelength can be directly computed from the frequency from the relationship $\lambda = c/f$, where f is the center frequency of the transmitting radar and c is the speed of light $(3 \times 10^8 \text{ m/s})$.]

Figure 5.14 collects the quoted manufacturers prices for a wide range of radars collected by the author over several years with all costs adjusted by the required annual inflation factor to today (2003) from the year of the quoted price. It is seen that there is a definite trend in that longer-detection-range radars cost more than shorter-range radars (in an almost linear form) and

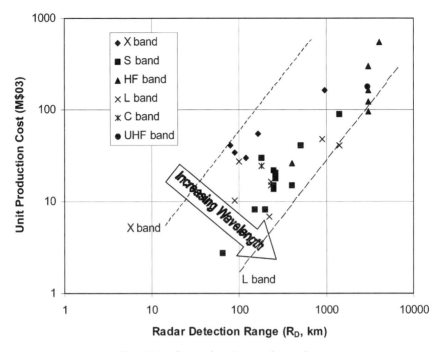

Fig. 5.14   Costs of early warning radars.

further that higher-frequency (shorter wavelength) transmitting radars cost more than the lower-frequency radars. The primary driver in both features is the power requirement P_{avg} and the size of the antenna A_e.

In this form the cost of a single radar can be expressed by

$$\$_{radar} = f(\lambda)R_D \tag{5.10}$$

In this cost relationship the cost is directly proportional to the detection range R_D but is modified by a function of the wavelength λ. From the radar equation one can see that this function $f(\lambda)$ is a complex function of several other interrelating parameters. Hence, $f(\lambda)$ cannot be explicitly related in a simple form because of these other factors in the radar equation such as antenna size, efficiency (antenna gain), and the varying design requirements from the customer or designer on the RCS value used.

Figure 5.14 has shown the costs for a range of radars that have the labels of their designated center frequency, which is another variable that is not always made obvious in the published descriptions of the radars. The radars have been labeled here as "early-warning radars" to emphasize that these values have been quoted assuming the main purpose of the radar is to see a target at some far distance. Some of these radars can also function as fire control radars for the interceptor system. This then introduces the discrimination feature that is a complicated function of the choice of center frequency f and size of target σ being detected. Only the early-warning feature is being included in this discussion. The literature is not always open about identifying the size of the target σ when quoting the detection range R_D, for obvious reasons. Such omissions also contribute to the spread of the data in Fig. 5.14.

It would appear at first glance that the unit cost of radars could be normalized by plotting cost against a parameter such as $\sqrt{\lambda R_D}$, because this parameter is proportional to the product of the power-aperture product $P_{avg}A_e$ and the antenna size A_e. Some cost estimators use the number of T/R modules as a key parameter because they represent both size and technological sophistication in the design of the radar. Unfortunately, such devices are difficult to collapse to such single parameters because of the many trade decisions of all parameters that comprise the final cost. Accordingly, the data available will be left in the form shown in Fig. 5.14 for the cost estimates used in the evaluation method in Chapter 8.

The radars in Fig. 5.14 have been identified by letters (X, S, L, etc.), and this requires some explanation. When radars were introduced during WWII, it was decided to identify the frequencies that were being used by letters, partly for convenience and partly for security reasons to confuse the enemy and to minimize the chance of having the radar jammed in that frequency. The confusion continues today for a different reason — over the years different groups have attempted to improve the lettering system for radar use and for electronic countermeasure (ECM) uses generally, and there are now several lettering systems being used in the user community, with some of the same letters being used for different frequencies! Also, some users also mix the lettering systems as they discuss both radars and ECM!

During WWII the lettering system was P, L, S, C, X, K, Q, V. After WWII the radar lettering system was VHF, UHF, L, S, C, X, K_u, K, K_a, Millimeter.[52] The ECM lettering system is A, B, C, D, E, F, G, H, I, J, K, L, M.[53]

These different lettering systems cover approximately the same range of frequencies from about 0.1–300 GHz. As can be seen, this causes confusion because the same letters appear in each grouping. Some designers and manufacturers use the post-WWII radar lettering system to describe their systems, and some use the ECM system. (Mercifully, the early WWII system of lettering has disappeared from the literature.) Figure 5.14 uses the post-WWII system because of its familiarity and more consistent use of radar frequencies. Table 5.2 provides a guide to the various ways of describing frequencies.

Some attempt has been made in Table 5.2 to line up frequencies between the two systems. It will be noticed that there is not an exact match between

Table 5.2 Comparison between radar and ECM lettering systems

Radar nominal frequency range f, GHz	Midfrequency range wavelength λ, cm	Radar (post-WWII) lettering system	ECM frequency range f, GHz	ECM lettering system
0.003–0.03	2000	HF	0–0.25	A
0.03–0.30	200	VHF	0.25–0.50	B
0.30–1.00	55	UHF	0.50–1.00	C
1.00–2.00	15	L	1.00–2.00	D
2.00–4.00	10	S	2.00–3.00	E
			3.00–4.00	F
4.00–8.00	5	C	4.00–6.00	G
			6.00–8.00	H
8.00–12.0	3	X	8.00–10.0	I
12.0–18.0	2	K_u	10.0–20.0	J
18.0–27.0	1.5	K		
27.0–40.0	1	Ka	20.0–40.0	K
			40.0–60.0	L
40.0–300	0.5	mm	60.0–100	M

the two systems. This difference causes more confusion and helps explain why some manufacturers still use the old (radar) system. One example is the X-band radar that is currently under development for both the upper-layer (now terminal-phase) interceptor system THAAD and the new missile defense system under consideration for homeland security in the United States. Most of the literature today still uses the radar labeling (where X band is referred to as 8–12 GHz), and it is rarely referred to as an I-band radar, which is in the 8–10 GHz band. Eventually, one hopes, these labeling systems will be stabilized to some commonly acceptable terminology for both radar and ECM use, but in the meantime Table 5.2 can be used as a guide if the reader finds different labeling in the literature.

The unit costs of radars with the appropriate lettering identification as used here will be as given in Fig. 5.14. As the radar designer seeks to get the "best" design possible for his particular application, he has the ability to adjust design parameters within limits to achieve his design goals. This can mean adjusting the center frequency, as just one example. One radar could be designed to operate at 3 GHz, and the other could be designed to operate at 3.3 GHz, but with the lettering system given in Table 5.2, they are both S-band radars, similarly for the other choices of center frequency for which there is some latitude for design optimization.

Table 5.2 provides a clue as to why it is difficult to collapse radar costs to simple parameters. The radar cross section σ varies with radar frequency and distance from the antenna. For the radar to discriminate the target, high frequencies are required. For simple detection, lower frequencies are best, especially if the wavelength approaches the size (length) of the target. For the radar enthusiasts this means moving away from the resonance region and into the Rayleigh region, where the wavelength is equal to or greater than the length of the target. Table 5.2 shows that this starts to occur in the UHF, VHF, and HF radar-frequency ranges for the most likely sizes of threat missiles (say, of the order of 1 m or less). Such a wide range of wavelengths contributes to significant changes in the type of antenna used, with the expected change in costs. Figure 5.15 shows some typical radars in some of the radar frequencies in use today.

The different antenna designs seen in Fig. 5.15 illustrate the difficulty of seeking a common cost equation for all types. In older missile defense systems it was normal to provide the interceptor system with two radars; one for surveillance (say, a larger S band) and one for fire control and discrimination (say, a smaller X band). There are significant departures today to this approach. The X-band radar (ground-based radar planned for use in different missions; GBR) example, already cited, moves away from this approach and uses a single radar of high power P_{avg} to reach the detection

THAAD radar (GBR)
X band (λ= 3 cm)

MESAR
S band (λ= 10 cm)

BMEWS RAF Fylingdales
UHF band (λ = 55 cm)

USAF OTH-B
(transmit antenna only)
HF band ($\lambda \approx$ 2000 cm)

Fig. 5.15   Some typical radars.

ranges R_D required but retains the high-frequency (X-band) feature for the fire control function. This gives the apparent out-of-trend look in the set of radars shown in Fig. 5.15. Both the UHF radar shown (Fylingdales) and the HF radar shown move into the realm where the dimensions approach real estate values rather than transportable battlefield direct antenna measurements.

These changes in basic construction illustrate how the cost of such radars is complicated in the setting up of common rules.

INTERCEPTOR MISSILE COSTS

Even though missiles have been part of the military arsenal since the earliest times (see Chapter 2), the relative sophistication of the modern-day missile is new, and there has not been an established pattern that can be used for discerning reliable cost trends. Figure 5.16 shows the general trend of cost of ballistic missiles including those that have the label "strategic" as well as those labeled "tactical." At this macro level it is seen that speed is a significant parameter in determining the cost.

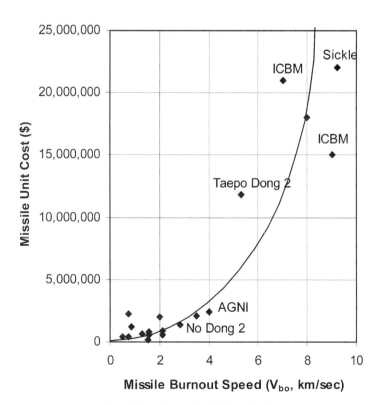

Fig. 5.16   Cost of ballistic missiles.

The cost data have been compiled from a variety of sources, most of which have ensured that the proviso of estimate is added to the entries. This would be especially true for the estimate of missiles built in nations not known for publishing reliable information. Figure 5.16 does show those missiles designed for extremely long ranges and with speeds approaching 10 km/s at burnout; the cost of a single missile can approach millions or even tens of millions of dollars.

If now a more focused look is made of those ballistic missiles designed more for the tactical arena (say, $V_{bo} < 6$ km/s), a slightly different perspective can be achieved even with the scant database of available missiles. Figure 5.17 shows these data collected from studies, and single prototype data, and missiles in production.

The trendlines added to Fig. 5.17 reflect the expected effect of both the numbers built as well as the effect of speed. Prototypes such as the HEDI, HOE, and ERIS missile programs are clearly more expensive than those missiles that have been designed for a production run of some amount, such as Patriot and Aster missiles. [Early prototypes that led the way in "hitting a bullet with a bullet" are the U.S. Army *High Endoatmospheric Defense Interceptor (HEDI)*, the U.S. Army *Homing Overlay Experiment (HOE)*, which had a successful hit-to-kill on 10 June 1984, and the U.S. Army *Exoatmospheric Reentry vehicle Interceptor Subsystem (ERIS)*, which had a successful hit-to-kill test in January 1991.] The general trend of cost vs both speed and quantity built can be seen by the shape of the curves. The data are still being refined as the missile manufacturers continue with their devel-

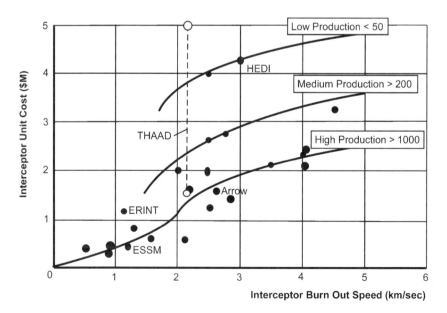

Fig. 5.17   Costs of interceptors as a function of speed and quantity built.

opment. An indication of the spread in the expected production cost can be ascertained by the vertical band shown for THAAD and PAC-3. This band represents the expected spread on both the THAAD and PAC 3 missile costs, which is reported to vary from $1.5 to $5 million per missile.[55]

Although at first glance the variation between $1.5 million per missile and $5 million per missile appears large, this is not incompatible with the values expected from the use of learning curves in the production run expected for such missiles. The concept of "learning curves" was first introduced by Wright[56] when he was evaluating the cost of airplanes during production runs leading up to WWII. As a result of Wright's work, it is now generally accepted that the cost of any product reduces by some constant factor in a learning process during production. Specifically, Wright postulated that as the total quantity of aircraft units produced is doubled, the average cost declines to 80% of the average cost of the previous batch.

Since that original work, it has been found that the concept holds true in most manufacturing processes but that the actual value of the learning-curve factor varies from 80% to other values depending on the product. Wright's work was based on the production of aircraft, and in ship construction, for example, the value of 90–95% is found to be more typical. Other factors apply in radar fabrication, electronics, and so on.

If this concept is applied to the production of missiles, the equation can be written as follows:

$$\$_Q = \$_1 \cdot Q^{[\ell n(lc)/\ell n 2]} \tag{5.11}$$

in which the key parameters are as follows:

$\$_Q$, cost of the Qth missile;
$\$_1$, cost of the first missile;
lc, learning-curve "slope"; and
ℓn, natural logarithm.

The slope of the curve of the unit production costs ($\$_Q$) vs quantity produced Q is not the mathematical slope of the curve, of course, because of the factor 2 assumed in this learning-curve relationship of price reduction for every *doubling* of the manufacturing batch. If the production cost run is expressed by

$$\$_Q = \$_1 \cdot Q^m \tag{5.12}$$

then, if the learning curve slope is, say, 80%, then the exponent m in Eq. (5.12) would be (-0.319). It is not unreasonable to assume because of the sophistication of missile fabrication that a learning-curve factor of 90% would be a good approximation ($m = -0.152$) to the reduction of costs as missiles are produced.

The planned production run for the THAAD missile is approximately 1300 missiles. Figure 5.18 shows a 90% learning curve from an initial cost of $5 million over a production run that includes the 1300 missile production value [using the published *Aviation Week* values (Ref. 55) for illustration]. It can be seen that reaching a goal of $1.5 million, all else being equal, is not an unreasonable value to expect.

The dramatic impact of the learning curve can be seen from Fig. 5.19, in which the same data are displayed with other possible values of learning from 80–95%. It is seen from Fig. 5.19 that if the learning curve were 80% (the original observation from Wright's work on aircraft) then the missile unit average cost for a quantity of 1300 missiles could drop to less than $500,000. Conversely, if the learning curve were only 95%, then the missile cost would only drop to about $3 million per missile.

Such considerations emphasize the importance of including the right key parameters when either making estimates or comparing data from different sources on missile systems that have reached different maturity levels in their production quantities.

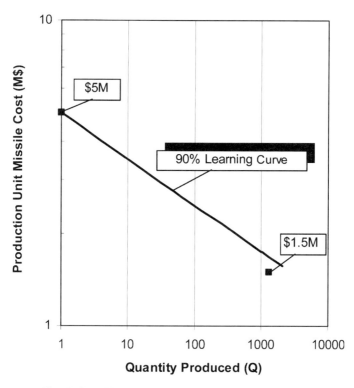

Fig. 5.18   Effect of learning curve on cost of missiles.

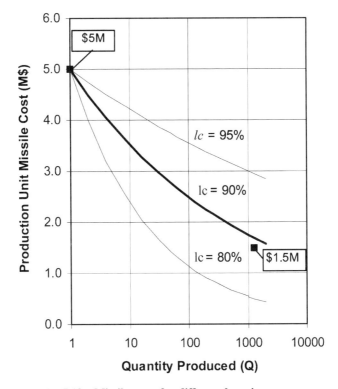

Fig. 5.19   Missile costs for different learning curves.

With the preceding results it becomes possible to represent the cost of the interceptor missiles through the following relationship:

$$\$_{\text{interceptor}} = \$_1 Q^{-0.152} f(V_{\text{bo}}) \tag{5.13}$$

in which, in a similar vein as the preceding production curves, $\$_{\text{interceptor}}$ is the expected cost of the interceptor after a production run of Q units that will have fallen from the cost of the first missile $\$_1$. The exponent (-0.152) reflects the assumed 90% learning curve, and V_{bo} is the burnout speed for the interceptor (see Chapter 4 for values of V_{bo}). The functional form $f(V_{\text{bo}})$ could be some polynomial, as shown in Fig. 5.17, but it is best left to graphical representation in this early stage of cost estimation pending more detailed information becoming available.

COST OF CRUISE MISSILES

The cost of cruise missiles is much less amenable to simple algorithms than even ballistic missiles because of the wide range of types extant in the world. Chapter 2 describes the many types under the broad definition of cruise missiles. Chapter 2, Fig. 2.12, for example, displays the weight and range characteristics for all of the known types of propulsion used in this class of missile. Because the cruise missile is an aerodynamic vehicle, weight is a key parameter that also shows up in the cost. The database on cruise missiles has grown over many years and the quantities Q built are quite wide ranging from the 20,000+ HARM built to 2500+ *Phoenix* missiles built. From the earlier discussion one would expect a significant impact on the unit costs with such variations in the numbers built. Figure 5.20 shows some available data (collected over a period of years from such sources as Refs. 57–59 and other references already cited) on unit costs of cruise missiles compiled as a function of the key parameter of weight.

Figure 5.20 shows that weight W is a key costing parameter, which should not be surprising given the aerodynamic nature of cruise missile design. Although there is a wide spread in the data, the trendline as a function of weight for the available data can be expressed by

$$\$_{CM} = 0.0043 \cdot W^{0.85} \tag{5.14}$$

where the unit cost of the missile $\$_{CM}$ is expressed in millions of dollars (in 2003 dollars) and the weight is expressed in kilograms.

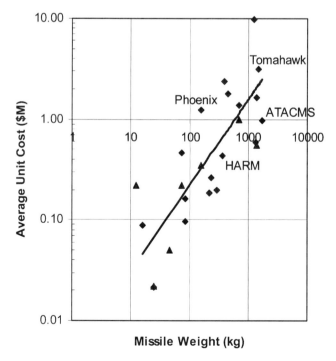

Fig. 5.20   Cost of aerodynamic missiles as function of weight.

A second key parameter would be the design speed of the missile, which is not as easy to identify as for a ballistic missile, where V_{bo} is a distinct characteristic. In the case of the cruise missile (or aerodynamic missile if a broader connotation is sought), the speed depends on the flight profile used, design altitude (if launched from an aircraft), and the type of propulsion used. In Chapter 2 the discussion is provided on the effect of different forms of propulsion and design speed. For example, one particular form of propulsion is the ramjet, which requires that the missile be first launched with some other form of propulsion to accelerate the missile to a speed where the main engine (ramjet) can best function, which is typically around Mach 3 or about 1 km/s (depending on the altitude). So which speed characterizes the missile and its cost is not obvious.

The speed V is not as readily available in the published literature, and the values used in compiling the costs are the published values of the quoted Mach number but expressed in terms of kilometers/second under the assumption that the quoted speeds are the launch speed from aircraft at 10-km altitude. This is a blanket assumption even though it is recognized that the *Tomahawk* (for example) is launched from aircraft, ships, and submarines with obviously different values of speed. Given these caveats, the unit costs of such missiles as a function of speed are displayed in Fig. 5.21.

Figure 5.21 gives some indication that speed is important but that there are other hidden factors not immediately apparent. The sophistication of the *Tomahawk* vs the relative simplicity of *Phoenix* for example shows in the different cost values of the missile even though the *Phoenix* is more than five times faster than the *Tomahawk*. It is not always clear in the literature if the quoted costs are indeed for the quantity produced at that time or are projected costs for some future production quantity design goal. Hence, corrections for quantity built have not been done in Figs. 5.20 or 5.21, which would account for some of the spread in the reported data.

Accordingly, one would expect that the cost of cruise missiles would be of a form such as

$$\$_{CM} = \$_1 Q^{-0.152} f(V) W^{0.85} \qquad (5.15)$$

where $\$_{CM}$ is the cost of the cruise missile after a production run of Q missiles and $\$_1$ is the unit cost of the first missile of the batch. For reference, an assumed learning curve of 90% has been used in Eq. (5.15). While not completely rigorous, it is interesting to note that the published data for the unit cost of the AMRAAM cruise missile follows the same trend in cost data as the other data used here for ballistic missiles. Figure 5.22 shows the recently published data for the cost of AMRAAM plotted as a function of the quantity Q produced (see Ref. 40 for quoted costs of the AMRAAM missile).

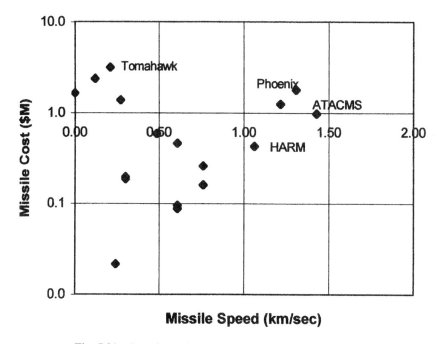

Fig. 5.21   Aerodynamic missile cost as a function of speed.

It is seen that the quoted average costs over a buy of 5000 missiles and the costs of the recent batch production follow approximately a 90% learning curve. This shows that the unit cost of the AMRAAM has dropped from about $1.24 million per missile during its initial production phase to about $304,000 per missile after a production run of 5000 missiles.

At this time, in Eq. (5.15) for CM costs, the speed function $f(V)$ must be considered as a "place holder" pending the availability of more detailed data that discriminate the various types discussed in Chapter 2 (such as ramjets, turbojets, etc.) and the degree of sophistication in the design together with the mission type. Weight, quantity, and speed are insufficient parameters to differentiate between, say, the *Tomahawk* mission from the *Global Hawk* or *Predator* mission.

For the purposes of this book, the end-cost values shown in Fig. 5.20 are sufficient to at least determine gross trends for early decision-making purposes.

COST OF SHIPS

In this book the cost of ships is not pursued in any detail, other than to show the general order of magnitude of costs for reference purposes. Also, ships have a unique characteristic in that they are usually multimission

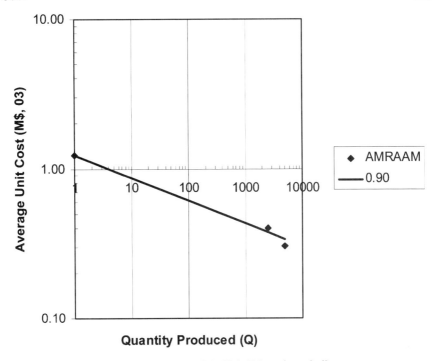

Fig. 5.22 Estimation of AMRAAM cruise missile cost.

capable, and in any cost of missile defense systems it would be unfair to burden the full costs of any ship against the cost of missile defense. There are ways of apportioning the cost to the mission, and, accordingly, Fig. 5.23 provides a basis from which allocated percentage costs can be ascertained.

Figure 5.23 collects the costs of the lead ship (first in the class) of both aircraft carriers (CV) and cruisers (CG) and destroyers (DD). For convenience, the costs have been plotted against the year of commission of each ship. The costs are quoted in 2003 dollars assuming a 4% inflation factor per year. The O&S costs for these ships over their projected 30-year life are approximately *double* these quoted procurement costs (using data from Ref. 41).

In Eq. (5.3), Ridell[44] had shown that weight and power was a significant cost driver for most vehicles, especially ships. In the case of ships, the effect of speed is minimal in that the speed has stayed relatively constant (25–30 kn design speed) for most of the ships displayed in Fig. 5.23, and weight is the dominant factor. The weight, power, and sophistication of the ships have increased considerably over the last half-century of ship data shown. In the case of the aircraft carrier (CV), only the cost of the platform itself is included. The cost of its payload, namely, a mixture of different

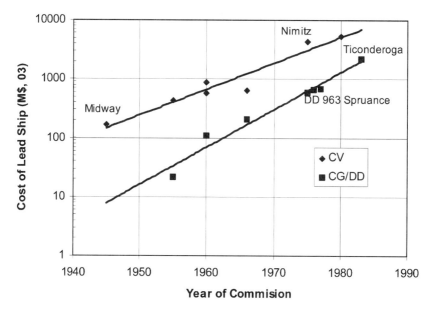

Fig. 5.23   Costs of CV and CG/DD.

aircraft types, is approximately equal to the cost of the CV itself, depending on the makeup of the 60–90 aircraft in the usual load-out, such that an aircraft carrier fully loaded with its full complement of aircraft can approach $4–5 billion each. The CG and DD today with their combat suites are approximately $1 billion each.

COST OF AIRCRAFT

As in the case for ships, the cost of aircraft is not pursued to any detail here, but some general data are provided as a point of reference. Like ships, the unit cost of aircraft has increased significantly each year since the early days of Orville and Wilbur Wright (1903), where speed, weight, and sophistication of mission and design have changed dramatically. It was shown earlier, that most CERs for aircraft show that the unit cost of an airplane is proportional to the product term $W^a S^b F^c$, where W is the weight, S is the design speed at altitude, and F is the thrust [for example, see Eq. (5.2) for one set of CER for fighter aircraft]. The exponents (a, b, c) vary for different type of aircraft such as fighter/attack aircraft, military transport, and civil aircraft, etc. Again, it is not the purpose of this book to develop the CER for aircraft other than to show the general trends with time, for ease of use in the management methods developed in Chapter 8. The user can substitute more detailed costs if available.

Figure 5.24 shows how the typical costs of fighter aircraft* and civil airliners have increased over the last half-century from about 1940–1990. Current costs appear to show no sign of reduction in unit costs or capability.

Figure 5.24 can only be used as a general indication of the costs of aircraft because of the many different types and missions, but it does provide an order-of-magnitude appreciation for the cost of aircraft in comparison with the other components of any missile defense system. Again, more detailed correlation between aircraft costs and the product terms involving key driving parameters of weight, speed, thrust, mission, and sophistication of design (e.g., inclusion of stealth features) need to be taken into account for an accurate representation of costs.

Armed with the component costs, either from accurate programmatic cost estimates or from the default values sketched out here in Figs. 5.12–5.24, it

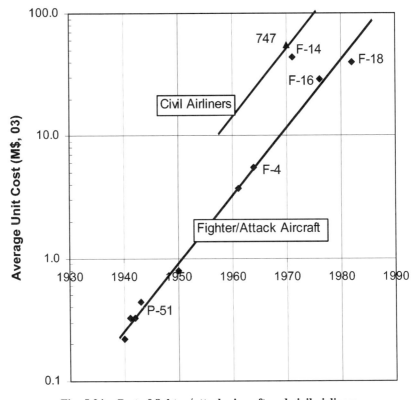

Fig. 5.24 Cost of fighter/attack aircraft and civil airliners.

*Most readers are familiar with the quote attributed to President Calvin Coolidge, who in 1928 when presented with a budget request for $25,000 for a *squadron* of aircraft, said: *"Buy one aircraft and let the aviators take turns in flying it!"* Many years later, Norman Augustine, in his *Augustine's Laws*,[62] suggests that this might have to occur in 2054!

becomes possible to build up the costs of various missile defense systems. The costs will depend quite strongly on the architectural arrangement of combinations of systems designed to defend against a set of perceived threats over different geographical areas.

ILLUSTRATIVE DEFENSE SYSTEM COSTS

Chapter 3 on Deciding on the Defense outlines possible goals that the defender might want to set up to protect different nations or a group of nations in some form of Alliance defense. These can vary quite widely. However, it is of value to provide some preliminary collection of components into illustrative defense system units. For illustration, just a few of the possible combinations are shown here to give an indication of the level of costs likely to be involved in any actual architectural arrangement.

EARLY-WARNING SATELLITE SYSTEM COSTS

Table 5.3 summarizes a possible set of early-warning satellite systems costs.

Hence, depending on the system used, the cost of a satellite early-warning system can cost from approximately $1–6 billion. The actual costs will vary from these representative or illustrative values depending on such factors as whether cruise missile defense is to be included or not, whether nations will donate or share costs of existing satellite systems or not, and similar variations. As just one example, the United States has offered to share early-warning data with its European Allies of any ballistic missile launches detected by the U.S. GEO satellite system (DSP and SBIRS). In this case all that the other nations would have as cost to a missile defense system would be the cost of the downlinks to their own or collective ground station.

LOWER-LAYER LAND-BASED INTERCEPTOR SYSTEM

In similar fashion as just shown for the illustrative early-warning satellite system, Table 5.4 displays an illustrative (i.e., no specific program) lower-layer land-based missile interceptor system, which can be considered as a representative battery that would be adapted depending on architectural scenarios in any particular planned conflict.

In Table 5.4, a nominal 80 interceptors have been assumed for this notional battery that has five launchers and one multifunctional fire control radar. This radar would typically be capable of receiving the early-warning data from the satellite system described in Table 5.3. There are normally various levels of communication required at the battery level, the battalion level, and to the command structure, in which case the indigenous BM/C^3 system shown as trucks will have this ability, and the cost is estimated for these various

Table 5.3 Illustrative early-warning satellite system

Picture	Component	Unit cost	Total costs
	GEO satellite primarily for ballistic missile detection	$500 million (see Fig 5.12)	$1000 million (two GEO satellites for stereo)
	LEO satellite primarily for cruise missile detection	$120 million (see Fig 5.12)	$4800 million (40 LEO satellites for Earth coverage)
	Ground station	GEO: $40 million LEO: $60 million	GEO: $40 million LEO: $60 million
	Launch costs	Per satellite (see Fig 5.13) GEO: $100 million LEO: $20 million	GEO: $200 million LEO: $800 million
—	—	—	**GEO (total): $1140 million**
—	—	—	**LEO (total): $5660 million**

functions. The cost of the national BM/C^3 network into which the individual systems connect is not included in the cost of this grouping (see later for cost of BM/C^3 systems). While clearly the actual makeup of the lower-layer interceptor systems will depend on the scenario and overall architecture of the defense, this particular illustrative system allows for 40 missiles being ready to fire (in five launchers) plus 40 interceptors ready for reload. This notional battery has been structured on the assumption that the threat is from short- to medium-range ballistic missiles. If the system is also to defend against CM at the same time, one possibility is that the number of missiles be retained at the same number (80), but an extra fire control radar could be added, with the boresight aligned to lower elevations for cruise missiles and other airbreathing threats, and the original FC radar's boresight angle aligned to the higher incoming angle of the ballistic missile. Such tables allow for rapid adjustments in possible system costs depending on the assumed threat.

Table 5.4 **Illustrative lower-layer land-based interceptor system**

Picture of similar system	Component	Approximate unit cost	Approximate total costs
	Interceptor (endo intercepts)	$1.3 million (see Fig 5.17)	$104 million (80interceptors)
	Launcher	$3 million (estimate)	$15 million (five launchers per system battery)
	Fire control radar	$30 million (see Fig 5.14)	$40 million (includes support)
	BM/C^3	$20 million (estimate)	$20 million (HUMVEE-like trucks)
——	——	——	**TOTAL: $179 million**

UPPER-LAYER LAND-BASED INTERCEPTOR SYSTEM

A similar construction can be put together for the upper-layer interceptor system such as seen in Table 5.5, again arranged for a notional battery for the defense.

Although photographs of the U.S. Army THAAD interceptor system are used in Table 5.5 as illustrations of the type of system that is designed for endo/exo-atmospheric intercepts, the values of costs shown are not official programmatic values for that program. The photos and sketches serve only as an indication of the type of component being discussed in this illustrative system. The costs are estimates for a similar system as taken from the generic cost trends shown in earlier figures.

SEA-BASED INTERCEPTOR SYSTEMS

There are many possibilities in sea-based missile defense, many of which have either not been tried yet or are in various stages of development today. The main

Table 5.5   Illustrative upper-layer land-based interceptor system

Picture of a similar system	Component	Approximate unit cost	Approximate total costs
	Interceptor (endo-exo intercepts)	$2.5 million (see Fig 5.17)	$100 million (40 interceptors)
	Launcher	$3 million (estimate)	$6 million (two launchers per system battery)
	Fire control radar	$150 million (see Fig 5.14)	$200 million (includes support, e.g., diesel generator)
	BM/C^3	$5 million (estimate)	$20 million (four HUMVEE-like trucks)
———	———	———	**Total: $326 million**

platforms under consideration are surface ships and submarines. Of the surface ships, both cruisers/destroyers (CG/DD) and aircraft carriers (CV) are being used as the launch platform. Both lower-layer and upper-layer interceptor systems can and have been installed in such platforms.* In the case of the submarine, in the latter half of the last century concentration was on submarines equipped with strategic (ballistic) missiles and some cruise missiles.

This is changing today, as the strategic submarines seek new missions after the Cold War, and modifications are already underway to convert such strategic submarines to carry cruise missiles.† This concept of converting

---

*In the United States, the lower-layer SM-2 (currently cancelled) and the upper-layer SM-3 are planned to be on the Aegis ships. In France, the aircraft carrier *Charles de Gaulle* is equipped with ASTER 15 (SAMP/N) missiles, and the ASTER 30 with its planned "TBM capable" upgrades is planned for various European navy frigates.

†Conversion is currently underway, beginning in 2003, in both Bremerton and Norfolk shipyards, to convert the four *Ohio* class strategic submarines to equip them with *Tomahawk* cruise missiles. The original 24 missile tubes that were for the strategic missiles (*Trident I & II*) will now hold up to 168 *Tomahawks* (current plan is for 154 missiles with remaining "holes" to be used for other purposes).

strategic submarines to carry *cruise missiles* could be expanded to converting strategic submarines to carry in addition to *cruise missiles*, both *upper layer interceptors* (such as the SM-3) and *lower layer interceptors* (various candidate systems available) for a total two layer active defense mobile platform to defend against the full range of missile threats. This would provide a stealthy, mobile defense that could extend missile defense to many trouble spots around the globe.

Because it is relatively early in the concept of such diverse sea-based defense, it is better to compile the possibilities rather than collect specific cases. Also, the issue of how much of the sea-based missile defense costs should be allocated to the total sea-based platform, given their multimission capability, belies the ability to neatly summarize the costs.

Figure 5.25 collects together the main elements in the cost of sea-based missile defense.

Aegis Cruisers Underway (Approx $1B each)

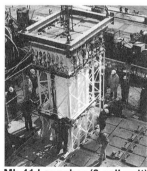

Mk 41 Launcher (8 cell unit)
(Approx $25M and 64 missiles)

BM/C^3 System
(Approx $20M)

SPY 1 Radar (3 faces)
(Approx $40M)

Mix of 128 Possible Missiles:
Lower Layer $1.5M ea
Upper Layer $2.5M ea
Cruise Missiles $2.0M ea

Fig. 5.25   Illustrative possible sea-based system.

Figure 5.25 shows a possible sea-based interceptor missile defense system, using the U.S. Aegis system as a baseline. In other non-U.S. ships the launcher could be either the same Mk 41 launcher (in its different BMD, Strike, and AAW variants) or the SYLVER [*système de lancement verticale* (vertical launch system)] launcher as planned for several European navies. The load-out of the missiles can be selected from a wide range, including SM-2 Blk IVA (or its replacement), SM-3, TACMS, Harpoon, ESSM, ASTER 15, ASTER 30, Tomahawk, and others.

Figure 5.25 serves as a collection file for later use as decisions are made as to what share of these costs should be allocated to the specific mission of missile defense.

AIR-BASED INTERCEPTOR SYSTEM COSTS

Of all of the various defense systems under development, the air-based interceptor system is the least developed. The aircraft can be used in a variety of ways and can provide both an upper-layer and a lower-layer defense capability depending on how it is used in the architecture and what type of interceptor is used. There are two concepts under consideration: 1) an aircraft plus interceptor combination and 2) an aircraft and directed-energy weapon (DEW) system. Of these, the aircraft and DEW combination has received the most attention and R&D funding because of two underlying beliefs: One, it is extremely difficult to develop an interceptor with a high enough speed V_{bo} to achieve the objectives, and two, the use of directed energy promises more shots per aircraft. The development of a high-speed interceptor is discussed in more detail in Chapter 4. It is also expected that a missile-carrying aircraft might be limited to about one or two shots before having to return to base for a reload. (This assumes the current fighter aircraft technology with the air-to-air missiles being carried externally on launcher rails. If, instead, the alternative concept of a possible fighter-bomber is used that can carry many missiles housed internally, this limitation of one or two missiles per aircraft might be alleviated. This would require new R&D.) The aircraft and DEW combination promises an airborne defense system with about 20–40 shots before having to return to base. Each concept has its advantages and disadvantages. Some comments on the possibilities for an airborne missile defense system and their likely costs follow.

AIRCRAFT AND INTERCEPTOR SYSTEM.   There is no airborne system (or any other system) today capable of providing BPI except perhaps in very limited scenarios. The air-to-air missile (interceptor) required has already been shown to be the result of intensive development. However, reasonable projections can be made based on the available data. Figure 5.26 summarizes the likely set of needed components for such an airborne system.

Obviously, the cost figures given in Fig. 5.26 are highly subjective at this stage because no system has reached any level of maturity for reliable use,

Fighter Cost: approx $40M

New improved on board
sensors and data links to
external Early Warning
sensors

High Speed AAM Cost: $5M each
(Assume 4 per aircraft)

(Cost per aircraft: $10M)

Approx "aircraft + interceptor system" cost: $70M
Number of aircraft required: TBD & scenario dependent

Fig. 5.26   BPI/API aircraft and missile system cost.

but, based on the typical values of the component costs already provided, they appear reasonable. The aircraft cost is consistent with the aircraft trend data provided in Fig. 5.24. The air-to-air missile (AAM) must be considered a "guesstimate," but from Fig. 5.17 for ballistic missiles it was shown that $5 million per missile was not an unreasonable assumption, especially if the speed requirement for BPI must be at the high end of the experience base at 6 km/s. Finally, in Fig. 5.26, it is shown that for any BPI aircraft to be able to conduct such a mission new onboard sensors need to be developed in order that timely links can be made from the off-board sensors (and early-warning sensors), which will be required to vector the aircraft to the correct position in the sky before the aircraft can launch its missile.

The number of aircraft required has two dimensions. In comparison to land-based systems and sea-based systems, the air-based system is essentially a "dynamic" system as distinct from the "static" nature of land-and sea-based systems. Hence, the procurement cost of an air-based system must include the cost of those aircraft required to keep one aircraft on station. A typical rule of thumb is to multiply the aircraft needed (say two in this case assuming that there are two aircraft in the CAP) by a factor of 3–4 to account for aircraft in transit plus those being readied to fly back at the air base. Such a calculation is quite distinct from the second dimension that would be comparable to the land-and sea-based systems to account for some O&S costs related to some agreed-upon operating life (say 20–30 years).

AIRCRAFT AND **DEW.**   In this particular case, for the aircraft plus DEW, there is a program already under development. It is the U.S. Air Force Air Borne Laser

(ABL) program. The development of lasers (acronym for light amplification by stimulated emission of radiation) has seen dramatic strides in development since the concept was first put forward by Albert Einstein in 1917. Laser technology is used in many fields of endeavor from meteorology to medicine and biological sciences to defense. Efforts were made in the late 1960s to develop many forms of directed energy with lasers receiving the greatest attention. It was found that the technology and the limitations of cooling systems produced designs that were too heavy for shipboard use even into the 1970s. In the 1980s, however, it was found possible with the use of a carbon-dioxide gas dynamic laser that airborne systems could be successfully developed. A prototype system on board a KC-135 aircraft successfully shot down a towed drone over the White Sands Missile Range on 2 May 1981. Further successes occurred with this program with another shooting down of a Sidewinder AAM on 26 July 1983. This successful test led to the inclusion of the airborne laser concept in the Strategic Defense Initiative in 1983.

Continued development of the laser brought in the newer technology of the chemical oxygen iodine laser (COIL) that is now being developed as part of the ABL program. This newer concept is lighter, more compact, and has the potential for greater ranges of lethal effectiveness. The current program is to place the COIL system on board a Boeing 747-400 freighter aircraft. This program started on 12 November 1996. The plan is to have an operational system by 2007. Seven aircraft are planned for a cost of $1.1 billion. Figure 5.27 summarizes a possible ABL aircraft system with very approximate costs.

As for the aircraft plus missile concept, these cost values are very approximate at this early stage of development because of the continued development and removal of risk from the concept. The high-power (approximately 5 MW) laser operating at 1.3 μm wavelength and a 1.5-m-diam telescope (compiled from public release sources such as *IEEE Spectrum*, *Defense News*, MDA Fact Sheets, and http://www.airbornelaser. com, etc.) is undergoing extensive R&D to remove weight problems and to increase the lethal range. The laser, during early tests, was exhibiting dissipation through the turbulent air with much reduced lethality much beyond 200 km until in 1991 after the Gulf War; a previously classified technique of adaptive optics was incorporated into the design. This gave an improved lethality range out to about 400 km.[63] This range is similar to that discussed earlier in the standoff distances required for a successful BPI without the aircraft being placed in harm's way by flying over enemy territory. This standoff distance allows for BPI of such missiles as SCUD, Al-Hussein, and Nodong.[64] One important difference between the concept of an aircraft plus missile combination and the ABL system is that the use of a large aircraft such as the Boeing 747-440F allows for the incorporation of its own onboard early-warning (AEW) system as part of the system package.

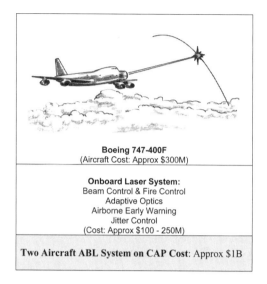

Fig. 5.27   ABL system costs.

Another important difference between the aircraft plus missile solution and the aircraft plus laser solution is the *cost per shot*. In the case of the aircraft and missile, each shot can cost up to $5–10 million, if a salvo-firing doctrine is used, whereas for the airborne laser solution the cost of a "laser burst" will cost somewhere between *hundreds and a thousand dollars*. (Public release handouts at AIAA conferences give the value of $1000 per engagement.) Each aircraft will have the capability for 20–40 shots before having to return to base for reload (or possibly replenished in flight by C-17 transport aircraft). This leads to comparing 1) the low cost of a fighter aircraft with a high-cost interceptor shot and 2) the high cost of the airborne laser aircraft system but with a very low-cost laser burst.

COST OF WEAPONS IN IRAQ WAR

The cost trends for missiles, ships, and aircraft of a missile defense system have been shown in the preceding trend charts. In addition to these trend charts, it is informative to collect together individual data points recently published as a result of the Iraq War that began 19 March 2003. The data are collected in the following three tables. These data have been compiled by the research staff of the Seattle Times,[45] from a variety of sources, including the U.S. Air Force, the U.S. Army, U.S. Navy, Jane's Information Group, Global Security, Boeing company, Raytheon, and Federation of American Scientists.

Table 5.6 gives the data for manned and unmanned aircraft. Table 5.7 gives the data for precision weapons, cruise missiles, and air defense

Table 5.6 Manned and unmanned aircraft in the Iraq War 2003

Aircraft	Mission	Range	Speed (Mach)	Unit cost, 2003 $
Aircraft				
A-10 Thunderbird II	CAS	1300 km	0.64	$9.8 million
AC-130 Gunship	CAS[a]	Unlimited[b]	0.45	$72 million
AV-8B Harrier II	Attack (day and night)	3888 km	0.95	$21.6 million
B-1B Lancer	Multirole bomber	Intercontinental[c]	1.2	$283 million
B-2 Spirit	Stealth bomber[d]	Intercontinental[c]	n/a[e]	$1.2 billion
E-3 Sentry (AWACS)	Surveillance, C3	>9250 km	0.55	$123.4 million
F-117A Nighthawk	Stealth fighter/attack	Unlimited[f]	n/a[e]	$45 million
F-16 Fighting Falcon	Multirole fighter	3200 km	2	$18.8 million
EA-6B Prowler	ECM	1850 km	0.77	$52 million
F-14 Tomcat	CV multirole fighter	2600 km	1.88	$38 million
F-15 Eagle	Tactical fighter	5550 km	2.5+	$28 million
F-15E Strike Eagle	A/G attack	3860 km	2.5+	$29 million
F/A 18E/F Super Hornet	Multirole fighter/attack	2360 km	1.8+	$60 million
F/A-18C/D Hornet	Multirole fighter/attack	2015 km	1.7	$40 million
Unmanned aircraft				
RQ-1A Predator	Surveillance, reconnaissance, target ID	730 km	0.18	$40 million
RQ-4A Global Hawk	Reconnaissance[g]	1930 km	0.60	$40+ million

[a]Includes close air support, air interdiction, force protection.
[b]Unlimited range with airborne refueling.   [c]Unrefueled.
[d]Multirole, heavy.   [e]High subsonic speed.
[f]With inflight refueling.
[g]Provides reconnaissance through high-resolution imagery in real time via satellite.

Table 5.7   Precision weapons, cruise missiles, and air defense

Weapon	Weight	Range	Speed (Mach)	Unit cost [a] 2003 $
JDAM[b]				
GBU-31	925 kg	Up to 24 km	Freefall[c]	$21,000
GBU-32	460 kg	Up to 24 km	Freefall	$21,000
AGM-65 Maverick	210–360 kg	Up to 24 km	$1 < M > 2$	$17–$110,000
AGM-114 K/M Hellfire	45 kg	∼8 km	1.1	$47,904
AGM-86C ALCM	1475 kg	2415+km	0.83	$1 million
Tomahawk Land Attack	1453 kg	925 km	0.83	$500,000
PAC-3	699 kg	117 km	5[d]	$2.5 million

[a]No cost–quantity relationships provided.
[b]Joint direct attack munition.
[c]Guidance tail kit gives accuracy to the freefall.
[d]Gives an approximate $V_{bo} = 1.7$ km/s.

Table 5.8   Aircraft carrier battle group

Ship	Length, m	No. of aircraft	Speed, kn	Unit cost, 2003 $
Aircraft carrier				
Kitty Hawk class	324	85	30+	$400 million[a]
Nimitz class	332	85	30+	$4.5 billion
Cruiser				
Ticonderoga class	173	2 helos	30+	$1 billion
Guided missile destroyer				
Arleigh Burke class	154	None	30+	$800 million
Frigate				
Oliver Perry class	137	2 SH-60 or 1 SH-2 helo	29	$200 million
Attack submarine				
Los Angeles class	110	Tomahawk missiles	20+[b]	$900 million[c]

[a]In 1961 dollars.
[b]Speed while underway submerged.
[c]In 1990 dollars.

(specifically the PAC-3 missile in the Patriot family but has a hit-to-kill capability). Table 5.8 collects the data on the makeup of a carrier battle group.

There is some spread in the data, but generally these specific data points fall within the trend lines given in the earlier charts for each of the type of systems being considered.

OPERATING AND SUPPORT SYSTEM COSTS

Except in specific instances where reliable data were available, this book has not tackled the second largest component of the LCC, which is the operating and support costs (O&S). This is because the O&S costs are very much scenario and mission dependent. Depending on the operating lifetime sought by the defender, say, 10, 20, or 30 years, the O&S costs can be at least equal to in magnitude the procurement costs shown in this chapter. The costs are not additive, of course, and come from different accounts in any defense establishment's budget (see earlier discussion in this chapter on LCC). For the purposes of this book, it is sufficient to highlight the existence of such costs that should be incorporated in any final analysis.

ARCHITECTURE COSTS

Armed with the preceding component and system costs, it is now possible to apply them in some form of defense architecture for the defense of a nation or a group of nations. Clearly, such an architecture will vary strongly from one nation to another and from one set of threat and geography scenarios to another. To place some degree of perspective on the possible costs of a defense architecture, consider a base case that can be used to both illustrate the type of costs and how they might vary from one case to another.

ILLUSTRATIVE BASE CASE

In Chapter 3, on deciding on the defense, it was shown that there was one common feature that serves to be representative of the possible architectures that might be used in Europe. That common feature was that most nations can be circumscribed by circles with a radius of about 500 km. Two nations, the United Kingdom and France, are used here to illustrate this feature. Figure 5.28 shows such a circle around the United Kingdom.

For this particular nation (United Kingdom) it is seen that because of the geographical shape of the land masses, there is a considerable amount of water that is also "protected" which is in and around Great Britain and Northern Ireland, which make up the United Kingdom. Ireland is also enveloped in this protective circle. As will be shown, it is possible to be more

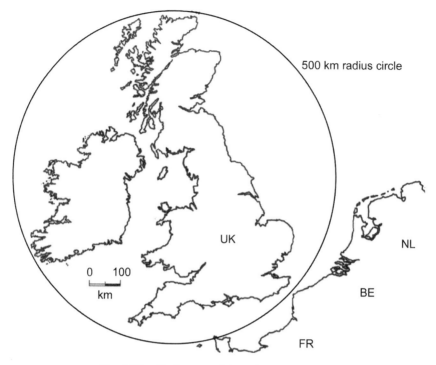

Fig. 5.28   Circle around UK of 500-km radius.

prescriptive on the covered areas once a selection is made on the defense systems.

A second nation, France, has a different geographical distribution of its land mass inside a 500-km-radius circle, as shown in Fig. 5.29.

In this second example, France, it is seen that the land mass is more evenly distributed throughout the circle. Any assets, including maritime assets operating in the littoral waters around France, would be included in this area. In the first example, United Kingdom, there are large areas of international waters that need not be included in any defended boundary. These two examples describe most of the possibilities for the nations of Europe. A similar approach can be used for the characterization of the needed defended areas of North America, but for simplification this has not been done here.

The next problem facing the decision maker is, what to protect within this 500-km-radius circle. Should all areas receive equal protection? Should the rural areas receive less protection than the more densely populated cities and towns? Do strategic military installations receive greater protection than civil centers (cities)? More than national capitals? Even the formulation of these questions is reminiscent of the "old" conventional weapon thinking. In this new era of WMD, it might be more strategically effective and sufficient

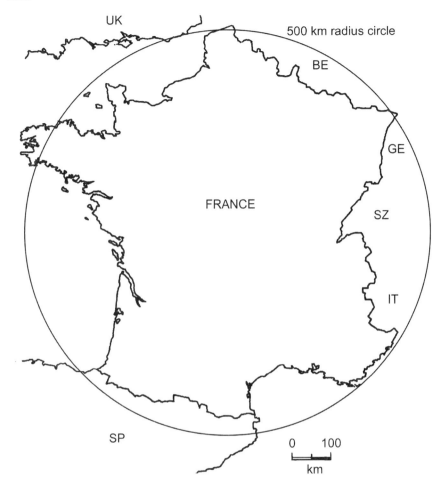

UK

500 km radius circle

BE

GE

FRANCE

SZ

IT

SP

0   100

km

Fig. 5.29   Circle around France of 500-km radius.

for the attacker to detonate his chemical and biological agents upwind of a
strategic or even tactical target and allow the local wind conditions or water
supply to carry the deadly material into cities without any further action from
the attacker.

The answers to these questions must be obtained in other forums than
those discussed in this book. However, regardless of how the decision is
made as to which specific areas to protect within the broader area encompassed within the 500-km-radius nation, eventually the problem reduces to
the costing of a specific number of defense systems. For the illustrative base
case a specific number of systems will be costed.

If the discussion is restricted to a layered defense, then the problem
reduces to the decision to protect the overall area (for example, as shown in

Figs. 5.28 and 5.29 for the United Kingdom and France) to 1) an agreed-upon level of acceptable leakage and then 2) to protect a specific number of sites or assets within this wide area, as just discussed, to some higher level of protection (i.e., lower leakage).

For the illustrative base case, the example of the United Kingdom will be used.

EXAMPLE PROBLEM 5.1—COST OF DEFENDING THE UNITED KINGDOM

It has been decided to defend the United Kingdom from missile attack using state-of-the-art missile systems. It has been taken that systems available within the next five years will be assumed. It has been decided to use a two-layer active defense system. It has been decided to protect the broad land mass and littoral waters with a single-layer wide-area defense system with a performance capability that will limit the leakage of any incoming ballistic missiles to less than 10%. Within this broad area it is further decided to protect 15 key assets with a higher level of performance (lower leakage) such that the leakage over these assets is reduced to less than 1%. Within this requirement of leakage rates (10% and 1%) is the implicit assumption that defense against WMD is not sought. It is expected that the threat ranges could reach as high as 3000 km.

For this illustrative base case the following two systems are used: *

1) *wide-area protection with*
    a) *an interceptor system with $V_{bo} = 2.0 \ km/s$ and*
    b) *a radar detection range, $R_D = 1000 \ km$; and*
2) *point-area protection with*
    a) *an interceptor system with $V_{bo} = 1.3 \ km/s$ and*
    b) *a radar detection range, $R_D = 200 \ km$.*

Because it is expected that the threat axis will vary from 1) threats coming from a large distance overland in one direction to 2) threats coming from a short distance anywhere over water from the surrounding international waters, it is decided to use the forward-edge radius R_{FE} as the representative net defended footprint from these systems (see Chapter 4, Fig. 4.38 for reduction of footprint size caused by variable threat axes).

In Chapter 4, Fig. 4.27, a set of forward-edge footprint radius R_{FE} charts are provided. The assumptions in the generation of these curves are provided in Chapter 4. For the case of a wide-area system ($V_{bo} = 2.0 \ km/s$; $R_D = 1000 \ km$), Fig. 4.27 showed that the defended footprint would have a radius that varied from a maximum of about 300 km for low values of threat range (less than 1000 km), then reduce to a value of about 165 km against a

---

*For the base case it will be taken that the radar detection ranges R_D can be achieved against the threat RCS signatures without the need for precise cueing from early-warning systems (satellite or otherwise). In a later sensitivity analysis the cost of any required early warning will be added to the cost of the defense.

3000-km-range threat. The minimum number of systems required to meet all threats would thus be the number that provided a defended footprint radius of 165 km. Figure 5.30 shows such a set of footprints superimposed on a map of the United Kingdom. As can be seen, a total of five wide-area systems will be required.

In any actual detailed analysis more complete calculations on the geometries of the defended footprints using the techniques outlined in Chapter 4 for R_{FE}, R_{BE}, and R_{SE} can be computed, but Fig. 5.30 illustrates the approximate number of systems that would be required. Note that more circles could be fitted within the 500-km radius for the United Kingdom but these would probably not be required. A certain amount of overlap, as shown, is also desirable for any actual lay-down for both redundancy and the possibility of defense systems becoming inoperable during the conflict. Also, as noted earlier, for the majority of expected threats with ranges less than 3000 km, the defended footprints would be much larger ($R_{FE} \rightarrow 300$ km) with a comfortable amount of overlap of defended areas.

For the purposes of this illustrative base case, it is taken that within these five wide-area defense footprints there would be 15 point defense (lower-layer) systems to protect the 15 key assets. For the purposes of this example, it is not important as to where these 15 key assets are specifically located,

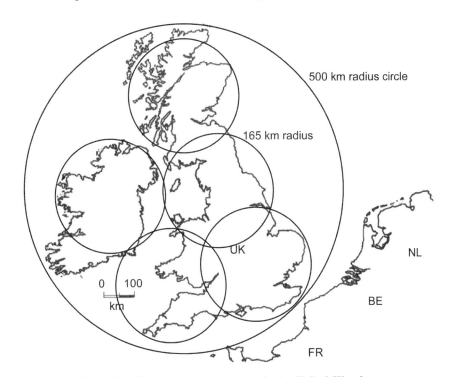

Fig. 5.30   Five upper layer systems in the United Kingdom.

provided that they are inside the five wide-area circles in order to receive the necessary two-layer protection and leakage containment (of less than 1% in this example case).

For the two-layer missile defense system the costs can be expressed in a simplified form for each of the two layers, as follows:

Upper (wide-area)-layer interceptor system:

$$Cost = (Site\ cost) \times (no.\ of\ sites) + (salvo) \times (no.\ of\ missiles\ in\ war)$$
$$\times (interceptor\ unit\ cost)$$

Lower (point-defense)-layer interceptor system:

$$Cost = (Site\ cost) \times (no.\ of\ sites) + (salvo)$$
$$\times (BM\ leakage\ + CMD\ share) \times (interceptor\ unit\ cost)$$

These are clearly not all of the costs but are representative of the procurement cost of the systems. Such O&S Costs, such as manpower, fuel, etc., are not included, which as seen earlier can double the costs shown here for a LCC evaluation. In this procurement-cost equation it has been assumed that the lower-layer system will also contribute to some share of the defense against the airbreathing threat, and specifically the CM. It is also taken that the lower layer will also defend against those BM that are not intercepted by the upper layer.

Finally, as some indication of the total procurement costs it is assumed that the two-layer defense system will be designed to intercept all enemy missiles launched during the conflict or war. For the purposes of this illustrative base case, it will be assumed that this is a defense against a 500-missile attack over some (unspecified) period of time. No attempt is made here to cost depot and supply line of interceptors or spare parts. In this sense the costs derived by the preceding equations must be considered as a low estimate or minimum cost expected for missile defense. In this example problem it is assumed that defense against the CM is handled by conventional air defense (AD, SHORAD) and is not part of BM defense (see example problem 5.2 for the impact of CMD defense). It is now possible to construct the likely costs of a two-layer system using the component and system costs derived earlier coupled with the scenario descriptions in this Example problem. Table 5.9 summarizes the site costs of the system.

Assume that the interceptor unit costs are (from Fig. 5.30) $2.5 million per interceptor for the upper-layer system and $1.5 million per interceptor for the lower-layer system. One further modification of the cost equations is the impact of the type of trajectories that might be used by the attacker and the

Table 5.9 Site costs of two-layer missile defense system

Component	Quantity per site	Unit cost, $ million	Site cost, $ million
Upper layer			
Radar and support	1	200	200
Launcher	2	3	6
BM/C^3 trucks	4	5	20
			226 (total)
Lower layer			
Radar and support	1	40	40
Launcher	5	3	15
BM/C^3 trucks	4	5	20
			75 (total)

effect on the lower-layer defended footprints, which from Chapter 4, Table 4.5, would be of the order of 25-km radius. Although not discussed in detail in Chapter 4, the effect of the use of lofted or depressed trajectories is to cause the defended footprints to be displaced from the center around the key asset. This could leave the key asset outside of the defended footprint. This effect can be alleviated by simply adding more lower-layer systems and placing them around the perimeter of the key asset. Although it would require detailed analysis in any particular case, it is not unreasonable to increase the number of lower-layer systems by some factor, say, 50%, to account for this effect. Thus the 15 key assets could be defended by 22 lower-layer systems. With these additional factors included, the costs of the two-layer missile defense system (BMD only) for a 500-missile war, would be as follows:

Upper layer:	*$3.63 billion*
Lower layer:	*$1.79 billion*
Total:	*$5.42 billion*

That is, for a cost of between $5 and $6 billion, it appears possible to provide the defense of the United Kingdom from a ballistic missile attack of about 500 missiles. Note that this provides defense against conventional warheads with a residual leakage into the key assets of about 0.81% or just over four incoming ballistic missiles. This leakage would be randomly distributed over the stated 15 key assets. (To keep the analyses within bounds, it has been taken in all that follows that all incoming missiles are targeting the key assets. This is a conservative approach, rather than pursue the complicated, but less insightful, analysis of randomly distributing the incoming missiles over the single, wide-area layer as well as the two-layer defense over the key assets.)

INTEGRATION OF CRUISE MISSILE DEFENSE

It is unlikely that any defense system would be put into place against ballistic missiles unless defense against the airbreathing threat and specifically CM was also included. It is a matter of military planning that determines which systems would be assigned which missions in the general defense both in the defense of forward-deployed troops and in the defense of the homeland. Based on operational analyses, it is reasonable to assume that in the preceeding two-layer missile defense system that the lower layer (which functions mainly within the atmosphere) would also be assigned to defend against the incoming cruise missile. A reasonable assignment might be that the missile defense system would also have to provide say 45% of its capability to defend against the CM.

EXAMPLE PROBLEM 5.2—COST OF ADDING CRUISE MISSILE DEFENSE

Using the same two-layer missile defense system as in example problem 5.1, it is decided that it shall also provide defense against the CM threat that has been estimated to be of the order of 250 missiles. How does this inclusion change the cost?

A simple substitution of these additional factors into the cost equations shown in Example Problem, with the assumption that the lower layer would now have to deal with not only the leakage of the BM from the upper layer but also 45% of the incoming CM, is made. This modifies the costs slightly to give the result

Upper layer:	*$3.63 billion*
Lower layer:	*$2.12 billion*
Total:	*$5.75 billion*

That is, compared to example problem 5.1, an additional $330 million is required to provide the additional defense. Hence, this simplified analysis indicates that for a cost of slightly under $6 billion, a defense can be obtained against an attack of 500 ballistic missiles and 250 cruise missiles. Of course, detailed analyses would have to be embarked upon to assess the ability of the various systems (upper and lower layers) to not be saturated with uneven raid sizes and rates of attack within the these total numbers of incoming missiles.

Further, the issues of early warning and detection and communication networks have not been addressed in this treatment.

EARLY-WARNING COSTS

These important topics have not been addressed in this book to any detailed degree. Early warning can be provided by satellites (GEO, LEO, etc.) as already shown. Early warning can also be provided by various forms of long-range radar (land or sea based). Early warning can also be provided by the many possibilities in airborne systems that include manned aircraft, unmanned aircraft, and lighter-than-air vehicles. The costs of such systems can vary widely as already discussed, and it is a matter of total system planning to determine how much of these (early warning) costs are to be apportioned to missile defense as they are in many cases multimission in nature.

COMMUNICATION NETWORK COSTS

In a similar vein to the early-warning costs, there is the general topic of how all systems "talk to each other" such that the defense can actually be performed. Missile defense brings in a new dimension to warfare that has not as yet been realized as to its transnational nature. NATO is developing its BM/C^3 network called Air Command and Control System (ACCS), part of the NATO Air Force Command and Control System, that will go into effect during this decade. Developments are also underway to extend the basic air defense ACCS into a system capable of handling the early-warning features and message traffic associated with the ballistic missile threat and the advancing cruise missile threat.

Additionally, there are efforts underway to further augment this NATO-wide communications network with a transatlantic link, so that the US satellite early-warning system (see earlier discussion on DSP and SBIRS) can be integrated into a global network for the Alliance. This could be accomplished within the decade. Although missile defense will be greatly enhanced by such developments, the costs should not be directly applied to missile defense systems. The linking of missile defense systems into this broader network are costly, and these costs have been included in the preceding example problems.

COST SENSITIVITY ANALYSES

It has already been shown that many of the elements in missile defense systems are still in their infancy of development and the costs are not as yet firm numbers. It is useful to the decision maker to see how variations in the cost of a particular component or system would affect the cost of the total two-layer defense system. Some variations are provided here to indicate likely trends.

INTERCEPTOR UNIT COST

In Fig. 5.17 and surrounding text, it was seen that the interceptor costs are still being tackled to bring them to lower values, but what is the effect if they vary from the expected values? Using the same equations as used in example problems 5.1 and 5.2, it is possible to see the effect if the interceptor unit costs increase by 25, 50, or even 100% from the nominal values assumed. Figure 5.31 shows the sensitivity of the total two-layer system cost to such variations in the interceptor unit cost.

In the illustrative base case for the defense of the United Kingdom against both BM and CM, it was found that a total system cost of $5.75 billion was representative of the procurement costs of such a system. It was assumed in that example that the upper-layer interceptor unit cost was $2.5 million, and that the lower-layer interceptor unit cost was $1.5 million. Figure 5.31 indicates how the cost of the total system would increase if these interceptor unit costs increased by the amounts shown. This shows that if the interceptor unit costs *doubled* (i.e., to $5 and $3 million per missile respectively) which is a cost increase of 100%, then the total system cost would increase by about 52% to $8.73 billion. Similarly, if the interceptor unit costs increased by 50%, it is expected that the system costs would increase by about 26%.

INTERCEPTOR EFFECTIVENESS IMPACT ON COST

The experience to date of the effectiveness of interceptor kills on dummy warheads and in the heat of war is well documented and does not need to be expanded upon here. Most of the analyses in the community assume certain effectiveness values for interceptors, whether they be of the fragmentation

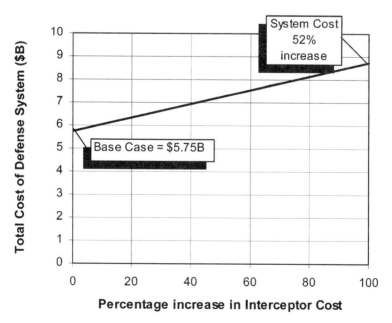

Fig. 5.31   System cost sensitivity to interceptor unit costs.

warhead design or the newer hit-to-kill variety. In the illustrative base cases used earlier, it was assumed that the typical value of SSPK = 0.70 would be a representative value. But suppose that such values are not attainable in battlefield conditions, what is the impact on cost and leakage in the defense?

There are several ways of obtaining the performance of interceptors, all else being equal, and that is by changing the SSPK or by changing the choice of firing doctrine with salvo firing. Chapter 4 provides a treatment of the effect on the defended footprint for different firing doctrines as they relate to salvo-look-salvo and the effect on the defended footprint area. Here, the impact on cost is treated.

As the SSPK of the interceptor falls to lower values than the assumed base case value of 0.70, then the leakage from the upper layer increases such that more lower-layer interceptors are required with the commensurate increase in system cost. Such an effect also has a negative effect on the number of incoming ballistic missiles that leak through to the key assets. This increase in leakage can be offset to varying degrees by increasing the salvo (from the base case of two) to larger values. Such an increase in salvo, of course, brings with it, increased costs of the system. Figure 5.32 shows these effects for various values of SSPK and for two salvo values (two and three). The base case is shown as a "star" on the diagram.

As seen in Figure 5.32, the illustrative base case for defense against both ballistic missiles and cruise missiles, the total system cost was $5.75 billion, and with an assumed SSPK for all interceptors the leakage was approxi-

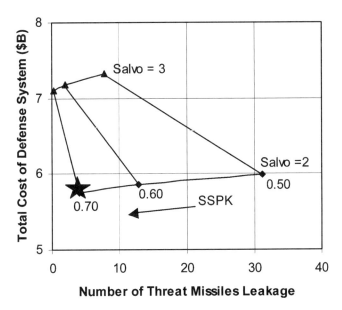

Fig. 5.32   Effect of interceptor effectiveness on system costs.

mately four ballistic missiles. If the interceptor individual SSPK drops to, say, 0.50, the total system costs only increase slightly to a value of $5.99 billion, which is a modest 4%, but, unfortunately, the leakage has increased dramatically by almost eight-fold to more than 31 missiles (spread over the 15 key assets). It is reasonable to assume that such a reduced defense would be unacceptable by the decision maker.

If, now, the alternative approach of increasing the salvo (from two to three) is used, then a significant improvement can be achieved in reducing the leakage over the key assets, but unfortunately at a large increase in cost. Figure 5.32 shows that if a salvo of three is used and the individual interceptor SSPK remains at 0.50, the leakage can be significantly reduced from 31 ballistic missiles to about eight missiles, which is double that of the base case. Unfortunately, the total system cost has risen to $7.32 billion (a 27% increase in cost). The decision maker must now decide if the 74% reduction in the number of missiles getting to target is worth the extra 27% increase in system costs—probably so.

This provides information to the decision maker on the relative merits of the various choices. For example, if it is determined that the R&D costs and lengthened schedule required to achieve a high-performing interceptor are unacceptable then an alternative is to accept the lower value of SSPK (=0.50) and simply change the firing doctrine from a salvo of two to a salvo of three and achieve a nine-missile leakage (which is 30% of the leakage of the alternative system) for an approximate 27% increase in total system cost (from $5.75 to 7.32 billion). Further, if an SSPK of 0.70 is achieved under battle conditions then a salvo of three would reduce the leakage from four missiles to one missile, which might be acceptable for the 23% increase in system cost (from $5.75 to 7.10 billion).

There are other variations that might be explored such as using a salvo of two in the upper layer and a salvo of three in the lower layer and other variations. These have not been costed here and are left as exercises for the reader. The technique shown here allows for such rapid evaluations.

COST OF PROTECTING MORE KEY ASSETS

In the illustrative base case it was taken that an arbitrary number of 15 key assets was selected that were to be protected. To account for the airbreathing threat such as the CM and also to account for the effect of "shifting defended footprints" because of the possibility of the BM threat being launched on lofted or depressed trajectories, a factor of 50% was incorporated to give 22 lower-layer interceptor systems needed. The 15 key assets would include the capital, key cities, seaports, industrial centers, and military facilities and other assets that the nation deemed important to protect.

If the number of key assets were increased, it is informative to see the effect on the total cost. This was done for the base case and the results shown in Fig. 5.33. It is seen that if the number of key assets were increased by a factor of two to give 30 key assets and that the number of lower-layer interceptor systems were proportionately increased to 45 systems, then it is seen that the 100% increase in the number of key assets to be protected would require an approximate 29% increase in total system costs.

It is possible that it might not be necessary to increase the number of lower-layer interceptor systems in direct proportion to the number of key assets in all cases. In any detailed treatment of the defense of a nation, it might be that the assumption of 50% would be modified to some other value, but the preceding treatment gives a good approximation for initial analyses.

COST OF PROTECTING EUROPE AND NORTH AMERICA

If the cost of protecting one nation (the United Kingdom in the illustrative base case) is of the order of $6–8 billion), then it is immediately noticed that this could lead to large costs to protect all Europe (16–25 nations) if a simple additive approach is taken. An alternative approach might be to assume a collective defense posture for the upper-layer systems and leave the lower-layer systems to be part of the individual national defense budgets. Because some of the nations have much smaller land mass then the illustrative base case that used a 500-km-radius circle (see Chapter 3), then it is possible that a bilateral defense might be considered for some of the nations. Hence, rather than show a blanket cost it is best left to individual analyses to assess the

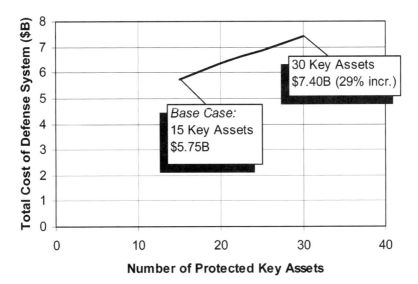

Fig. 5.33   Cost of protecting more key assets.

costs for protecting all nations using various combinations of shared Alliance costs and national costs. Similar techniques could be used with the preceding approach for the cost of the defense of North America and then the total cost for protecting the Alliance as an entity. Such calculations are easy to do using these techniques and left for later analyses.

COST OF PROTECTING AGAINST WMD

Most of the preceding analyses and discussions have the implicit assumption that the threat, whether it be ballistic missile or cruise missile, have conventional warheads and that some combination of firing doctrine, SSPK, and defended footprint will result in the desired protection and minimum leakage of incoming missiles over the key assets. Such an assumption embodies the additional assumption that a hit is a kill. Unfortunately, as shown in more detail in Chapter 3, for the case of chemical and biological weapons, such an assumption can be woefully off the mark. While intensive R&D continues in seeking and improving interceptors with much higher SSPK, there is the possibility that an interceptor with an SSPK that approaches the values needed to ensure a high level of success against all submunitions containing chemical or biological agents with high lethality might not be attainable.

If specific program information on costs is not available to the reader, then the various cost values for components, systems, and architectures can be used as default values for cost for use as inputs into Chapter 8.

EXAMPLE PROBLEMS

5.1   In example problem 5.2, in the protection of the United Kingdom from both ballistic missile and cruise missile attack, assume that the performance capability of the interceptors in both layers of defense has been improved to give SSPK = 0.80. What is the improved leakage (in number of missiles) over the key assets, and what is the increase in cost of the total system?

5.2   If, in addition to the improved SSPK as given in example problem 5.1, the firing doctrine is changed to a salvo of three interceptors instead of two, what is the improved leakage, and what is the increased cost?

5.3   If, now, it is learned that 50% of the incoming missiles contain chemical and biological warheads and that a survival probability of 99.9934% is deemed acceptable, what is the impact on the defense in terms of how many missiles get through, and what is the cost of the total defense system, assuming that a two-layer defense is used with interceptors?

5.4   Using the default cost figures provided in this chapter, compare the costs of using either of the two airborne defense concepts, that is, 1) aircraft + interceptor or 2) aircraft + laser. What is the crossover point in cost between the two concepts?

5.5   If, in example problem 5.1, a higher-speed interceptor is used to give an improved value of $V_{bo} = 3\ km/s$, how many upper-layer systems would be required, and what is the cost of the total (two-layer) defense system?

5.6   Assuming each nation in NATO Europe uses the same two-layer defense system as given in example problem 5.1, what would be the cost to protect all NATO Europe? Assume each nation has 15 key assets that they need to protect.

Chapter 6

SCHEDULE

> *Remember that time is money.*
> Benjamin Franklin, 1748

Like "cost" (see Chapter 5), the word "schedule" can be deceptively simple. From a defense decision-maker's viewpoint, the subject of schedule usually refers to the question *"when can the proposed defense system be ready for use?"* There are, of course, immediate subsidiary questions, such as the following: *When does the clock start ticking on such a schedule? How does the system under consideration fit in the (ongoing) acquisition cycle?* and other related and important questions.

There are two main and overlapping time streams of interest to the consideration of any new defense system. The first relates to the *Acquisition Program* structure, and the second relates to any required *Research & Development (R&D)* that might be needed before any consideration of approved acquisition can proceed. In most nations, in the structure of the defense budgets it is taken that before any R&D for military purposes can be approved and placed in the defense budgets that there is at least a general belief that the approved R&D is for some military purpose with some military need focus. This ensures that *"research for research's sake,"* which is more the purview of university scientific research, is not part of the defense R&D budget. Because of this, there is automatically a link between R&D and Acquisition. As will be shown later in this chapter, the linkage between the *preacquisition phase* (which includes the bulk of R&D) of any defense system and the *acquisition phase* and their *overlapping schedules* has been the subject of debate for at least the last 33 years.

BRIEF OVERVIEW OF ACQUISITION POLICY

It is informative when examining the schedule of defense systems to examine briefly the various Acquisition Policies used by various nations. Such a subject deserves a much more detailed treatment than is possible in a book of this type, but an overview can provide the needed appreciation of the question, *"when can the defense system be made available?"* First, a synopsis of the main features in the U.S. system is given, followed by those used by NATO, and then some individual nations of the Alliance, specifically France, Germany, and the United Kingdom.

ACQUISITION POLICY IN THE UNITED STATES

As of April 2002, a new set of policies[66,67] has been issued that has not yet been tested and is already being modified by decision authorities in missile defense, but it is provided here as the "new" thinking, especially as it relates to missile defense.

The form of the schedules for the total acquisition process has actually not changed in substance too much over the last 33 years, although their timescales certainly have changed and in most cases lengthened. All management actions on the setting up of the approved schedules for acquisition have been directed at reducing the complexity and to reduce the time it takes to get a new system fielded. The main differences have appeared as to when decisions are to be and by whom. There are three revisions to the acquisition schedule of importance: 1) the original process initiated in 1969, 2) the first major revision to the process in 1981, and 3) the current process instituted in 2000–2002. These three acquisition processes and the relevant schedule milestones are shown in Fig. 6.1.

After the Vietnam conflict the management of the U.S. Defense Budget came under scrutiny and Deputy Secretary of Defense David Packard issued a set of instructions that established the first Defense Systems Acquisition Review Council (DSARC).[68] This schedule for the process to acquire defense systems became known as the DSARC process. The top schedule shown in Fig. 6.1 is a brief outline of the main features of this original process. It shows the unfolding of the approval of any given defense system from the first recognition of a formal requirement for a new capability, expressed in a *mission element need statement* (MENS), through to the actual deployment of a system. It is important to note that the chart as shown does not highlight where in this schedule does the requirement for a system enter the R&D cycle. For example, if the proposed system utilizes a technology that has been in Basic Research for some time prior to the issuance of the formal (military) requirement, this is not considered part of the schedule. As will be shown later, acquisition in a formal sense does not begin its acquisition schedule until *sometime into the R&D phase*. The 1969 DSARC process required all milestone decisions to be made at the Secretary-of-Defense level.

The middle schedule shown in Fig. 6.1 was introduced in 1981[69,70] by then Deputy Secretary of Defense Frank Carlucci in an effort to further improve efficiency and competition over the previous 1969 DSARC process.

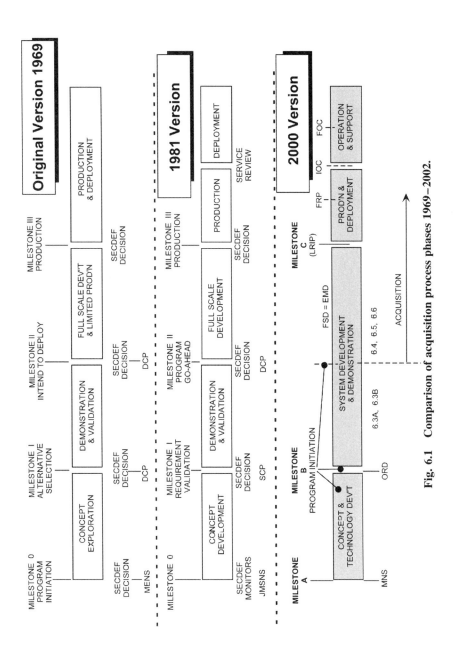

Fig. 6.1 Comparison of acquisition process phases 1969–2002.

It is seen that one of the changes was to place more of the decisions into the hands of the services responsible for the program.

Finally, the lower schedule in Fig. 6.1 was issued in 2000 (with instructions issued in 2002) by Under Secretary of Defense Pete Aldridge, as a further refinement of the original schedule produced in 1969 and is extracted from the current cited DoDD 5000.1 and DoDI 5000.2 documents. One of the main differences between the current acquisition process and that of the earlier versions is the flexibility of the Milestones, especially the new Milestone B that can occur at different points (as shown) in the development depending upon the maturity of the technology being used in the defense system.

Basically, the decision milestones in the overall schedule remained intact for the first two acquisition schedules shown in Fig. 6.1 but were revised in today's version. It is useful to compare the descriptions of the milestones because today depending on where ongoing individual defense systems under development are in the acquisition process determines which acquisition policy (old or new) they are to follow. Today is a transition period in the approval cycle improvement process. Table 6.1 describes the "old" milestones, generally applicable to the original (1969) and the

Table 6.1   Description of old milestones (1981)

Milestone	Description[a]
Milestone O	*Mission Need Determination* Once a mission need is determined and reconciled with other Department of Defense (DoD) capabilities and priorities, a program initiation is approved with a previously developed justification for a major system new start (JMSNS) requirement document.
Milestone I	*Demonstration and Validation* After completion of competitive exploration of single or alternative systems concepts to the point where selected alternatives warrant system demonstration, approval is given to proceed into a demonstration and validation effort. This begins with the recommendations documented in a system concept paper (SCP).
Milestone II	*Full-Scale Development* When the demonstration and validation activity is complete and deemed satisfactory, approval is given to proceed into full-scale Development of the system or systems. The recommendations at this milestone are documented in a decision coordinating paper (DCP).
Milestone III	*Production and Deployment* With the necessary updates of the DCP, the system is approved to enter the production phase and ultimate deployment as an operational defense system.

[a]Paraphrased from the official documents for brevity.

improved (1981) DSARC process. A new acronym has entered the process with the creation of the *Milestone Decision Authority* [*MDA*, not to be confused with the Missile Defense Agency (MDA), created on 1 January 2002] which varies, as before with the deemed importance or concern over the program. The MDA might be Secretary of Defense or Service Chief or Program Manager within the Service. Such identification has been left off Fig. 6.1 for clarity as it varies widely depending on whether the program is certified (by the Secretary of Defense) as an Acquisition program (ACAT) or a nonAcquisition program (non-ACAT).

In the preceding description of the Milestones, much official detail has been removed to capture just the essential elements. For example, variations of the Milestones were allowed to accommodate the realities of any particular acquisition. In 1983, as just one example, the Milestone III was broken into Milestone IIIA to allow an initial low-rate production (LRIP) followed by Milestone IIIB to begin full-rate production (FRP) with a suitable spread in the schedule to allow for lessons learned during initial production and initial deployment. Such a split of Milestone III also allows the new defense capability to be fielded earlier, albeit with a smaller number of units. The FRP decisions did not have formal "Milestone" designation but were a key program decision requirement. In Table 6.1 the decision maker at each level has been not identified for clarity. The decision maker might be at the Secretary-of-Defense level or within the individual services depending on certain criteria established within the Office of Secretary of Defense (OSD) (see later).

The level of the decision maker indicates the degree of importance or concern that the Defense Department has about the program under consideration. If the program carries a large amount of risk (see Chapter 7 for a more in-depth treatment of risk and R&D expenditure) and is considered of prime importance to the makeup of defense, then this raises the level of decision maker. This is handled in the accounting system by the establishment of Acquisition Categories (ACAT). Those programs containing a large risk (large dollar amount of R&D) and a significant percentage of the Procurement Budget are called ACAT I, with lesser risk programs given designations of ACAT II, III, and IV, respectively. These ACAT Programs are handled within each of the pertinent services again at different levels of decision authority. A description of these ACAT categories and the cognizant MDA is given and expanded upon in Chapter 7.

Ongoing defense systems that have not yet been approved as acquisition programs, which means that they are still in the component R&D stage of development and thus prior to Milestone B in the new system, will be required to adhere to the new (2002) Acquisition Process milestones, which are defined in Table 6.2.

Table 6.2   Description of new milestones (post-2000)

Milestone	Description[a]
Milestone A	*Begin Concept and Technology Development* This milestone serves as the old Milestone O. It signals the start of Concept Studies and development of strategies on how the mission need statement (MNS) will be evaluated. This is a PreAcquisition milestone.
Milestone B	*Program Initiation (Normally)* This Milestone signals approval to enter System Development, and Demonstration and Program Initiation normally begins here, but the system allows Program Initiation to begin earlier during the Concept and Technology Development Phase (specifically during the Component and Advanced Development stage) if the technology is deemed mature enough. Thus under "normal" procedure Milestone B is equivalent to Milestone II. **This Milestone signals the start of the Acquisition Phase of the program**. This Milestone also requires that a firm Operational Requirement Document (ORD) has been established.
Milestone C	*Start Low-Rate Initial Production* This Milestone starts LRIP for major systems and full production for non-major systems. FRP for major programs begins during the Production and Deployment phase after a Full-Rate Production Design review is successfully passed. Such a decision is not marked by a formal (letter designated) Milestone.

[a]Unfortunately, the new DoDI 5000.2 dated 5 April 2002[7] does not give names to these new milestones (only a letter designation is given), and one must interpret their meaning from the acquisition phase that begins following approval of that milestone. The names used in this table are indicative only.

Table 6.2 shows that considerable flexibility has been built into the process to accommodate a variety of types of programs that do not all fit into one mold. It does make it difficult, however, to collect the data on when programs start, or which part of R&D is included in Acquisition Cost and other important aspects of acquisition.

As already mentioned, this new policy for acquisition is in a stage of transition and testing. In the case of *missile defense acquisition*, for example, the requirement for a formal Operational Requirement Document (ORD) has already been waived so that a change of focus to a capabilities-driven acquisition process rather than a requirements-driven process has been given to the program [see testimony given before the Senate Armed Services Committee (SASC) by Secretary Pete Aldridge, Under Secretary of Defense (AT&L)[71]]. The strategy evoked in that policy is that it is better to get a system in place that can handle a large number of threats rather than insist that development continue to meet the threat preordained in an ORD that had defined a specific set of threats that might or might not come into existence.

The approach would then be to improve the missile defense system later through what is being called *evolutionary acquisition.* A formal definition of evolutionary acquisition appears in DoDD 5000.1. The accompanying R&D program that supports this approach is called *spiral development.* Prior to the recent changes (2000–2002), the acquisition approach that followed this practice was first introduced in 1981 as *preplanned product improvement (P3I).* It is left to other writings to expound on all of the different features of the various Acquisition Policies that have been introduced over time. It is sufficient for the purposes of this book to recognize these variations and how they change the milestones and schedules that will bring a missile defense system into operational use.

R&D SCHEDULE IN THE UNITED STATES

To determine when a defense system can be introduced in the field, it is necessary to first outline the preacquisition phase, which is intimately tied to the R&D phase. Chapter 7 in the treatment of Risk details the history of how R&D is structured in the United States and how it has changed in make-up; here today's form of R&D accounts will be used to illustrate how the schedule is influenced by the type of R&D.

The Research, Development, Test, and Evaluation (RDT&E) Categories and the corresponding Budget Activities (BA) have not changed substantively since their introduction in 1962 (see Chapter 7). However, because there have been subtle changes in language and now the Milestones have changed names, it is useful to use the latest text taken from the DoD Financial Management Regulation,[72] which is provided in Table 6.3.

The FYDP is approved as part of the DoD portion of the President's Budget. It must go through a cyclic approval process that typically takes 14–16 months in three interrelated phases. This is part of the Planning, Programming, and Budgeting system (PPBS)[73] that was introduced in 1962 to bring together the complete decision process on policies, strategies, and the development of defense capabilities to meet mission requirements.

The reference to program elements in Table 6.3 is a reference to the smallest subdivisions of the RDT&E programs used for accounting. Each Program Element contains an identifying number that identifies the developing activity, the RDT&E category, the functional area (ASW, etc), project number, and other identifying codes. It provides an immediate way of identifying where in the acquisition process the particular RDT&E effort belongs. Assuming broadly that RDT&E categories 6.4 and above belong in the Acquisition phase, Table 6.4 provides the official descriptions taken again from the DoD Financial Management Regulations cited earlier.

Because the acquisition policies are quite new, it is not surprising that some of the terminology still reflects the old system's terminology. Such

Table 6.3   Definition of Pre-Acquisition RDT&E Categories

RDT&E[14] Budget Activity/Category	Description
BA 1: Category 6.1	*Basic Research* Basic research is a systematic study directed toward greater knowledge or understanding of the fundamental aspects of phenomena and of observable facts without specific applications towards processes or products in mind. It includes all scientific study and experimentation directed toward increasing fundamental knowledge and understanding in those fields of the physical, engineering, environmental, and life sciences related to long-term national security needs. It is farsighted high-payoff research that provides the basis for technological progress. Basic research can lead to a) subsequent applied research and advanced technology developments in defense-related technologies and b) new and improved military functional capabilities in areas such as communications, detection, tracking, surveillance, propulsion, mobility, guidance and control, navigation, energy conversion, materials and structures, and personnel support. Program elements in this category involve pre-Milestone A efforts.
BA 2: Category 6.2	*Applied Research* Applied research is a systematic study to understand the means to meet a recognized and specific national security requirement. It is a systematic application of knowledge to develop useful materials, devices, and systems or methods. It can include design, development, and improvement of prototypes and new processes to meet general mission-area requirements. Applied research translates promising basic research into solutions for broadly defined military needs, short of system development. This type of effort can vary from systematic mission-directed research beyond that in BA 1 to sophisticated breadboard hardware, study, programming and planning efforts that establish the initial feasibility and practicality of proposed solutions to technological challenges. It includes studies, investigations, and nonsystem-specific technology efforts. The dominant characteristic is that applied research is directed toward general military needs with a view toward developing and evaluating the feasibility and practicality of proposed solutions and determining their parameters. Applied research precedes system specific research. Program control of the applied-research program element is normally exercised by general level of effort. Program elements in this category involve pre-Milestone B efforts, also known as concept-and-technology-development phase tasks, such as concept-exploration efforts and paper studies of alternative concepts for meeting a mission need.

BA 3: Category 6.3A   *Advanced Technology Development (ATD)*   This budget activity includes development of subsystems and components and efforts to integrate subsystems and components into system prototypes for field experiments and/or tests in a simulated environment. ATD includes concept-and-technology demonstrations of components and subsystems or system models. The models can be form, fit, and function prototypes or scaled models that serve the same demonstration purpose. The results of this type of effort are proof of technological feasibility and assessment of subsystem and component operability and producibility rather than the development of hardware for service use. Projects in this category have a direct relevance to identified military needs. ATD demonstrates the general military utility or cost reduction potential of technology when applied to different types of military equipment or techniques. Program elements in this category involve pre-Milestone B efforts, such as system concept demonstration, joint and service-specific experiments, or technology demonstrations. Projects in this category do not necessarily lead to subsequent development or procurement phases.

BA 4: Category 6.3B   *Advanced Component Development and Prototypes (ACD&P)*   Efforts necessary to evaluate integrated technologies, representative modes, or prototype systems in a high-fidelity and realistic operating environment are funded in this budget activity. The ACD&P phase includes system-specific efforts that help expedite technology transition from the laboratory to operational use. Emphasis is on proving component and subsystem maturity prior to integration in major and complex systems and can involve risk-reduction initiatives. Program elements in this category involve efforts prior to Milestone B and are referred to as advanced component development activities and include technology demonstration. Completion of technology readiness levels 6 and 7 should be achieved for major programs. Program control is exercised at the program and project level. A logical progression of program phases and development and/or production funding must be evident in the five-year defense plan (FYDP).

Table 6.4   Definition of Pre-Acquisition RDT&E Categories

RDT&E[14] Budget Activity/Category	Description
BA 5: Category 6.4	*System Development and Demonstration (SDD)* SDD programs have passed Milestone B approval and are conducting EMD tasks aimed at meeting validated requirements prior to full rate production. This budget activity is characterized by major line item projects, and program control is exercised by review of individual programs and projects. Prototype performance is near or at planned operational system levels. Characteristics of this budget activity involve mature system development, integration and demonstration to support Milestone C decisions, and conducting live-fire test and evaluation (LFT&E), and IOT&E of production representative articles. A logical progression of program phases and development and production funding must be evident in the FYDP consistent with the department's full funding policy
BA 6: Category 6.5	*RDT&E Management Support* This budget activity includes research, development, test, and evaluation efforts and funds to sustain and/or modernize the installations or operations required for general research, development, test, and evaluation. Test ranges, military construction, maintenance support of laboratories, operation and maintenance of test aircraft and ships, and studies and analyses in support of the RDT&E program are funded in this budget activity. Costs of laboratory personnel, either in-house or contractor operated, would be assigned to appropriate projects or as a line item in the basic research, applied research, or ATD program areas, as appropriate. Military construction costs directly related to major development programs are included.
BA 7: Category 6.6	*Operational System Development* This budget activity includes development efforts to upgrade systems that have been fielded or have received approval for full rate production and anticipate production funding in the current or subsequent fiscal year. All items are major line item projects that appear as RDT&E costs of weapon system elements in other programs. Program control is exercised by review of individual projects. Programs in this category involve systems that have received Milestone C approval. A logical progression of programs and development and production funding must be evident in the FYDP, consistent with the department's full funding policy

inconsistencies do not cause problems provided the intent is clear. Under the ATD or the "old" 6.3A programs for example, as shown in the list with categories 6.1–6.3, the category actually groups the ATD and the related Advanced Concept Technology Demonstration (ACTD). Using the language of the 2002 issued DoDI 5000.2,[67] these are now defined as follows: the ATD to "demonstrate the maturity and potential of advanced technologies for enhanced military operational capability or cost effectiveness," and the ACTD is to "determine military utility of proven technology and to develop the concept of operations that will optimize effectiveness." Such considerations influence where the dividing line is between RDT&E activities and acquisition activities and who is the Milestone Decision Authority.

Although not defined in the new acquisition process documents (5000.1 and 5000.2), certain key functions are still retained. For example, in BA 5 (RDT&E category 6.4), the key development phase called *engineering and manufacturing development* (*EMD*) is an essential part of SDD. The equally important functions of initial operational test and evaluation (IOT&E), as required by law, and any required follow-on operational test and evaluation (FOT&E) are included in the latest 5000.1 and 5000.2 document.[66,67]

SO WHEN DOES THE SCHEDULE BEGIN?

It has already been shown that the start of the schedule, defined as the *approval to proceed with some defense system*, occurs somewhere during the RDT&E program. The new Milestone B and its flexibility as to where it is in the process makes it difficult to precisely identify a point in the acquisition process as a prescriptive requirement for all programs. However, it is useful to examine a "normal" program, and Fig. 6.2 shows the approximate connection between RDT&E and acquisition.

For a normal program, which is any program that has not been decided at the Secretary of Defense level to receive special treatment, approval to proceed or program initiation begins at the start of the System Development and Demonstration Phase (see Table 6.2 and Ref. 67). Prior to this point, the process is moving out of the Tech Base* (=6.1 + 6.2) and into advanced Development category (6.3) to explore alternative solutions to meeting a mission need. As has already been stated, if the technology is deemed sufficiently mature and the risk is low the program could start during the latter stages of category 6.2. On the other hand, if a need is seen for technology demonstration before a decision can be made then Milestone B slips to after any category 6.3A activity and at the beginning of the category 6.4 or SDD of the selected defense system.

---

*It is convenient to use some of the old terminology here as it is believed to be useful in communicating the basic ideas.

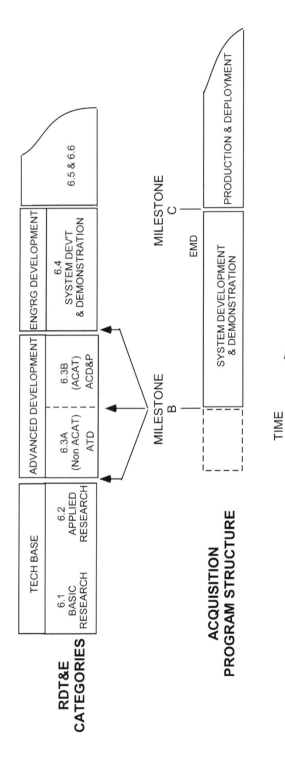

Fig. 6.2   Approximate connection between RDT&E and acquisition.

These three choices for milestone B that signal the start of acquisition are shown in Fig. 6.2. Note that these choices change the dividing line between RDT&E and Acquisition, and thus care should be taken in interpreting the dividing line between the list of categories 6.1–6.6 and Fig. 6.2 in the compilation of the RDT&E Categories. As the Milestone B changes, depending on the program, the cost of the program phases also change.

How Long Is the Schedule?

It is taken that the acquisition decision-maker's interest in the schedule is the determination of how long it will take once the program is underway (post-Milestone B) to introduce the defense system into the force structure. This means that the schedule should be measured from Milestone B up to the introduction into service use. This is usually measured by the date of Initial Operating Capability (IOC), but it should be recognized that this date only reflects the *first* unit or batch. If in the Performance analysis (see Chapter 4) it is determined that a set of systems is required, then the date of all units called the Full Operational Capability (FOC)* must be used. The difference between these dates can be significant. For example, suppose it is determined that using five typical ballistic missile (TBM) capable ships satisfies the mission requirement and that a particular IOC date has been determined for the first ship, then assuming one ship per year can be launched from the shipyard, the full operational capability (FOC) would become available in IOC plus four years. The decision maker must ask: *Is a stop-gap defense needed during this time between IOC and FOC? Will the defense be adequate during those intervening four years?*

Some idea of the length of time it takes to develop programs can be seen from historical data. Figure 6.3 shows data taken from several sources. It shows the length of time from Program Concept to the start of Full-Scale Development for several missile programs. There was concern expressed in the late 1970s about the perceived increase that was occurring in many defense programs in the time taken to develop them and to get the systems to the operating forces. Figure 6.3 confirms this increase in time. The data shown have been complied from a variety of sources. The data labeled "RAND data"[74] were done for the Office of the Under Secretary of Defense in response to concern about the lengthening acquisition process. The data on more recent missile programs shown have been taken from published program files. Additionally, the Defense Science Board (DSB)[75] conducted an analysis of many different major acquisition programs that were subject to the new (at that time) DSARC process and came to the same conclusion.

---

*This shows the problem with acronyms. In shipbuilding parlance it is common to use the FOC acronym to mean "first of the class," which would be the same as IOC.

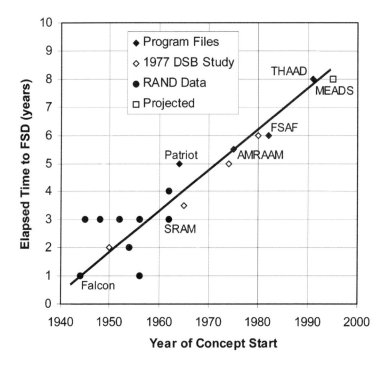

Fig. 6.3   Time from program concept to start of full-scale development (FSD).

Those data are also shown on Fig. 6.3. It would appear from the more recent missile program data taken from publicly available data that in the 1990s and beyond this trend is still evident. It remains for historical perspective to see if the new acquisition policies initiated in 2000–2002 will reverse this trend.

Figure 6.3 essentially shows the program development up to Milestone B, in today's acquisition parlance. The next period of time in the acquisition process, is from Milestone B to the start of LRIP, which today is called Milestone C and corresponds to the start of the System Development and Demonstration Phase. Here, it was found that the data were considerably more scattered. Figure 6.4 shows the available data. The historical data for the missile programs were again taken from the RAND report. The projected data for the missile programs shown have been compiled from recent Congressional reports and similar documents. The *downward* trend for aircraft programs (over 30 programs in the database) is also taken from the RAND report for reference to show what could be the trend. However, the RAND Report goes on to show that if the trend curve were shown for the time to get to the 200th aircraft (and not the IOC of the first aircraft) the data would exhibit an *upward* trend. It is suggested by RAND in that data that

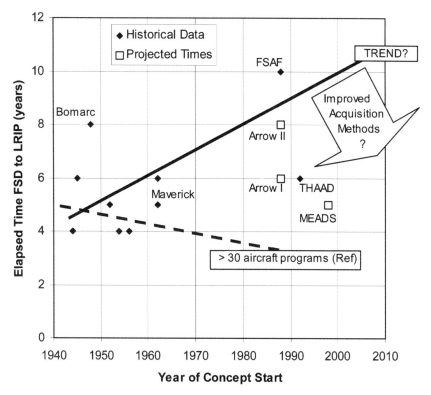

Fig. 6.4   Time from start of full-scale development to LRIP (Milestone B to Milestone C).

such a trend is incurred by the frequent reprogramming as budgets wax and wane and government reviews expand. It is possible that similar factors are driving the current trends in development times for the ongoing missile programs. As one example, the Theater High-Attitude Defense (THAAD) program recently entered its EMD phase (after an eight-year concept-and-technology-development phase) 23 June 2000 for a 6-year EMD phase with a development schedule that gives milestone C (for LRIP) in 2007 and with FRP in 2010 (or possibly 2008). From the available data it appears that such elapsed times are quite typical of today's programs.

The trend line shown in Fig. 6.4 should be regarded as a place holder until the historical data become available and new acquisition policies and methods take hold. Some insight into the reasons for the scatter of data in Fig. 6.4 can be ascertained by examining some ongoing missile development programs both in the United States and internationally. A brief overview of these programs is given here.

ASSESSMENT OF SOME OF TODAY'S MISSILE PROGRAMS

Table 6.5 shows the key dates for the ongoing missile defense programs, namely, THAAD (U.S. Army upper-layer land system), Patriot (operational family of U.S. Army integrated extended air defense systems), FSAF [*famille de systems anti-ariens futur* (family of future anti air systems), French air-defense systems under development for both land and naval use], and Arrow (cooperative U.S. Israeli Anti-Tactical Ballistic Missile system).

As can be seen from Table 6.5, there is always some confusion as to when a program actually starts, but over the life of the development of any particular program the times tend to even out to a statistical average that does not vary significantly from one program to the next. For example, despite the wide variations between them at the start of their development all of the programs in Table 6.5 have required approximately 14 years to reach IOC. The U.S. Patriot is the main missile defense system in the world today. It was the first to be tested and combat proven during Desert Storm in 1991 and was used against threats for which it was not originally designed. It first intercepted a SCUD TBM on 18 January 1991 (at 0428 local time in Saudi Arabia). (It is left to other writers to continue their analyses of the various successes and failures of these first combat-tested interceptors against the new threat of TBM that was introduced in the Gulf War.)

It is possible to misread the column on Patriot in Table 6.5, in that after the introduction of Patriot a series of improvements or "advanced capability" with the Patriot system has been instituted. Advanced-capability (PAC) versions with PAC-1, PAC-2, and PAC-3 have appeared on the scene. These could be read as P3I versions of Patriot, and thus the quoted dates of IOC merely indicate the later introductions of the advanced-capability version of the Patriot rather than any representation of the time it takes to develop each one. For example, PAC-1 and PAC-2 are derivatives of the basic Patriot with additional improvements of low noise receivers, addition of remote launch capability, and other P3I improvements. In 2002 a new technology was integrated into the PAC-3 configuration to give the Patriot a hit-to-kill (HTK) capability as an improvement for ATBM use. The HTK technology was introduced into the Patriot from previous activities from other development programs, such as the Homing Overlay Experiment (HOE, which had a HTK success on 10 June 1984) and Exoatmospheric Reentry-Vehicle Interceptor Subsystem (ERIS, which had a HTK success in January 1991) programs. So the development schedule should really be measured from these programs (not shown). With this proviso the dates for the Patriot and its advanced capability versions are correct. Other versions of the Patriot, not shown, as the development continues include such features as guidance enhancement for improved acquisition and tracking of low radar-cross-section targets operating in a clutter to produce the guidance enhancement missile (GEM).

Table 6.5 Elapsed times in development of today's missile defense programs

Milestones	THAAD	Patriot	FSAF	Arrow
Milestone 0	1990 (3 studies)	1965 (3 studies on XMIM 104)	1982 (2 studies on SYRINX)	1986 (U.S. Israeli studies)
Milestone I	1992 Program start	1967 Program start	1968 Program start	1988 Program start
Milestone I to Milestone II	8 years	9 years	0 years	7 years
Milestone II (start EMD)	23 June 2000	1976	1988; ASTER 15 1990; ASTER 30	1994; Arrow I 1997; Arrow II
Milestone III (start LRIP)	2006	1981	1998; ASTER 15 and 30	1999
From Milestone I to Milestone III	6 years	5 years	10 years	5 yrs for Arrow I 7 yrs for Arrow II
IOC	2007 (40 missiles)	1984 1988 PAC-1 1990 PAC-2 2002 PAC-3	1999; ASTER 15 2002; ASTER 30	2000
FOC	2010 (>1200 missiles)	—	2006	—
Program start to IOC	14 years	14 years Patriot 18 years PAC-1 20 years PAC-2 32 years PAC-3 (HTK)	11 years ASTER 15 14 years ASTER 30	12 years
Program start to FOC	17 years	Various	18 years	n.a.
From concept to FOC	19 years	Various	24 years	n.a.

The French missile system has gone through several iterations. In 1982 the intent was to develop a *"système rapide interarmés à base d'engins et fonctionnant en bande X (SYRINX)"* or an "X-band fast tri-service missile system." There were so-called predevelopment contracts awarded in 1984, and the program, when started in 1988, immediately launched into full development. This would mean by the usual planning system that the standard concept-and-technology-development phase had been bypassed (hence the "0 years" listed). Obviously, the design work had been moved into the 1988 SDD phase, which would account for the apparent increased elapsed time shown. However, the end result of 14 years to IOC for the ASTER 30 missile system matches that of the other programs shown. Further improvements to ASTER 30 not shown will add an ATBM capability with later milestone dates out to post-2010.

The Arrow program is a cooperative venture between the United States and Israel. The projected costs ($2.2 billion) are divided evenly between the United States and Israel. Three Arrow batteries are planned, with one in place at time of writing.

SOME ACCELERATED INTERNATIONAL MISSILE DEFENSE PROGRAMS

There are two other ongoing missile defense programs that are important to consider because the approach being taken in these programs will greatly affect their schedules. They are international in content and contain some aggressive acquisition methods worth commenting on. The programs are the *Medium Extended Air-Defense System (MEADS,* intended to replace the Patriot in the U.S. Army and to be the mobile air-defense system for Germany and Italy in 2015) and the NATO Active-Layered TBM Defense system (under preacquisition planning today) programs. The schedules for these two activities are shown in Fig. 6.5.

Without getting into the details of these two programs, there are some interesting aspects of acquisition that could be instrumental in how missile acquisition can improve in the coming years as ways are explored to reduce Risk and Schedule to develop such systems. Perhaps, these techniques can reverse the trend shown in Figs. 6.3 and 6.4.

The first program, MEADS, designed to replace both the HAWK and Patriot batteries as protection for a maneuvering force was initiated on 28 May 1996 and contained an aggressive schedule to achieve the IOC. One of the reasons for such confidence was the planned use of existing missile systems rather than develop a new missile. It is planned to use the PAC-3. Even so, it was decided in November 1999 to insert a risk-reduction effort, as shown in Fig. 6.5, in the normal NATO schedule (see later) between the Project Definition Phase and the Design and Development Phase. This

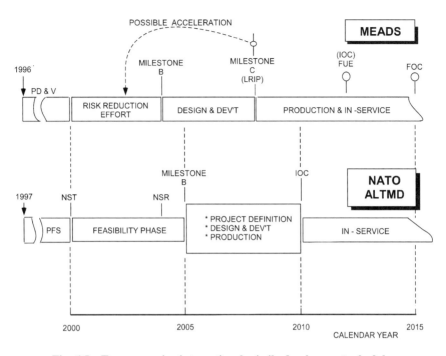

Fig. 6.5   Two aggressive international missile development schedules.

actually lengthened the schedule to give a projected IOC or First User Equipment (FUE) in 2012, some 16 years after start of project definition. There are planning activities underway to recover this time and provide an IOC date of 2007 as shown. These are still projections at this stage.

In the second example a similar aggressive schedule is shown (in recognition of the perceived threats expected by 2010) in the NATO pre-acquisition work to provide NATO with a layered defense (see Chapters 3 and 4) against TBM attacks. As stated in all of the preliminary work and planning documents, such an aggressive schedule is predicated on the maximum use of existing systems. This eliminates the need for protracted development (and reduced Risk), and effort can be concentrated on the integration features of existing systems into a new defense architecture. The necessary requirements document (issued in 1999) is the NATO Staff Target (NST), which has allowed the Feasibility Phase to begin, with a planned NATO Staff Requirement (NSR) (similar to the MNS referred to earlier in the U.S. System) to be issued in 2004, such that a decision for a Program Start can be made in 2005.

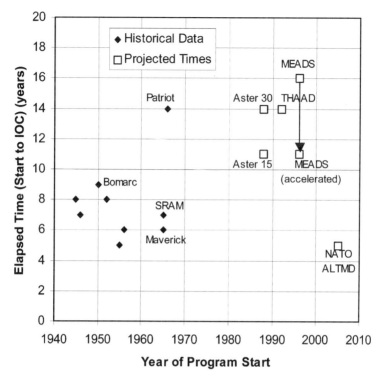

Fig. 6.6   Time from program start to IOC of current programs.

The key feature of both of these international programs is that maximum use is made of either existing systems or ongoing developments such that an introduction can be made of such systems sometime around 2010+. If the IOC of these programs are displayed together with the ongoing missile defense programs already discussed, an appreciation can be made as to the likely trend in the future. Figure 6.6 shows both the historical data base and these new developments. No trend line has been added because of the wide disparity of the elapsed time to IOC. The reasons have been discussed here, and the numerical display of such dates is given in Fig. 6.6. It remains for historical viewpoint to determine if such aggressive acquisition methods will have borne fruit.

SCHEDULE TO BRING A DEFENSE SYSTEM INTO BEING

Clearly, it is not possible to be too prescriptive on the total length of a defense system schedule because of all of the factors already discussed, but Table 6.6 gives a broad outline of these various contributors to the schedule based on the data presented in this chapter. A total elapsed time of 10-20

Table 6.6 Approximate total timelines for missile system acquisition

Starting milestone	Phase	Approximate elapsed time
A	Concept and technology development	6 years (see Fig. 6.3)
B	System development and demonstration	
	System integration	2 years (see Fig. 6.4)
	EMD	7 years (see Fig. 6.4)
C	Production and deployment	
	LRIP	2 years
	Block production	5 years
	Operating and support	10–15 years (see the following)
Totals	Concept to IOC	10–20 years[a] (see Fig. 6.6)
	IOC to disposal	10–15 years[b]
		20–35 years (totals)

[a]Depending on how far back into basic research (6.1, 6.2) it is required to capture the key technology.
[b]Would be extended with P3I and SLEP programs.

years is not an unreasonable assumption for risk removal and with a total "cradle-to-grave" time of 20–35 years.

It is suggested that the lengthening of the acquisition schedule is not all caused by the increased complexity or risk of the defense systems. It is difficult to reach too far back in time for evidence on this point because of the lack of documented proof, but it is interesting to note that over the last 200 years the time taken to develop major systems from the point of Program Start to Start of Full-Scale Development has not increased dramatically (being relatively constant of the order of 3–4 years)[76] and the lengthening of the acquisition cycle began at about the time of the increased review process that was created to reduce Risk and Schedule! Some of today's actions in missile defense acquisition and already discussed promise to reverse this trend. Indeed, the scatter of data shown in Fig. 6.4 emphasizes that in the critical period between Milestone B and Milestone C, many factors that cannot be related to analysis contribute to delays and increased costs. To illustrate this point, an example not related to missile defense will be used (see Fig. 6.7). It concerns the introduction of the submarine into the U.S. Navy that began with a concept introduced to the Navy in 1875 by Holland.[76]

The story behind this effort characterizes the various stumbling blocks to the introduction of a new capability that can easily occur and is not amenable to analytical treatment. The *Holland* submarine story shows the problems of budget issues, impact of changing administrations, stubbornness, technology

1875 John Holland presents his designs to Capt Simpson at Naval War College. Receives "lunatic" comment and no action is taken	**1893** Navy issues third RFP. Eleven bids received. Baker's bids and all others, except John Holland's were rejected on technical grounds. Submarine Board recommends moving ahead with Holland's design and dispense with penalties. Adequate design margins were incorporated.
1877 Group of Irish rebels in New York, *The Fenians,* advance Holland funds to build a model. Their plan was to build such a craft for ramming the British Fleet to bring about the overthrow of British Rule in Ireland.	**1893** New Chief of BuOrd, Admiral Sampson disagrees with findings and requested a study be done on the technical differences between Holland and Baker designs.
1881 After 2 years of construction, the full scale vehicle, *"Fenian's Ram",* is ready for testing in New York harbor. Several test failures, with craft becoming swamped.	**1894** Congress supports submarine idea and appropriates in its Bill H.R. 5445, $200,000 for submarine construction and test.
1882 Lt Edward Zalinski, US Army, becomes intrigued with Holland submarine and builds similar craft. Craft designed to approach within one mile of enemy with conning tower awash, and fire gun. Zalinski boat struck a piling and sank.	**1894** Holland submits design to BuOrd. Adm Sampson still unconvinced.
1887 Adm. Montgomery Sicard, Chief of BuOrd, impressed by Zalinski boat and alerts Secretary of Navy Whitney to pursue idea.	**1895** BuOrd awards contract to Holland's company for $150,000 and $50,000 to be retained for Navy tests. Contract insists on delivery in 12 months or suffer penalties. Submarine to be called *"The Plunger".*
1887 Navy issues RFP with stringent requirements. Certified cheque of 5 per cent of bid to accompany bid and upon acceptance by Government, plus 60 per cent for performance bond. Also, performance specified as 15 knots on surface for 30 hours; 8 knots submerged for 2 hours. Other detailed requirements included. Successful bidder must meet performance or forfeit bond.	**1895** John Holland becomes ill and unable to monitor design. Navy Department continues to "improve" the design, adding among other things, Baker's down haul propellers.
1888 Two bids received: one from John Holland's company, *"Nautilus Torpedo Boat Co.",* and one from *Cramp's Shipbuilding Company.* Holland's design came closest but neither company would agree to the performance bond. Navy rejected bids.	**1897** Well past the one year contract time, *"The Plunger"* undergoes abortive dock trials. Navy makes many changes throughout 1897-1902. Many criticisms abound on poor craftsmanship. Secretary Herbert testifies to general problem in Navy of poor ship construction.
1888 Second RFP issued. Two bidders, Holland and a George Baker of Iowa, provided designs. Baker's dismissed because of no guarantees. Holland offers limited financial liability. Admiral Sicard recommends moving forward with Holland's approach.	**1898** John Holland, having witnessed the corruption of his design of the *"The Plunger",* had moved ahead to design and build a submarine on his own. This was called *"The Holland"* and was launched in 1898. The Navy took cognizance shortly after launch and received favorable initial trial reports.
1888 Secretary Tracy replaces Secretary Whitney and submarine appropriations diverted to other use.	**1899** *"The Holland"* satisfactorily completes "OPEVAL-like" trials, firing torpedoes at full speed while submerged and on the surface.
1890 Baker constructs crude submarine using down haul propellers.	**1900** The US Navy purchases *"The Holland"* and places orders for six boats of *"The Adder"* Class, which is an improved and slightly larger submarine than *"The Holland".*
1892 Cdr Converse witnesses trials of Baker's boat on Lake Michigan. Despite problems, Cdr Converse recommends Navy pursue the design. Baker boat claimed only 9 knots speed and 4 hours endurance on the surface. A Submarine Board earlier in 1888 had rejected the Baker design.	**1902** *"The Plunger"* finally completed.

Fig. 6.7 *Holland* **submarine story.**

development issues, and other factors all too familiar in the RDT&E process of today's programs.

ACQUISITION POLICY IN NATO

NATO acquisition policies have not been subjected to the same scrutiny as U.S. programs, and consequently less data are available. Also, in many cases a NATO program is followed according to national policies if the decision has been made to do so. For those programs that do adhere to NATO Acquisition Policy, the "rules" are much more flexible and are usually provided as guidance only. This results in considerable variations in the various NATO programs. Nevertheless, it is useful to compile the same schedule and Milestone data for NATO as was done earlier in this chapter for U.S. programs.

The essential difference between the U.S. system and that of NATO is that the procedures have been prepared under the assumption that they will be cooperative programs and based on harmonized military requirements. Because of these needs, there are more milestones and review points among the nations of NATO, to ensure that all national concerns have been integrated into any common program.

The governing guidance document is the Phased Armaments Programming System (PAPS) Handbook.[77] In the NATO PAPS process there are seven defined phases and eight distinct milestones as shown in Fig. 6.8.

Although an exact correlation does not exist between the Milestones in the U.S. system and the Milestones used in the NATO system, there is a sufficient overlap that provides useful insight as they apply to missile defense. An approximate correlation between the two systems is shown in Fig. 6.9.

Although an exact correlation does not exist between the U.S. and NATO Acquisition processes, there are sufficient similarities that allow for comparison of the Milestones. Figure 6.9 shows the connection between the two systems. In the NATO acquisition system the main preacquisition phases are the PreFeasibility Study (PFS) phase and the Feasibility Study (FS) phase. These phases end, respectively, as shown, with the NST and the NSR, which are similar documents but with more detailed requirements and alternative system design information in the NSR. This NSR, which is the equivalent of the MNS in the current U.S. acquisition policy (see Table 6.7), is the required document for an acquisition program start (equivalent to Milestone II or B) and the start of Project Definition of the approved program.

It should be emphasized that the flexibility of the process both in the NATO PAPS and in the new U.S. Acquisition Policy program start can be approved to occur at earlier or later phases, depending on technology maturity level and national security needs. These are for guidance in planning and not rigid control.

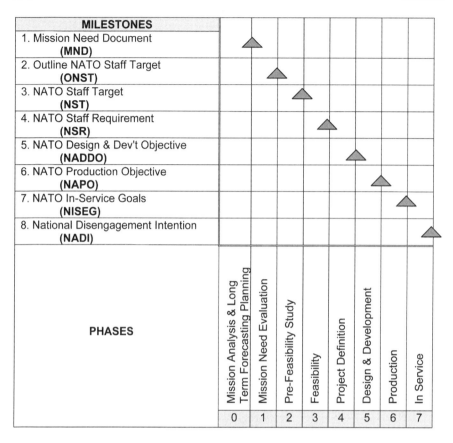

Fig. 6.8   NATO PAPS phases and milestones.

In like fashion to the U.S. system, the NATO preacquisition process phases are dealing with concepts and studies of alternatives up to and including the Feasibility Phase, at which point a NSR is established and approval given (or not given) at the *Conference of National Armament Directors* (CNAD), which is a body of senior officials responsible for armaments procurement within their nations of NATO that reports directly to the Secretary General).

ACQUISITION SCHEDULES IN FRANCE, GERMANY AND THE UNITED KINGDOM

There are many similarities and some differences in each of the 19 nations of the NATO Alliance in the schedules and milestones for defense system program acquisition. Figure 6.10 summarizes the main phases of acquisition in three nations (France, Germany, and the United Kingdom) as compared

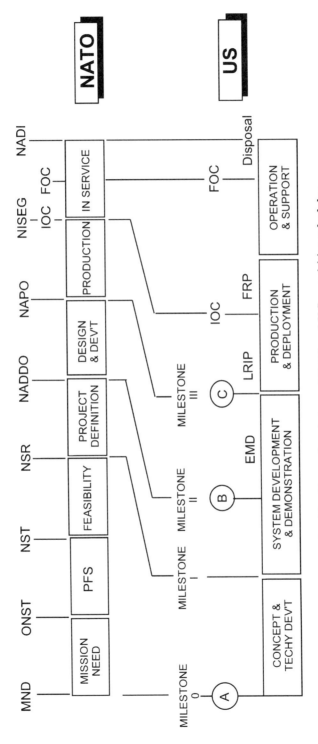

Fig. 6.9 Comparison between NATO and U.S. acquisition schedules.

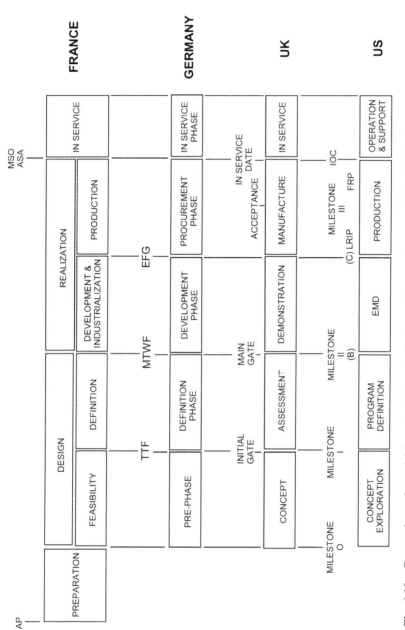

Fig. 6.10 Comparison of acquisition schedule and milestones (France, Germany, United Kingdom and United States).

Table 6.7   NATO and U.S. acquisition cycle milestones

NATO Milestone	Acquisition Phase	Approximate equivalent U.S. Milestone
Mission Need Document (MND)	*Mission Need Evaluation* Establishing common Alliance or multinational military needs	0
Outline NATO Staff Target (ONST)	*Prefeasibility Phase* Studies of alternative concepts and system designs (noncompetitive)	—
NATO Staff Target (NST)	*Feasibility Phase* More detailed studies on alternative system designs (competitive)	I or A
NATO Staff Requirement (NSR)	*Project Definition* Preliminary design of selected system	II or B
NATO Design and Development Objective (NADDO)	*Design and Development* Detail design and prototype fabrication and test	—
NATO Production Objective (NAPO)	*Production* System manufacture	III or C
NATO In-Service Goals (NISEG)	*In-Service* Operation of the system, Initial and Full Deployment	—
National Disengagement Intention (NADI)	Scrap or disengage	—

with the U.S. system.* In most cases it is simply a name change for each phase and milestone, and it is fairly easy to ascertain which phase is being discussed, especially if particular program schedules of various nations are to be compared and used in the database as used in this book. Most of the differences occur at the beginning of the acquisition schedule as each nation tackles the start of any program in a slightly different manner in terms of level of approval required and the allocation of defense budget funds.

---

*This construction of the nations' acquisition schedules is the result of discussions held by the author with national representatives to the NATO Industrial Advisory Group as part of the prefeasibility studies (PFS) into missile defense. They might not represent the latest versions of the official documents issued by the respective government offices. Note that the U.S. system has changed somewhat since 2000. They are presented here as good indications of the various schedules and milestones sufficient to make assessments of international program schedules.

Table 6.8   Translation of milestones

Milestone	Description	English translation
France		
AP	Agrément Préalable	Preliminary Approval
MSO	Mise en Service Opérationnel	Launching of Operational Service
ASA	Admission en Service Actif	Admission to Active Service
Germany		
TTF	Taktisch/Technische Forderung	Tactical/Technical Requirement
MTWF	Militarisch Technisch Wirtschaftliche Forderung	Military Technical Economic Requirement
EFG	Einführungsgenehmigung	Approval for Production

One important difference between the nations is which Milestone is considered the *start of the program*. In France the decision to move from Feasibility to Definition in the Design Phase is made by the Ministry of Defense, and this is taken to be the Program Start. For the United States, Program Start is not considered until Milestone B, which is the start of System Development and validation, and specifically the start of EMD. Similarly, in the German system of acquisition the decision to include the program in the Federal Armed Forces Plan (Budeswehrplanung) occurs once the TTF milestone (see Table 6.7) has been reached and approved. In the United Kingdom program start occurs at the Milestone Main Gate, which signals the start of the Demonstration Phase and is similar to the U.S. system. Such differences suggest care in comparing schedules from different nations for their program start dates.

The Milestones shown in Fig. 6.10 use the acronyms in the language of the particular nation identified. Table 6.8 provides an English translation of each Milestone.

EXAMPLE PROBLEMS

6.1 If a new missile development program were to start today and it followed the current trend in missile defense programs,

a) How long would it take from initiating Concept Studies to start of Program?

b) How long would it take from Program Start (milestone B) to the start of LRIP at Milestone C?

c) How long from program start to IOC?

6.2 If aggressive acquisition methods similar to those being used in today's international programs were implemented, how much shorter would the schedule be compared to those in Exercise 6.1? Is it a uniform improvement for Milestones A, B, and C?

6.3 What is the underlying reason for the lengthening of the acquisition schedule?

6.4 What changes should be made to current Acquisition policies that would shorten the schedule?

Chapter 7

RISK

Great deeds are usually wrought at great risks.
Herodotus, c 425 B.C.

Take calculated risks. That is quite different from being rash.
General George S. Patton, 6 June 1944

Risk management is one of the driving factors in the acquisition of any defense system. Indeed, the management of risk is the cornerstone of the research, development, test, and evaluation (RDT&E) programs within all national defense budgets. If there were no risk, the customer would move directly to the purchase of the defense system and immediately add the system to the force structure without further schedule delay. In reality, of course, all new systems embody some feature that requires study or research or technology exploration or even tested prototypes to ensure that the defense system(s), as designed, will function as planned.

The intent of RDT&E is to gradually eliminate or at least to reduce the risk to acceptable levels as the defense system progresses through its research and development into production and ultimately enters the military force structure, that is, as it approaches its Initial-Operating-Capability (IOC) date. Figure 7.1 illustrates the intent within each of the various phases of Acquisition (see Chapter 6) to lower the risk from high to moderate or medium to low and even no risk, as the IOC date approaches.

In Fig. 7.1 the horizontal axis is easily understood. It is simply the structured time-phased segments of the Acquisition Cycle (discussed in Chapter 6). It is the vertical axis, risk, that causes the main difficulty. Some planners operate on a direct probability assumption, which really translates to a feeling on risk with such descriptions as "*95% probability of not functioning as planned,*" meaning *high risk*, and, say, "*5% probability of not functioning as planned*" for *low risk*. Such descriptions, however, are not satisfying or convincing, especially if large costs are involved in bringing the systems into being. How were those probabilities derived?

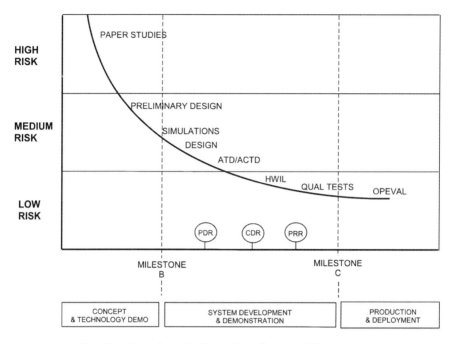

Fig. 7.1   Reducing risk throughout the acquisition process.

RISK-MEASUREMENT MODELS

The importance of Risk Management has resulted in many methods to measure risk. These methods have been published either widely in the public domain or have seen more restrictive use within defense establishments for defense program use. It is not the purpose here to provide a summary of all of these methods and their applicability or even validity. It is useful, however, to indicate the nature of these models to provide a perspective on this key topic which can spell the success or failure of any system.

Typically, most models tackle the subject from an analysis of the component parts of the system, then apply probability analyses to the likelihood of failure caused by some (postulated) form of failure and its effects, and then sum these effects to the total system and express the results in some form of cost increase that must be added to the basic estimated system cost to account for risk.

Figure 7.2 is a representation of most methods that illustrate this component approach. It shows that the system is expressed in a tree in terms of a Work Breakdown Structure (WBS). Under each WBS (and sub-WBS if necessary) a set of possible failure modes is listed (e.g., missile seeker window will not cool sufficiently). A probability of occurrence is estimated (usually subjectively), a cost impact is then estimated, and thus an expected

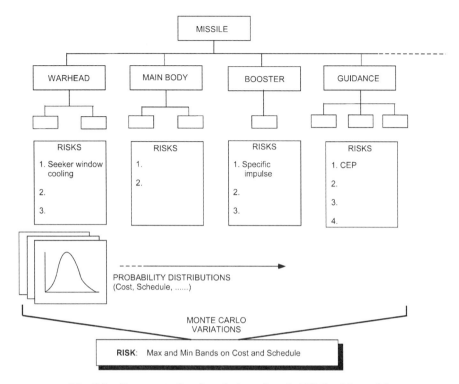

Fig. 7.2   Representative description of probabilistic risk models.

value (of cost) is computed. There are various forms of this basic method used. In some models the probabilities are estimated as single entities; in other models a chain of events is estimated using probability theory of interdependent events. In some models the cost impact is taken as a single value within each WBS, in others the cost is treated in a similar manner to the 1960s-developed program evaluation review technique (PERT) networks, which allows for "least likely, most likely" type cost and schedule variations to be included. Usually, if such complications are added into the model then Monte Carlo techniques are incorporated to compute the most likely distribution of risk impact, expressed as a *cost impact* and *schedule impact* on the program. These methods have enjoyed different acceptance levels within the community. First, the entire method relies on risk factors that are subjective and difficult to quantify. Second, as any program manager will testify, any identification of *cost additions* to account for risk highlights them such that they are the first items on the chopping block by such oversight committees (Congress et al.) when it comes time to cut the budget (leaving the program manager with the same risk but now without the funds!) Some examples of Risk-Analysis methods as used in the Services in managing

RDT&E programs are the U.S. Army-developed Total Risk-Assessing Cost Estimate (TRACE)[78] and Risk Information System and Network Evaluation Technique (RISNET 1979) and its derivatives as used in the U.S. Navy and the U.S. Air Force. The Air Force also uses a similar method RISK.[79] The reader is encouraged to explore these methods in the literature to understand in more detail such Risk-Analysis techniques.

ALTERNATIVE APPROACH TO RISK MEASUREMENT

The method presented here is not meant to supplant the previous methods, which represent a discipline for any program manager who must put forward a sound business plan to ensure that risk is being properly taken care of in his program as it progresses through the RDT&E phases. Rather a method is presented that removes the subjective estimates of risk factors and utilizes the *measurable* quantities of *RDT&E cost* and *RDT&E schedule* that each program manager must prepare in order to pass the designated Milestones. Such measures allow for decision makers to gain an appreciation of how risk influences the choice to be made between systems (see Chapter 8), using familiar parameters of cost and schedule.

It is generally agreed that complex systems are riskier than simple systems. It is also agreed that if the development is rushed the risk is higher. Good examples of this can be found in the work by Bearden[80] and the earlier work by Sarsfield,[81] who examined the reasons behind the failure of NASA spacecraft that were rushed through their planned RDT&E. Bearden devised, based on Sarsfield's work, a Complexity Index that captures the many features of the spacecraft (mass, power, design life, battery type, pointing accuracy, etc.) and plotted the results against both time (development schedule) and cost. The results show quite convincingly that moving too quickly through the RDT&E schedule with inadequate cost contributed significantly to the failure rate of the spacecraft.

Such reasoning is intuitively correct and is reflected in both *RDT&E Cost* and *Schedule*. One can also argue in reverse. If technical experts, after proper examination of the problem to be solved (to get the desired defense system to function), determine that it will take a large sum of RDT&E dollars and a long time to develop to achieve a reliable and acceptably low-risk defense system, then it is reasonable to mark this as high risk. If it is determined that a particular defense system will only take a small sum of RDT&E dollars and a short time to develop, then it is reasonable to state that that system is Low Risk. And accordingly, Moderate or Medium Risk would be some middle combination of RDT&E dollars and schedule, such as large RDT&E dollars, short time to develop and small RDT&E dollars and a long time to develop.

Simply stated, after an assessment of all of the technical details required have been analyzed and incorporated in the necessary RDT&E cost and schedule to reduce risk then risk can be *measured* by

$$\text{Risk} = (\text{RDT\&E cost}) \times (\text{RDT\&E schedule})$$

Although certainly simple in form, such a calculation of risk brings to the decision maker most of the factors that he needs to consider in a *measurable manner* and readily calculate and further can be computed from information that is already at hand when the decisions need to be made. R&D planners are responsible people who will want to do the best job possible, and thus the detailed estimates and review processes that they must go through to generate the needed plans for RDT&E (whether at the Basic Research level or at the Engineering and Manufacturing Development level) will give a realistic understanding of the risk involved in bringing the system to IOC and to a relatively risk-free operational status.

To be fair, those programs that have been in development for some time and have only a short time left before IOC should not be penalized (i.e., given a High Risk rating) when compared to another system that has just started and has a small amount of RDT&E costs and schedule to incur. This can be handled by simply making sure that the calculation (cost × schedule) is from *today* (or whatever date is to be used in the comparison of systems) forward. This is best illustrated by example.

In the first example consider two systems: System A and System B, as shown in Fig. 7.3. Both systems (in this example) start their preacquisition phases at about the same time. System A has a greater RDT&E cost and a longer schedule than System B. Neither system has passed their Milestones B and C. Clearly, System A has *more* risk than System B.

Now consider a second example, which compares the same two systems (A and B), but in this second example the more complex System A has been in development for some time and already has successfully passed its Milestone B and is close to completing Milestone C. System B, on the other hand, is as in the first example just beginning its development. This is shown in Fig. 7.4.

Clearly, with its milestones successfully completed and significant RDT&E cost and schedule already expended, System A now would be viewed as having *less* risk than System B.

So, in this formulation of risk the formula should be restated as

$$\text{Risk} = [(\text{RDT\&E cost}) \times (\text{RDT\&E schedule})]_{\text{from evaluation date}}$$

which is to remind the evaluator that the product of cost and schedule must

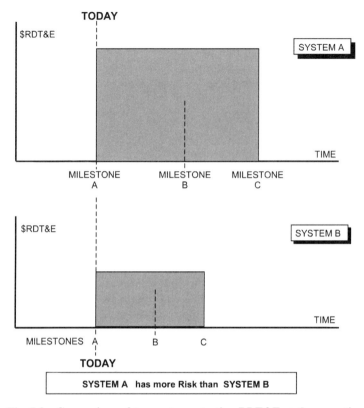

Fig. 7.3   Comparison of two systems starting RDT&E at the same time.

be computed on a common timescale and from a common starting point so as to properly reflect any sunk RDT&E cost and time and thus eliminated risk of the systems to be compared.

Before examples of this technique are presented, consider the guidelines that already exist in the defense establishments that indicate "thresholds" of concern about defense systems. Because of the importance of certain programs and the concern about their meeting the stated goals, this has prompted high-level decision makers in the defense establishment to set guidelines of when they should be involved and how detailed any RDT&E review to eliminate or reduce risk must be. Because of the public scrutiny in the U.S. defense budget process, such considerations are paramount, and this has given rise to the establishment of Acquisition Category (ACAT) programs. This is a way of providing visibility to those programs for national security that are of *concern—and risk—if not successful*. Table 7.1

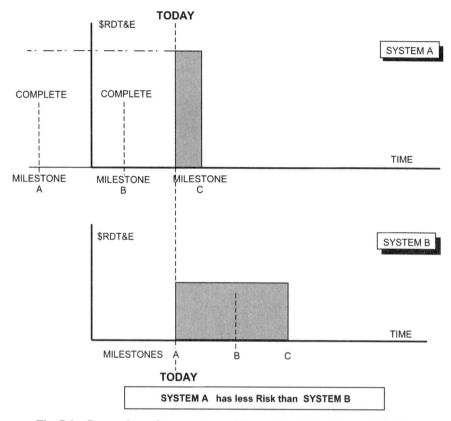

Fig. 7.4   Comparison of two systems with one already in advanced RDT&E.

shows the definitions and guidelines for the ACAT category of U.S. programs (compiled from data given in Ref. 67, where more details can be found).

It is seen from Table 7.1 that RDT&E Cost is a key driver of concern. The higher the cost of RDT&E, then the higher level of decision-maker review. It is probably unfortunate that guidance on schedule is not included in such U.S. Department of Defense (DoD) classification. It is usually the result of Congressional action to stop risky programs from continuing in time if they are not meeting mission needs. Such classification (shown in Table 7.1) demonstrates the threshold level as viewed in the decision chain. As a point of comparison, it is noted that other nations follow similar categorization of the importance of acquisition programs and the criteria placed on R&D (and Procurement), which establishes the thresholds perceived by the national

Table 7.1 Definition of ACAT categories

ACAT Category	Guidelines[a]			Milestone Decision Authority
	RDT&E	Procurement	LCC	
ACAT I				
D[b]	$365 million	$2.19 billion	—	USD(AT&L)[c]
C[b]	$365 million	$2.19 billion	—	CAE[d]
ACAT I				
AM[b]	—	$126 million or $32 million/yr	$378 million	ASD(C3I)[e]
AC[b]	—	$126 million or $32 million/yr	$378 million	CCIC[f]
ACAT II	$140 million	$660 million	—	CAE[d]
ACAT III (except for Army, Navy[g])	—	—	—	Lowest appropriate level
ACAT IV (Army, Navy[g])	—	—	—	Lowest appropriate level

[a]The apparent precision in the ACAT I dollar values given in this table appears to be the result of applying inflation factors to the earlier (1981) approximate values of $200 million RDT&E and $1 billion procurement. Such levels are meant to be general guidance and not precise values.
[b]The letters D, C, AM, and AC refer to the internal systems used for review, (e.g., D stands for Defense Acquisition Board). See Refs. 66 and 67 for details.
[c]Under Secretary of Defense (Acquisition, Technology, and Logistics).
[d]Component Acquisition Executive.
[e]Assistant Secretary of Defense (Command, Control, Communications, and Intelligence).
[f]Component Chief Information Officer.
[g]Includes U.S. Marine Corps.

Table 7.2 Classification of German Acquisition Programs

Category	Guidelines[a]		Decision Authority
	R&D	Procurement	
1	$9 million	$25 million	Bundestag[b]
2	$1–9 million	$2.5–25 million	Armed Service Command Chiefs
3	<$1 million	<$2.5 million	Individual Service Program Managers

[a]Using current exchange rates.
[b]Lower House of the Federal Parliament.

Table 7.3   RDT&E budgets for 2000

Nation	Total RDT&E Budget, billion	Approximate ratio European nation/U.S. RDT&E Budget
United States	$39.10 billion	1:1
France	$3.05 billion	1:13
Germany	$1.30 billion	1:30
Italy	$0.22 billion	1:178
Netherlands	$0.07 billion	1:559
Spain	$0.18 billion	1:217
United Kingdom	$4.03 billion	1:10

decision makers to be the dividing line between those programs that will receive more scrutiny than the others. As one example, Table 7.2 shows the makeup for German acquisition programs.

Note that because of the large difference in defense budgets between the United States and Germany (see Chapter 5) the threshold levels of concern are much lower in the German acquisition system, but the driving principle of review to reduce risk is retained. Some idea of the relative magnitudes of RDT&E between the United States and nations in Europe can be seen in Table 7.3.

Such wide differences in the funding available in RDT&E to remove or reduce risk will naturally cause differences in the relative levels of decision makers and in the amounts to be used for such evaluation in the acquisition process.

SUGGESTED RISK CATEGORIES

Using the U.S. values for ACAT levels of RDT&E limits as guidelines and the format for risk as just suggested, it is possible to structure a format for risk categories that accommodates both *cost and time* considerations. Assume that RDT&E costs expressed in $million is called $(R\&D)_{cost}$ and the time from point of evaluation until Milestone C expressed in years is called $(Sched)_{milestone\ C}$, then *Risk, as defined here*, is given as follows:

Low Risk:

$$R\&D_{cost} \times (Sched)_{milestone\ C} < 1000$$

Medium Risk:

$$1000 < (\text{R\&D})_{\text{cost}} \times (\text{Sched})_{\text{milestone C}} < 2000$$

High Risk:

$$(\text{R\&D})_{\text{cost}} \times (\text{Sched})_{\text{milestone C}} > 2000$$

in which $(\text{R\&D})_{\text{cost}}$ is the RDT&E cost ($million) from today's date or evaluation date and $(\text{Sched})_{\text{milestone C}}$ is the RDT&E schedule (years) from today's date or evaluation date until milestone C. (It is reasonable to debate whether RDT&E categories 6.5 and 6.6 should be included in this term.) This is shown graphically in Fig. 7.5, in which the U.S. RDT&E levels of $365 million for ACAT I programs and $140 million for ACAT II programs have been superimposed.

The ACAT values in DoDD 5000.1 are guidance values, and so for the sake of simplicity in the equation for risk as just displayed, the values for the

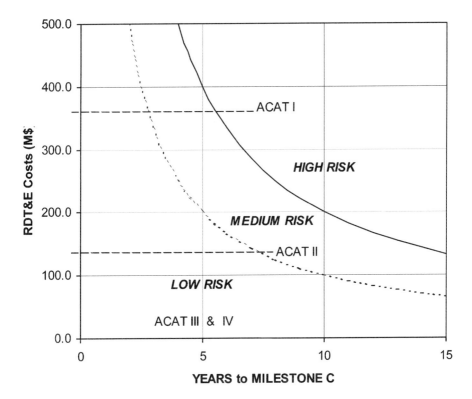

Fig. 7.5   Suggested risk categories with superimposed ACAT values.

risk boundaries of 1000 and 2000 million dollar-years have been rounded such that one could restate the ACAT guidance values, when the consideration of schedule is included, as follows:

ACAT I:

$365M RDT&E costs and approx 6 years to Milestone C

ACAT II:

$140M RDT&E costs and approx $7\frac{1}{2}$ years to Milestone C

ACAT III and IV:

Lesser costs and schedule than ACAT I or II

Such schedule values appear reasonable and tend to match the historical database for defense programs as given in Chapter 6 (see Fig. 6.4).

Alternatively, one might consider keeping the schedule common, say at five years, in an attempt to streamline the lengthening acquisition schedule, and provide the guidance as follows:

ACAT I:

$400M RDT&E costs and five years to Milestone C (*High Risk*)

ACAT II:

$200M RDT&E costs and five years to Milestone C (*Medium Risk*)

ACAT III and IV:

Programs with lower values than for ACAT I and II (*Low Risk*)

which keeps the numbers simple, but this would require a revision to DoDD 5000.1!

Because other nations have different levels of RDT&E funding before certain Milestone Decision Authorities (MDA) require review, then it would be expected that similar plots to Fig. 7.5 could be constructed. The intent here is to indicate how the current levels of review can be integrated into an assessment of risk, taking into account the two important factors of *RDT&E Cost* and *Schedule*.

RDT&E BUDGET ACCOUNTS

How much is spent on RDT&E both in absolute terms and in percentage of the total amount provided each service, known as the total obligational authority (TOA), provides a measure of the adequacy of RDT&E to eliminate or reduce risk. The structure of the R&D budget has gone through several iterations since the formation of the account after WWII. Fig. 7.6 shows how the budget has varied in amount and form since 1945 (compiled by the author from historical records). (The data from 1945–1961 were reconstructed from historical Navy Comptroller documents.[82] The data from 1962–1982 were taken from Navy Comptroller Information System Historical File dated 23 February 1984.[83]) These example data are for U.S. Navy RDT&E, but the data are similar for the other services.

Between the years 1948 and 1954, there were only three categories of Research and Development: Systems, Exploratory Development, and Research. From 1958 to 1962, the accounting of RDT&E entered a so-called transition stage, and in 1962 the current system of RDT&E categories was formalized as part of DoD initiatives to streamline the entire defense budgeting process in response to Congressional scrutiny after the end of U.S. involvement in Southeast Asia. One observation from Fig. 7.6 is that the form and trend of RDT&E over time was largely a matter of accounting as

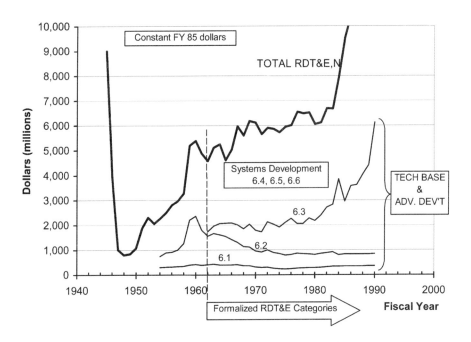

Fig. 7.6   Historical RDT&E funding for the U.S. Navy.

programs were being reclassified as "research," "technology," or "advanced development" as a result of the change from the previous 1948 planning system. The trends were due more to this reclassification than to any fundamental change in treatment of the subjects being pursued.

For example, there was concern expressed in the 1970s about the decline of the Tech Base (especially category 6.2), when in reality it was simply the reclassification of the RDT&E accounts, as the all-important Technology Demonstrators were being introduced into the system as an integral method of reducing risk by testing systems early in an operational environment. Technology Demonstrations increased dramatically around 1984 as such programs as the Trident II missile program and the High-Energy Laser (HEL) projects entered their Advanced Development RDT&E stages.

Historically, Basic Research (category 6.1) has enjoyed a relatively constant funding level. The division between Technology (category 6.2) and Advanced Development (category 6.3) has been the area of debate among R&D planners and decision makers for some time. Not shown in Fig. 7.6 is the introduction in 1970 of the Advanced Technology Demonstrator (ATD), called 6.3A (so as not to introduce a completely new accounting system). The ATD was introduced to facilitate the transfer of technology from research and exploratory development stages into systems development and to begin the process of subjecting R&D projects to demonstration of technological innovations in a real-world environment. The idea was to reduce risk early before transferring the R&D and preacquisition efforts into an approved Acquisition Program.

As might be imagined, such transfers caused some problems as decisions were made as to how ready was the concept for real-world testing. Each service has historically treated such Technology Demonstrations differently. As shown in Fig. 7.7, the U.S. Army tend to push its projects earlier into the ATD stage than either the U.S. Navy or the U.S. Air Force. Historically, both the Army and the Air Force expend more on ATD than the Navy.

Such trends in RDT&E between the three services emphasize the need to pay attention as to how the risk is reduced with the use of Advanced Technology Demonstrators, which in turn determines how much of the RDT&E budget is directed at such activities.

Although not shown here, the amount of the total budget that is allocated to all of the service accounts tends to reflect the values found in the Life-Cycle Cost (LCC) of the systems. For all three services the RDT&E budget is typically 10% of the total budget. The Procurement and Operations and Support budgets tend to be about the same (approx 40% or more). The difference, to make up to 100%, is assigned to military construction (MILCON) and military personnel.

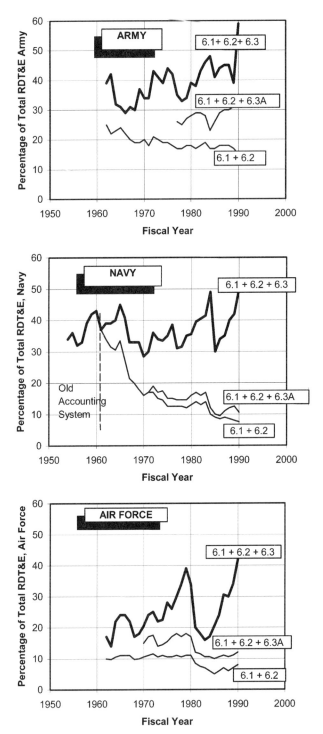

Fig. 7.7   U.S. Army, Navy, and Air Force trends in pre-acquisition RDT&E.

The preceding representation of risk has sought to place some measurable yardstick that can be used by all decision makers to a common language so as to eliminate the subjective nature that is usually applied to risk.

In Chapter 8, where the decision aids and methods are presented, the reader can insert his own measures of risk or use the default values presented here.

EXAMPLE PROBLEMS

7.1 A set of early-warning systems is being presented to the decision maker, with the intent to discuss which might have the lowest risk. These systems have the following characteristics:

System	RDT&E Costs to Go	Years to Milestone C
Nation A satellite	$4 billion	6
Nation B satellite	$2 billion	10
Long-range X-band radar	$400 million	5
Long-range UHF-band radar	$500 million	6
Long-range S-band radar	$300 million	7
Airborne infrared system	$200 million	10

Display these candidate systems on the Risk Chart (Figure 7-9) to indicate the relative risk categories. Which are the lowest risk systems?

7.2 In the same set as in Exercise 7.1, it is now learned that nation A is willing to share the early-warning system with all of the other nations and states so that the RDT&E costs will be absorbed by nation A. A nominal charge of $200 million is asked for this system How does this new proposal look on the same risk chart as used in Exercise 7.1?

Chapter 8

EVALUATION

> *The best is the enemy of the good.*
> Voltaire, *Dictionnaire Philosophique*, 1764

When evaluating different defense systems, most decision makers will say that they are seeking the "*best system*" or, expressed in another way, one that offers the "*best value.*" Intuitively, this feels "correct" but as any mathematician will immediately note, such a statement is incomplete. What is missing is the rest of the sentence, which should read something like "best value *with respect to* _____." Fill in the blank. The probability that any single system would be best in all features, under all circumstances, all of the time is very unlikely. This then forces the decision maker to make *choices*, or more often than not his staff must analyze the options and present *recommendations* to the decision maker for the final choices. Even if such a procedure is followed logically, properly, and without bias, the decision maker must still, when all of the recommendations are in, *make a decision.* How should he do that? How should he or his staff conduct the evaluation so as to make the decision that gives this best value?

FORMULATE THE RIGHT QUESTION BEFORE RUNNING THE COMPUTER

There are many decision aids on the market. Some rely on asking experts and taking the majority votes (Delphi technique), some rely on rigorous algorithms expressing the various features and calculating the lowest (or highest) combined score, yet others statistically compare each feature with some preordained set of criteria and generate the rms or some other average of the results. Some require the decision maker to stay in the loop as the decision unfolds, and some do not involve the decision maker until after all factors have been subjected to algorithmic manipulation and the final computed result is placed before him.

Unfortunately, there is no panacea. Only a magical computer program will take hundreds of factors as *inputs* and generate a unique best-value solution as an *output* if the *question has not been formulated properly*. As in all decisions in life, the decision maker must stay involved in the process

299

because he will probably change his mind as the process unfolds and the impact of incremental decisions are made known to him. As one simple example, suppose the decision maker sets the goal that he wants the *most cost-effective weapon system* out of a set of candidates. But then after the initial analyses he finds that the most cost-effective defense system is not the one with the lowest cost and is unfortunately beyond the set budget. The decision maker will now (correctly) change the criteria to be the *most cost-effective but within the available budget.* This would probably yield another choice from *the same set of defense system candidates.*

In the case of defense systems, one is often dealing with high technology with several unknowns. Suppose the most cost-effective system that is within the budget can handle most of the threat but not the particular case of chemical weapon fallout. Here the decision maker might be faced with intelligence information that is not firmly established. The chemical weapon threat might not materialize (in this hypothetical example), and a good enough solution (for the conventional threat) can rapidly move forward. How does the decision maker hedge his bets, if the intelligence information proves to be correct and the chemical threat could arrive before the good enough solution can be fielded? Because of the long development times of most weapon systems (see Chapter 6), this is a real possibility. By this time the decision maker has realized that he must formulate the questions a little differently.

APPROACH

What is proposed here is a controlled decision aid approach method that keeps the factors to be considered in manageable pieces which can be reevaluated as the process unfolds. There is a tendency in some circles to keep collecting data in the hope that large enough *data* banks will guarantee the needed *information.* All too often, however, data become confused with information. Indeed, in many instances extraneous data can actually become disinformation and distort the decision process. The issue facing the decision maker (and those setting up the computer programs) is how to isolate the *relevant data* from the *extraneous data* that could, if all data are treated equally, outweigh the correct data (and the correct choice).

Frequently, when important decisions are to be made there are not much data available, especially early in the acquisition phases of any defense system program (see Chapter 6). Hence, it is the aim of this book to seek the *lowest* number of pieces of information to make intelligent decisions. Then, as time progresses and more data and information become available it can be *inserted* into the decision process to *refine* the early decisions. The decisions should only be of sufficient detail to permit the development of the candidate systems to proceed (or not proceed) to the next phase in the acquisition program.

TOP-LEVEL CONSIDERATIONS

It would be naïve to believe that only technical, cost and programmatic information contribute to the selection of defense systems. Frequently, decisions are ultimately politically driven by such considerations as jobs in country, jobs in certain political districts, or national pride, or other intangible but nevertheless real considerations. But it is believed that any such decision is even better handled if the decision maker *also* knew the considerations of which systems provide the best solutions (with the provisos already mentioned about such a descriptor). It is just possible that a firm desire to have, say, one political factor as paramount might be assuaged if the second choice (from the technical evaluation) provided a significant amount of more defense against some very undesirable threat. The decision maker needs to remain in the loop as the information unfolds.

To this end, it is believed that the decision making needs to be conducted in at least two distinct and separate steps: first, at the technical, cost, and programmatic level and second, at the political level. These two steps probably should not be combined or analyzed by the same group.

DECISIONS AT THE TECHNICAL, COST, AND PROGRAMMATIC LEVEL

There are four top-level considerations for any defense system; they are as follows:

1) *Performance*: Can it do the job asked of it?
2) *Cost*: Can it be afforded (when compared to some prescribed budget)?
3) *Schedule*: Can it be fielded in time (to meet some predicted threat)?
4) *Risk*: Can all of the preceding be accomplished with high confidence?

How these four top-level factors are "combined" or treated in the formulation of the questions to be asked and choices to be made is key to making intelligent decisions. Each of the preceding four considerations or factors will have more detail to describe them but ultimately the decision maker wants to hear the word "yes" to each of these questions before he proceeds and folds in his own requirements or modifications to the analysis. Such additional considerations could spawn another round of evaluations, but now it will be a controlled and understandable and reproducible process.

The basic concept in the evaluation method as developed here is illustrated in Fig. 8.1. The basic concept, as shown in Fig. 8.1, is applicable to the user's own set of performance, cost, schedule, and risk parameters, but, as shown, default values can be taken from the other chapters in this book. Each of

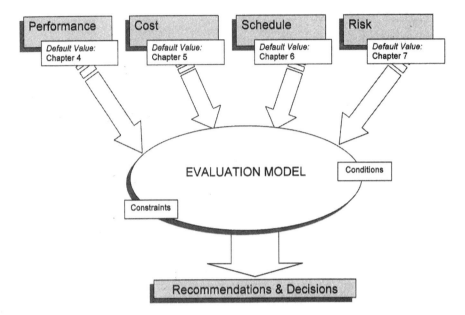

Fig. 8.1   Basic concept of evaluation model.

these four factors or parameters is given their own treatment in separate chapters in this book. They are performance (Chapter 4), cost (Chapter 5), schedule (Chapter 6), and risk (Chapter 7). This chapter on evaluation takes the end results of the four major considerations (user or default values) and subjects them to a consistent analysis in the decision making.

To emphasize how important it is to first express these (or any other factors) in a proper question format, the following illustrative combinations are presented. It is sometimes convenient to display such factors in a hierarchy or tree such as that shown in Fig. 8.2.

Figure 8.2 is a convenient way to "collect" all of the factors into a form for viewing, but it is also dangerous in that it provides no indication of the relative merits of each of the contributing factors or how they might vary in some manner with other parameters in some (possibly) nonlinear way. In some decision-analysis methods a "score" is attached to each of the subfactors or "children of the parents" (shown as blanks in Fig. 8.2) and then combined in an additive manner to get the score for the parent or higher level in tree. Clearly this could lead, in some instances, to spurious results. For example, system A could have zero score for performance but attractive scores for cost, schedule, and risk. This would then be compared to system B with some acceptable score for performance but with higher cost, a longer schedule, and more risk than system A. This could lead to the silly

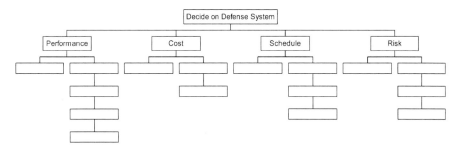

Fig. 8.2   Display of factors contributing to decision.

conclusion of deciding on system A because its additive score was favorable—even though it could not do the job!

What is missing from Fig. 8.2 is the set of boundary conditions. For example, the performance must be above some specified level, cost must be below some specific budget figure, etc. Further, Fig. 8.2 has not yet established the problem statement. That is, for example, is the decision maker looking for a system that considers each of the preceding (four) factors at a *common level*—even if they are weighted? Or is he looking for some other combination, such as was already suggested at the beginning of this chapter, such as cost-effectiveness? If this were the case, then the hierarchy should look like Fig. 8.3. This has now changed the problem statement to a three-factor consideration and would produce different results from the *same set of systems* arranged as in Fig. 8.2. This simple example has served to illustrate that simply adding up scores of all the factors will not satisfy logic even if they are weighted in some manner. An important part of decision making is to *ask the right questions first*. Simply putting all hundreds of factors (even if weighted) into a computer and trust that such an inanimate object will make the right decision is not logical, without first formulating the questions of interest to the decision maker.

Another example is the consideration of cost. As described in Chapter 5 and used extensively in the defense industry, cost is usually treated in

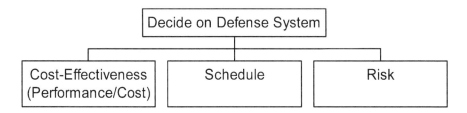

Fig. 8.3   Alternative display of contributing factors.

three major components that add up to what is called life-cycle cost (LCC), that is,

$$LCC = R\&D \text{ Cost} + \text{Procurement Cost}$$

$$+ \text{Operating and Support Cost} \qquad (8.1)$$

In the commercial world a fourth component is included, called the "scrap" or "residual value" that reduces the LCC at the end of the system's life, but this will not be treated here. It is very difficult to visualize an "LCC dollar." It cannot be spent or put in the bank. The defense community deals quite differently with each of the components of LCC, and each of the cost elements of LCC have quite different accounts (see Chapters 5 and 6 for further discussion on cost accounts used for the different contributors to LCC) in the defense budgets. (Just to keep NATO planners on their toes, it should be remembered that not all nations of the Alliance keep the costs of defense systems in the same accounts! This complicates cost databases and the comparison of system costs among nations in the Alliance.) Each of these accounts has its own set of decision makers, and the perception of each of these cost elements is different in each defense community. Further, when the subject of cost-effectiveness is treated it is usually just the procurement cost component of LCC that is considered in the analysis. Also, as discussed in Chapter 7 on risk, because the research and design (R&D) cost is part of the method of eliminating or reducing risk it would not be unreasonable to include the R&D cost as part of the risk consideration. Hence, another variation on the hierarchy might be as shown in Fig. 8.4, in which each of the three components of LCC have been placed in their respective parts of the decision tree.

This particular approach would emphasize the impact of the (operational) cost to the user as separate from the investment (procurement) cost and, further, separate from the cost of development (R&D cost) to eliminate or reduce risk. In this example, it is seen that now the operating and support cost

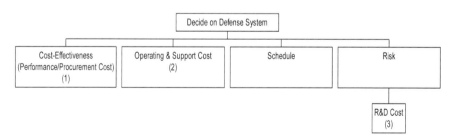

Fig. 8.4   Treating life-cycle cost components differently in the decision process.

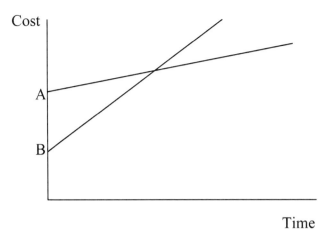

Fig. 8.5   Two systems costs vs time.

has become a significant item and on the same level as the other considerations. Such a formulation highlights that the indirect factor of time has
entered the picture in an important way. In the original setting of the
question(s), schedule was highlighted as to represent when will the system
become available? This translates into when is the date of the initial
operational capability (IOC, for convenience, U.S. terminology is being used
in this discussion), but the display in Fig. 8.4 points out that there
is another impact of time; that is, what is the cost incurred during the
operational life of the defense system *after* IOC? Over a typical operating life
of, say, 20 years, the operating and support costs can be as much as or greater
than the procurement cost.

By way of example of this impact of time, consider two systems A and
B. If system A and system B have two different procurement and operational
and support costs, then a different set of questions needs to be asked. Figure
8.5 shows a possible comparison for these two hypothetical systems.

In this hypothetical example system A has a high initial procurement cost
but is cheaper to operate over a long period of time. System B has a lower
procurement cost, but as a result of, say, a manpower-intensive operating
feature, its operating costs outweigh this advantage after some "break-even
point," which might well be inside the time frame of interest to the decision
maker. This does not automatically mean that system B should be decided
against because it might have the advantage of getting into the field ahead of
system A (that is, it might have an earlier IOC) and handle the threat
adequately and be capable of upgrading through a planned program product
improvement (P3I) to meet the new threat within the same time period. This
constructed example shows how important it is to first frame the questions,

and to keep the decision maker in the loop as the analysis unfolds. It is unlikely that all such features and even nuances would be obvious during the initial setting up of the problem.

Other variations must be considered when choosing a defense system for a particular mission or set of missions.

The purpose of the preceding discussion is to emphasize the importance of setting up the decision process or asking the right questions *first* before any computerized analysis goes into effect.

DECISIONS AT THE POLITICAL AND NATIONAL LEVEL

In some sense such considerations are easier to treat than the technical issues because they frequently encompass "go-no-go" decisions that can be visualized on an evaluation scale. For example, restricting the analysis (in this book) to just the factors already cited as typical then the political factors are relatively easy to estimate.

By way of example and for illustration only, the political criteria might be expressed in the following decision factors:

1) *Jobs in country*—the percentage of the production cost of the system to be in the nation of interest.
2) *Use of national systems*—percentage of subsystems or systems that are owned by the nation of interest.
3) *Interoperability*—percentage of systems that operate seamlessly with the nation's basic defense structure.

Clearly, this list of considerations can include other items of interest to the decision maker and must be decided upon before any analysis begins. This will be returned to later after the mathematical method has been outlined.

MATHEMATICAL TREATMENT

Before treating the problem of rearranging the four factors in the correct formulation as just discussed, first consider the mathematical problem of solving the basic equations. Assume the four factors are p, c, s, and r to represent *performance*, *cost*, *schedule*, and *risk*, respectively. Also, before rearranging these in some different combinations assume that these factors are treated at the same level (in the hierarchy tree) but have different relative importance to the decision maker as w_1, w_2, w_3, and w_4. These are the weightings of each of the considerations p, c, s, r. Further, assume that there are n candidates to evaluate, and the systems can be identified by $i = 1, 2, 3, \ldots, n$.

In this first combination of considerations, the problem statement becomes a standard one of solving a set of nonhomogenous linear equations. This approach is similar to that followed by Miller (first at Massachusetts Institute of Technology, Cambridge, Massachusetts, and subsequently at RAND)[84] and Raiffa[85] and others. The linear equations are

$$w_1 p_1 + w_2 c_1 + w_3 s_1 + w_4 r_1 = k_1$$
$$w_1 p_2 + w_2 c_2 + w_3 s_2 + w_4 r_2 = k_2$$
$$w_1 p_3 + w_2 c_3 + w_3 s_3 + w_4 r_3 = k_3$$
$$w_1 p_4 + w_2 c_4 + w_3 s_4 + w_4 r_4 = k_4$$
$$\cdots\cdots\cdots\cdots\cdots\cdots\cdots\cdots\cdots\cdots$$
$$w_1 p_n + w_2 c_n + w_3 s_n + w_4 r_n = k_n \tag{8.2}$$

The reader is cautioned here that addition of the weighted considerations $(w_i p_i)$ here in this manner is for illustration only. It will be shown later, based on the comments at the beginning of this chapter, that the factors p_i, c_i, s_i, and r_i should be *combined in other ways* depending on the objectives of the decision maker. The main issue for illustration here is the necessary algebraic manipulation required *once* the combination decisions have been made. The solution to the set of equations (8.2) is determined from matrix algebra, such that the solutions can be expressed by

$$
\begin{pmatrix} p \\ c \\ s \\ r \end{pmatrix} =
\begin{pmatrix}
1 & w_2/w_1 & w_3/w_1 & w_4/w_1 & \cdots & w_n/w_1 \\
w_1/w_2 & 1 & w_3/w_2 & w_4/w_2 & \cdots & w_n/w_2 \\
w_1/w_3 & w_2/w_3 & 1 & w_4/w_3 & \cdots & w_n/w_3 \\
w_1/w_4 & w_2/w_4 & w_3/w_4 & 1 & \cdots & w_n/w_4 \\
\cdots & \cdots & \cdots & \cdots & \cdots & \cdots \\
w_1/w_n & w_2/w_n & w_3/w_n & w_4/w_n & \cdots & 1
\end{pmatrix}^{-1}
\begin{pmatrix} k'_1 \\ k'_2 \\ k'_3 \\ k'_4 \\ \cdots \\ k'_n \end{pmatrix}
\tag{8.3}
$$

in which the constants k'_i are the result of dividing the constants in Eq. (8.2) by w_i. The inverse matrix made up of the ratio of the individual weighting factors (w_i/w_n) is the source of much debate within the mathematics community as to its characteristics for the proper solution of this set of equations. Saaty[86] uses these pair-wise comparisons as the basis for his approach to solving these linear equations. The solution to the preceding nonhomogeneous equations, when a particular subset of n equations in n unknowns, is easily solved by standard methods (e.g., Cramer's Rule) provided certain conditions exist in the determinant of the coefficients [formed from the square matrix in Eq. (8.3)].

One of the questions that arises from Eq. 8.3 is whether or not the solutions are unique or consistent. That is, are the solutions to one equation applicable to the other equations such that they are linearly dependent? This gives rise to the examination of the characteristic equation of the preceding set of equations (8.3), the roots of which are called the "latent roots" or "characteristic values" of the matrix. These characteristic values are sometimes called *eigenvalues* (a familiar corruption of the German word *eigenwerte* meaning *characteristic* values). The eigenvalues are important in the constraints on the solutions to these equations. Such values determines whether there are 1) no solutions, 2) a unique set of solutions, or 3) an infinite number of solutions. It is quite possible that a certain set of values of the parameters p, c, s, and r (or whichever other functions are chosen) for quite different defense systems could render the preceding equations *not* solvable. This problem in linear programming* must be evaluated before embarking on the mathematical solutions. This emphasizes again that the setting up of the problem is extremely important or spurious results can appear. The reader is referred to standard mathematical textbooks to avoid such pitfalls.

Fortunately, in the basic concept of hierarchy or tree treatments, it is not necessary to solve these equations explicitly because all terms are known. Values of p, c, s, and r are known from analysis of each of the candidate systems *performance*, *cost*, *schedule*, and *risks*. The values of the *weighting factors* are known from the (initial) interests of the decision maker. In some cases it is possible to develop the weighting factors with a little more discipline by using relative importance discussions with the cognizant experts. Three examples of this might be Albert,[87] Ulvila and Brown,[88] and Saaty.[86] The issue, which will be returned to later, is how these four factors are combined in the first place. The reader should be aware that there are mathematical pitfalls associated with incorrect application of the details of solving these linear equations as treated in detail by Barzalai.[89] Some of these pitfalls have already been mentioned. As Barzalai points out, care must be taken in comparing systems on either an absolute or ratio basis. The placement of absolute zero is can affect the outcome of any comparison.

Fortunately, when solving the linear equations, whether one uses the algebraic approach shown and developed here, or by the rating approach of Ulvila and Brown, or by using the *pair-wise comparison* technique for the weighting factors (per Saaty), one gets *identical results*, which should not be surprising because all of the methods quoted have as their basis the solution to a set of nonhomogeneous linear equations for which the standard mathematical texts have provided the form of solution.

---

*At this time it has been assumed that the problem can indeed be set up under the rules of linear programming. The treatment of nonlinear programming that relates different functional forms between the defense system characteristics, than the inherently assumed linear relationships, is an area beyond the scope of this book and the subject of research in the area of computer codes and nonlinear programming.

IMPORTANCE OF THE INITIAL SETUP OF THE PROBLEM

The problem caused by an incorrect setting up of the questions to be asked when determining *what is the best system* is best illustrated by examples. Also, to keep the mathematical problems to a minimum the examples chosen will keep the examples in the realm of an *equal number of equations as unknowns*, which in this case is four equations in four unknowns. The problem can be solved with more equations (i.e., more candidate systems) than the four unknowns, but the complications are mathematical in nature only and do not aid in the understanding of how to compare systems. In any particular case, however, in which a square matrix is not the norm, special treatment is required, and the reader is encouraged to read the cited references for more clarification. Alternatively, simply grouping the candidate systems into analyzable groups (a common technique when setting up competition in tournaments for elimination) will accomplish the same purpose without the need for advanced mathematical treatments.

Before identifying the characteristics of the four systems to be evaluated, it is important to establish another rule in this method, as used in this book, which is to avoid the need for *inversion* of the various parameters. Such inversion is frequently used in hierarchical analyses but can incur mathematical anomalies (and singularities) and should be avoided. This arises from the inherent problem of comparing two parameters when a high value is considered "good" for one parameter and a low value is considered "good" in another parameter for the same evaluation. Analysts usually solve this by *inverting* the parameter as part of the analysis. This can be a problem, especially if one of the parameters is close to zero, which on inversion would produce an abnormally high factor in the matrix, that would disproportionately distort the results. It can, in fact, invalidate the results.

This potential mathematical anomaly can be avoided in the initial setting up of the problem. In this instance it is suggested that in the definition of "performance" that the parameter sought would be best when low. This seems correct in the case of defense that the best-performing defense system is the one that *minimizes* casualties or *minimizes* leakage, etc. (This could lead to a discussion where in some treatments high values of "kill" in determining performance are sought. For defense it is suggested seeking no casualties is more correct.) What is desired is a system that for

performance,	seeks the *lowest* leakage,
cost,	seeks the *lowest* cost,
schedule,	seeks the *shortest time* (*lowest number of years*) to IOC, and
risk,	seeks the *lowest risk*.

It is noticed that in the (unlikely) event that any of these parameters is zero the singularities that occur through inversion are avoided. Not only does this approach avoid the problem of singularities, but it also emphasizes that the aim is to seek the minimum (optimum) of all attributes in a consistent manner. This does *not* mean that any sublevel component of the main factors needs to follow the same rules. For example, if a *high* interceptor speed or a *high* (meaning *long*) radar detection range is required to achieve *low* leakage then the hierarchy can be set up this way. One must be aware of anomalies occurring in the matrix if an *inversion command* is included.

Consider an example problem of deciding between candidate defense systems to meet some defense requirement. This example problem (and others to follow) will be used to illustrate various issues and also to illustrate how a general solution can be structured.

EXAMPLE PROBLEM 8.1—DIRECT VALUATION OF PERFORMANCE, COST, SCHEDULE AND RISK

Four candidate systems have been proposed to defend a nation from ballistic missile (BM) attack They all cover the defended nation's geographical area and provide different degrees of protection around the identified key assets. For this example no considerations to other missions are to be considered. The candidate systems are as follows:

1) *System 1—a system of only lower-layer, land-based defense systems made up of early-warning sensors, interceptors, indigenous battle management, command, control, and communication (BM/C^3) networks, operational support systems, launchers, and radars.*

2) *System 2—a system similar in makeup to system 1 but with upper-layer, land-based defense systems only.*

3) *System 3—a two-layer missile defense system made up of upper-layer, land-based defense systems that cover the area of the nation plus lower-layer land-based systems to provide extra protection around the key assets.*

4) *System 4—a system similar to system 3 but with a slightly less capable set of systems (with the commensurate lower cost, risk, and shorter schedule).*

For this example the risk can be determined from the following information provided for the R&D programs on the four candidates. Each of the systems to be evaluated has been in development for some time. The evaluation is to be made from today, and the remaining R&D and time to the end of development (milestone III, called milestone C for later programs; see chapter 6 for new designations on acquisition milestones) is as given in Table 8.1.

A visual representation of the risk in this example problem is shown in Fig. 8.6 using the risk categories developed in Chapter 7 (see Fig. 7.5).

From the data supplied, in this example problem it is seen that system 1 is deemed to be a low-risk system, system 2 and 3 are high risk, and system 4 is

Table 8.1   Computation of risk

Candidates	Remaining $R&D to milestone III, $million year	Remaining time to milestone III, years	Risk, $billion × years
System 1	200	1	0.20
System 2	500	2	1.00
System 3	700	5	3.50
System 4	300	2	0.60

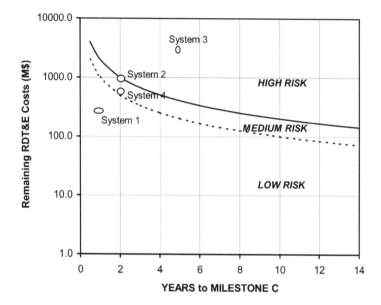

Fig. 8.6   Example risk categories displayed on general risk chart.

a medium-risk candidate. These values of risk are then included with the other key factors on the candidate defense choices to give the following set of parameters in Table 8.2.

Further, it has been decided that the following "weights" or relative importance be retained in the selection process:
Performance:

$$w_1 = 50\%$$

Cost:

$$w_2 = 40\%$$

Table 8.2 The four characteristics for the four candidate systems

Candidates	Performance (leakage), %	Cost (procurement), $billion	Schedule, years to IOC	Risk, $billion × years
System 1	9	3.3	4	0.20
System 2	9	2.1	9	1.00
System 3	0.81	4.6	12	3.50
System 4	5	4	6	0.60

Schedule:

$$w_3 = 5\%$$

Risk:

$$w_4 = 5\%$$

Note that system 1 and system 2 are single-layer defense systems with a planned leakage of 9%, which corresponds to a system with an interceptor salvo-firing doctrine of two per threat missile and an individual interceptor performance of SSPK = 0.70. System 3 is a two-layer defense system with the corresponding leakage of 0.81%. These values have been taken from Chapter 3, Table 3.8. Finally, system 4 is a lesser or in-between capable system with an arbitrary planned leakage of 5%.

SOLUTION

If each of these characterizations of performance, cost, schedule, and risk are treated at the same level (as shown in Fig. 8.1), then the equation set (8.2) becomes

$$\begin{bmatrix} 0.378 & 0.236 & 0.129 & 0.038 \\ 0.378 & 0.150 & 0.290 & 0.189 \\ 0.034 & 0.329 & 0.387 & 0.660 \\ 0.210 & 0.286 & 0.194 & 0.113 \end{bmatrix} \begin{bmatrix} 0.50 \\ 0.40 \\ 0.05 \\ 0.05 \end{bmatrix} = \begin{bmatrix} 0.292 \\ 0.273 \\ 0.201 \\ 0.235 \end{bmatrix} \qquad (8.4)$$

Before summing the weighted values, it was first necessary to "normalize" each factor to remove the scale distortion caused by the choice of units for each. Such a normalization also removes the problem of the equations being dimensionally incorrect (of attempting to add dollars to years). For example, if the cost had been quoted in lire instead of dollars, the cost term would have

Table 8.3   Ranking of system candidates

System	Ranking	Ratio of results
System 3	1	1.0
System 4	2	0.855
System 2	3	0.736
System 1	4	0.688

been approximately 1000 times larger and distorted the relative effect in the matrix. A similar distortion would have occurred in the other factors if different choices had been made on the units (decimals instead of %; months instead of years, etc.). This shows that the best system would be system 3, and relative to the other systems the results are as shown in Table 8.3.

That is, if system 3 is the best at treating all factors of performance, cost, schedule, and risk on a common level, then system 4 is 86% as good, and system 1 and system 2 are approximately 69–74% as good. Upon examination of the basic characteristics in the simplified example problem, the results are not too surprising when one considers that system 3 had a very good performance (a leakage of 0.81%, which is considerably lower than the other candidates), and managerially, performance was viewed as dominant with a weighting of 50%. The No. 2 choice, system 4, was approximately 86% as good, and thus it is possible that the ranking could be reversed when some of the political and national issues are considered. This will be returned to in Example 4. In this simplified example most of the relative rankings could be determined by inspection, but the purpose here is to demonstrate the method that can be used for less obvious comparisons of candidate systems. Figure 8.7 displays each of the factors of performance, cost, schedule, and risk taken from the preceding results.

In this (constructed) example problem it is seen that system 3 is the most expensive, takes the longest to field, and has the highest risk. Yet because it offers the lowest leakage against the perceived threat and because performance has been given such a high priority (50%), it is the preferred system. The net result of the analysis is displayed in Fig. 8.8.

If the performance weighting of 50% had been lower, one of the other systems would have become the selected system. Here is where the decision maker must decide how important each factor (performance, cost, schedule, and risk) must be in the decision process.

Consider now the more likely case, where the decision maker has some constraints *on the problem. Specifically, in addition to the four factors being* weighted *as shown, they are now to exhibit other characteristics, such as*

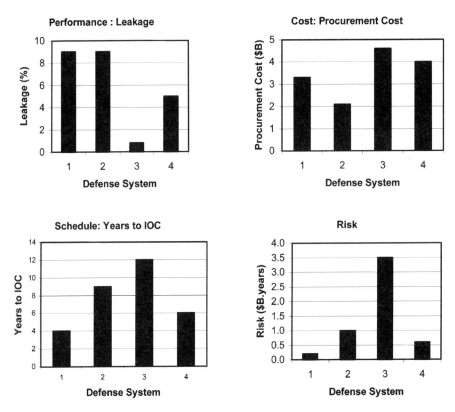

Fig. 8.7   Separate characteristics of each candidate system (example).

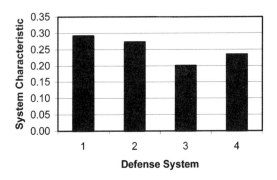

Fig. 8.8   Overall comparison of example candidate systems.

which is the most cost effective*? or other criteria such* as which is the most cost effective for a given cost*?* or risk, or schedule*? Such questions cannot be readily answered by the preceding equations, and different formulation must be set up. This is best illustrated by an example.*

EXAMPLE PROBLEM 8.2—INCLUSION OF COST-EFFECTIVENESS IN COMPARISON

In this example problem it is taken that cost effectiveness is more important than just performance. Using the same candidate systems as in example problem 8.1, decide which of the candidate systems is the most cost effective and which system is the best choice overall, taking into account the additional concerns of cost, schedule, and risk.

From Example Problem 8.1, the system characteristics shown in Table 8.4 can be constructed.

In a similar vein to that taken in example problem 8.1, assume that the relative importance of the attributes to be considered are as follows:

Cost effectiveness (performance/cost):

$$w_1 = 50\%$$

Cost:

$$w_2 = 40\%$$

Schedule:

$$w_3 = 5\%$$

Risk:

$$w_4 = 5\%$$

SOLUTION

Following similar normalization and weighting as before, the results shown in Table 8.5 are found for each of the candidate systems.

In this set of candidates, it is seen that system 3 is the most cost effective in that it has the lowest leakage per dollar and by a significant margin. If the other attributes of cost, schedule, and risk are included, the results shown in Table 8.6 are found for all candidate systems.

Table 8.4 Introducing cost-effectiveness to the evaluation

System	Cost effectiveness, % leakage/$billion	Cost ($billion, procurement)	Schedule, years to IOC	Risk, $billion × years
System 1	2.73	3.3	4	0.20
System 2	4.29	2.1	9	1.00
System 3	0.18	4.6	12	3.50
System 4	1.25	4	6	0.60

Table 8.5   Ranking of cost-effectiveness

System	Ranking	p/c	Ratio
System 3	1	0.18	1.0
System 4	2	1.25	0.144
System 1	3	2.73	0.066
System 2	4	4.29	0.042

Table 8.6   Overall ranking of systems

System	Ranking	Value	Ratio
System 3	1	0.194	1.0
System 4	2	0.204	0.951
System 1	3	0.264	0.735
System 2	4	0.338	0.574

In this case, it is seen that some changes have occurred from the results of example problem 8.1. System 3 is still the best overall but with a smaller margin over the second choice of system 4. System 4 that was 86% as good has closed the gap considerably to 95% as good. Also, the order has been changed in that now system 1 is in third place and system 2 has slipped to fourth place.

COMMENT ON COST EFFECTIVENESS

In example problem 8.2, there is an issue of logic in the determination of the familiar term *cost effectiveness*. This label is usually used in the defense community to mean *effectiveness per unit cost*, where "effectiveness" is some attribute of performance that is to be maximized. In example problem 8.1, for a quick and familiar comparison the ratio of "% leakage per $billion" was used as a recognizable parameter. However, in the development of this comparison method it should be remembered that *minimizing* the parameters would be in line with the consistency requirement in the equations. Clearly, dividing leakage by cost would not be the required approach if a consistent set of comparisons is to be used in the evaluation (otherwise increasing cost would be seen as one desired approach). In the terminology of this approach, the following is more correct:

Maximize (Cost Effectiveness) ≡ Minimize (Leakage × Cost)   (8.5)

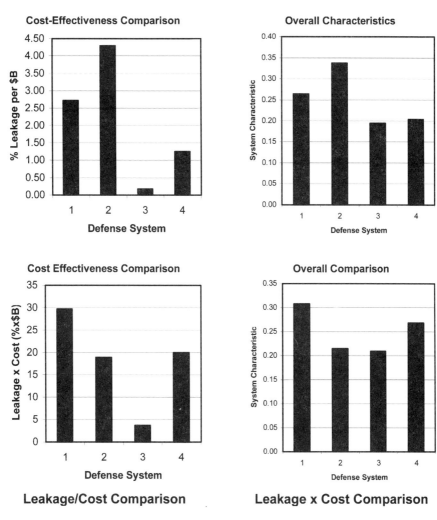

Fig. 8.9 Cost-effectiveness representation effect.

That is to say, if maximizing cost effectiveness is the aim then one should seek system solutions that satisfy the condition $p_i.c_i =$ minimum, and not $p_i/c_i =$ minimum.

Figure 8.9 shown the results in example problem 8.1 where the two versions of cost effectiveness are used (i.e. p/c and $p.c$). The upper part of Fig. 8.9 shows where the familiar (p/c) has been used in the matrix equations, first on the left-hand side showing a direct comparison of leakage per unit cost and the final result on the right-hand side on the four candidate system comparison. The lower part of Fig. 8.9 shows the comparisons done with cost effectiveness expressed as $(p.c)$. The left-hand side of Fig. 8.9

shows the product "leakage × cost," and the effect on the overall comparison is shown on the right-hand side. In this particular (constructed) example problem of the choice system 3 is still the best, but that is a fortuitous result for the particular set of numbers used and is not a general result. Again, the setting up of the problem is paramount if logical results are to be expected.

This emphasizes yet again that the relative rankings of all candidate systems depend entirely on how the problem is set up. There is no unique solution for a set of candidate systems as to which is "best" unless one completes the sentence such that the answer (solution) is given to *best with respect to some attribute of concern*. In the preceding examples the attributes include, but are not limited to, *cost-effective, cost effective within budget, cost effective with least risk*, etc. As discussed in Chapter 4 on performance, this will include additional attributes, such as *best with respect to conventional weapons* and *best with respect to weapons of mass distruction* (WWD), among other issues of interest to the decision maker.

BOUNDARY CONDITIONS AND SPECIFICATION REQUIREMENTS

The preceding linear equation approach or multiattribute approach is a good way to determine the *relative merits* of candidate systems. However, it is not capable of determining if candidate systems satisfy *specific criteria*, which are usually established as limits and goals and included in any specification or *requirements documents*. Such determination is determined by separate analysis. For example, it is usually required that the candidate systems meet some performance level (expressed either as minimum, desired, goal, or other limit); the cost might be expressed as to be below some budget level and similarly for the other attributes, the schedule and risk. This can be approached as an exercise in linear programming. For example, the problem can be set up for linear programming as follows:
Minimize:

$$w_1 p_1 + w_2 c_1 + w_3 s_1 + w_4 r_1 \tag{8.6}$$

Subject to the constraints:

$$p_1 < P_1 \quad \text{(say, some minimum leakage requirement)}$$
$$c_1 < C_1 \quad \text{(say, some specific budget figure)}$$
$$s_1 < S_1 \quad \text{(say, within some stated time frame)}$$
$$r_1 < R_1 \quad \text{(say, some low value of risk)}$$

This linear programming would be repeated for all of the candidates. If the basic system attribute equation (8.6) is set up differently, as suggested in the beginning of this chapter, in other combinations such as performance cost, procurement cost, operating cost, development time, operating time, etc., then the constraints (P_1, C_1, S_1, R_1) could be more complex such as "*R&D cost should be no more than 10% of procurement cost,*" as just one example. This is left as an exercise for the reader, and a slightly different approach will be developed here to illustrate some features of interest to the decision maker. An example using the same candidate systems as before will be used.

EXAMPLE PROBLEM 8.3—COMPARISON USING GRAPHICAL TECHNIQUES
Given the same candidate systems as in example problem 8.1 and retaining the same considerations of cost effectiveness as in example problem 8.2, which system offers the best system that can be fielded in the next 10 years, has no more than 5% leakage, and has the minimum risk?

SOLUTION
In this case the problem can be solved easily by inspection or more informatively for more general use where more candidate systems might be under evaluation, in a graphical format, as shown in Figs. 8.10 and 8.11.

Figure 8.11 now shows that all systems except system 3 (the originally selected system) meet the 10-year goal. This might cause a reevaluation of the requirement or acceptance of a longer schedule if the better cost effectiveness and lower leakage are required. Fortunately, in this case the No. 2 choice (system 4) does indeed meet the required 10-year IOC date (in this example it actually is predicting 6 years) and also has the acceptable leakage rate of 5% as shown in Fig. 8.10. The risk values are shown in both

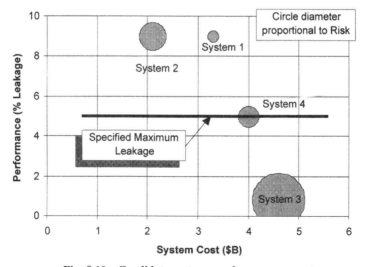

Fig. 8.10   Candidate systems performance vs cost.

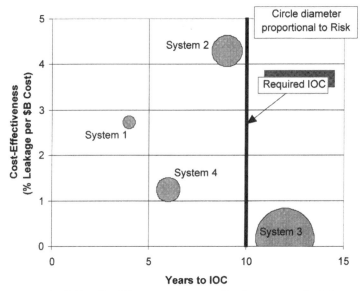

Fig. 8.11   Candidate systems cost-effectiveness vs schedule.

figures, where the diameter of the circles is proportional to the risk (in $ billion × years).

These charts show the boundary conditions or limits or requirements in an easy to understand format and provide the required information for the decision maker to readjust some requirements if necessary. In the following sections the next stage of evaluation is to bring in the other attributes such as the political and national interests that will effect the final selection.

POLITICAL AND NATIONAL INTERESTS

Once armed with the technical, cost, and programmatic recommendations, attention can now be turned to the political and national interests to see whether or not adjustments should be made in the selections. Again, this is best demonstrated through the use of examples. The preceding four candidate systems will be used.

Consider first that a particular defensive missile battery is made up its major subsystems. Figure 8.12 shows a typical missile (interceptor) battery made up of the set of missiles themselves, their launchers, fire control radar and its support (diesel generators, etc.), and the indigenous BM/C^3 network that allows all subsystems to function. This BM/C^3 subsystem would then link the battery to its own command structure and into any national or international network.

For the purposes of illustration, assume now that the question of international co-operation is being considered when considering the relative merits of the candidate defense systems. Assume that a nation has developed

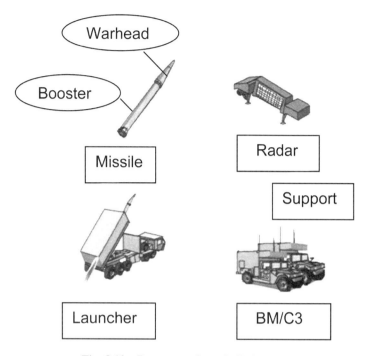

Fig. 8.12   **Representative missile battery.**

its own radar, launcher, and BM/C^3 systems but that the missile has been the result of a previous international cooperative effort where the booster has been developed and produced "in country" but that the warhead has some unique high technology features that it has collaborated with from another nation.

This means that the illustrated defensive missile system has a combination of 1) in-country development and 2) multinational development, which will incur a ratio of jobs in country to the total costs. Or, as an additional possibility, a complete subsystem, say, the radar, is procured from another nation. These two considerations would be important to a nation that is deciding when selecting a defensive missile system how much the national need to develop its own defensive systems vs, say, an economical procurement (or earlier delivery) from another nation might be important. These considerations might outweigh the technical, cost, and programmatic evaluations considered earlier. This is best illustrated by an example.

EXAMPLE PROBLEM 8.4—POLITICAL AND NATIONAL FACTORS OF THE FOUR
CANDIDATES
The decision maker has decided to hold off making a decision based on the results of example problem 8.3 until he has heard the results of the

Table 8.7   Statistics of political factors for each candidate system

System	Jobs in country, %	Use of national systems, %	Interoperability, %
System 1	100	100	100
System 2	80	60	100
System 3	85	75	80
System 4	90	75	80

implications of jobs in country and the need to develop his nation's own capability and industrial base.

An analysis has been made of the four systems, and results are shown in Table 8.7.

In the case of system 1, all systems and subsystems are 100% nationally owned and developed. For system 2, an upper-layer missile system, it was found that it was economical to develop and produce all systems in country except for the high-technology warhead, which is available from another nation that has the technology already developed. The measure of interoperability in this hypothetical example is simply the percentage of systems that have been designed to have the ability to operate with the Alliance systems [covering such features as electrical interfaces (power and frequency); BMC^3I message traffic protocol; mechanical interfaces, etc.]

As discussed in Chapter 5, such a high-technology warhead constitutes a high percentage of the systems costs (of the order of 40% in this case) and, further, required 20% of the manpower costs of the system to be allocated to the other country for such a development. For systems 3 and 4 the lower-layer missile systems were nationally owned and developed, and a portion of the upper-layer systems were from the other nation, as in system 2. This gave the values of the three attributes as just shown.

The economic situation in the nation was such that a high "weighting" was given to jobs in country and the use of national systems, such that the weightings of the three attributes were chosen to be

$$w_a = 70\%; \quad w_b = 25\%; \quad w_c = 5\%$$

The technical, cost, and programmatic values in example problem 8.2 must now be changed because the nation is deciding to procure certain subsystems from another nation and thus will not carry the risk for those subsystems. It is assumed (for this example) that the schedule to IOC remains unchanged.

The new values for the characteristics of the candidate systems now, would be as shown in Table 8.8.

Table 8.8 Reassessment of characteristics with political factors

System	Cost effectiveness % leakage/$billion	Cost, $billion, procurement	Schedule, years to IOC	Risk, $billion × years
System 1	2.73	3.3	4	0.20
System 2	4.29	2.1	9	**0.80**
System 3	0.18	4.6	12	**1.50**
System 4	1.25	4	6	**0.30**

[a]For the values in bold it has been taken that the risk would be reduced considerably for the nation because it is now only developing certain portions of the upper-layer system. The reduced values used here are arbitrary, but estimates would be inserted for any specific example in a real case. Also in this example, to keep the variations in bound it has been assumed that the cost of the system is not different in each nation and that the schedule remains the same. In an actual case one would have to analyze the different labor rates and schedules for the cooperating nations. However, such variations are not germane to the illustration of the method and have been omitted here for clarity.

Table 8.9 New rankings through adjustment in risk

System	Ranking	Value	Approximate ratio
System 3	1	0.188	1
System 4	2	0.203	93%
System 1	3	0.266	71%
System 2	4	0.343	55%

SOLUTION

First, by using the same technique as before, the technical, cost, and programmatic rankings of the four candidate systems (as a result of the reduced risk) are shown in Table 8.9.

This result is not surprising, in that the ranking has not changed and only minor variations have occurred in the relative values. This is because the risk consideration was only weighted at 5% in the overall evaluation factors.

Second, following the same approach for the political and national factors as provided in the example problem statement, the relative rankings are shown in Table 8.10.

This result summarizes the expected results that system 1 is now the most desirable from a national and job viewpoint because it utilizes 100% of national systems and that system 2 uses the least (although all systems were designed to be interoperable). Also, systems 3 and 4 are at the 82–85% ranking because of the fairly large national systems contribution in the two-layer missile defense system. Then, combining all factors (technical, cost, programmatic, political, and national factors) the candidate systems show the rankings as shown in Table 8.11.

Table 8.10   Relative rankings based on political and
national factors

System	Ranking	Value	Approximate ratio
System 1	1	0.212	1
System 4	2	0.248	85%
System 3	3	0.258	82%
System 2	4	0.282	75%

Table 8.11   Overall rankings with both technical and
political factors

System	Ranking	Value	Ratio
System 3	1	0.191	1
System 4	2	0.200	96%
System 1	3	0.224	85%
System 2	4	0.385	50%

When all factors are combined, the rankings do no change significantly, but there are significant conclusions to be drawn from the results when the impact on national defense are considered.

IMPACT ON NATIONAL DEFENSE

The preceding set of simplified results serves to illustrate how different factors must come into the decision process. Example problems 3 and 4 help to illustrate the choices that face the decision maker. Table 8.12 shows the results of these two examples.

Given the preceding results, the decision maker must consider the following: in this hypothetical example system 3 clearly had the best defense

Table 8.12   Comparison of example problem results

System	Before political considerations		After political considerations	
	Value	Ratio	Value	Ratio
System 3	0.188	1	0.191	1
System 4	0.203	93%	0.200	96%
System 1	0.266	71%	0.224	85%
System 2	0.343	55%	0.385	50%

in that it offered a minimum leakage (and a minimum leakage per unit cost) but did not meet the 10 year to IOC requirement. System 4, on the other hand, improved its ranking (over its competitors systems 1 and 2) upon inclusion of the political and national factors because of its greater use of national systems and opportunity to provide jobs in country. System 4 also met all requirements on leakage and schedule.

The decision maker must now decide whether 1) to stay with the (specified) 5% leakage and a "96% solution" or 2) to wait the extra two years and achieve the much improved 0.81% leakage. These results and recommendations become "go-no-go"-type decisions.

Such decisions now require human judgment, and the computer can only be an aid! As will be shown later in this chapter, the subject of leakage can be a major deciding factor.

LESS PRECISE DECISION MAKING

Although the preceding method does allow for the key system characteristics ($p, c, \ldots$) to be treated in a logical and consistent manner, it does require specific values of inputs without any reference to goals, limits, or, more likely, "*desirable values.*" In many instances the decision maker is more likely to say something like, "we need the most cost-effective solution, provided it does not get too expensive or is not too far away from the schedule that we are thinking of." Such direction from the decision maker expresses his desires, and he is saying implicitly, "*help me here in making a decision.*"

Such lack of precision is more normal than not, and any method that can indeed help or aid the decision maker would be useful. There are various methods on the market that tackle such imprecision in decision making, and they will be referred to as the method is developed here. In the preceding sections it was stated that the main purpose was twofold: first, to seek solutions that *minimize* the function

$$f(p, c, s, r) = w_{i1}p_i + w_{i2}c_i + w_{i3}s_i + w_{i4}r_i \tag{8.7}$$

where w_{ij} are the weighting factors for each of the system characteristics (p, c, s, r), and second, to seek which has the *lowest* value of $f(p, c, s, r)$ out of a set of system candidates. Note, as before, that for the purposes of illustration, it is taken that the system characteristics of interest are indeed p, c, s, r. The method is equally applicable to some other preselected set of system characteristics, such as cost-effectiveness, procurement cost, operating cost, etc. (see earlier discussion).

One means of treating the imprecision, or *provisional decisions*, in the questions asked by the decision maker referred to earlier is to modify Eq. (8.7) and express it as

$$f(p, c, s, r) = w_{i1}.v(p).p_i + w_{i2}.v(c).c_i + w_{i3}.v(s).s_i + w_{i4}.v(r).r_i \quad (8.8)$$

where now a *value function* $v(x)$ has been inserted into the equation to modify each term. (The method here is an adaptation of several earlier authors' approaches, most notably that given in Ref. 90.) In this formulation $x = p, c, s, r$. Such an insertion does not invalidate the equation, and it is still required to have the original weighting factors (w_{ij}) sum to unity:

$$w_1 + w_2 + w_3 + w_4 = 1 \quad (8.9)$$

with similar equations for the remaining system candidates.

There are various ways to express the value function, but algebraically and expectedly the exponential form is most common;

$$v(x) = \frac{1 - e^{-(x_e - x_d)/\alpha}}{1 - e^{-(x_m - x_d)/\alpha}} \quad (8.10)$$

In this exponential expression for the *value function*, the following key parameters to describe the constraints are as follows:

1) x is the system characteristic (e.g., $p, c, \ldots$)
2) x_m is the *maximum acceptable limit* of the system characteristic,
3) x_d is the *desired goal* of the system characteristic, and
4) α is a shaping constant for the exponential representation ($\alpha \neq 0$).

Such a representation of the value function allows for two things to be accomplished; 1) it allows "goals" and "limits" to be added into the consideration, and 2) it allows how any individual system characteristic is viewed by the decision maker. For example, if the decision maker wants to keep his options open, he can say that he would *prefer* to keep the performance requirement at say 50% (as used in the preceding examples), but would not want to do this stubbornly if the cost were too high. The *degree of preference* can then be handled through such value functions, by setting the limits x_m, x_d and the shaping parameter α. Kirkwood[90] provides equations for both increasing and decreasing preferences and also for assuming that the *maximum* value is sought by the decision maker. In the context of this book, the exponential equation has been modified to seek the *minimum* value in all parameters and further to handle how the characteristics behave as they

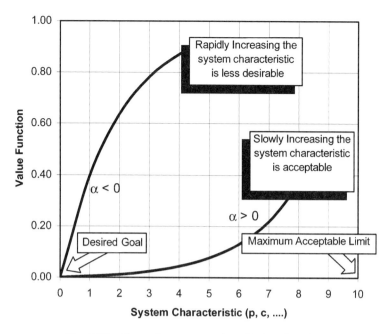

Fig. 8.13 General representation of value function.

depart from this *minimum* desired goal. Figure 8.13 expresses all of these features in a diagrammatic fashion. In Fig. 8.13 several germane features of this approach can be described.

SYSTEM CHARACTERISTIC

This is simply the system characteristic x that is to be minimized. In this approach it is the performance, cost, schedule, or risk as used heretofore.

MAXIMUM ACCEPTABLE LIMIT

This value x_m is the limit that the decision maker would not like to go beyond. For example, it might be *seek a solution that can be available within 10 years* or perhaps *cost less than $5 billion* or some other value that can be specific or approximate.

DESIRED GOAL

This value x_d is the value that the decision maker would like to achieve, even if no candidate can quite make that value. For example, the desired goal might be *select a candidate that can achieve a leakage value of 1% or less.*

SHAPING CONSTANT

This factor α is the most subjective, in that it expresses how strongly the decision maker feels about some system characteristic. For example, if the decision maker would like to achieve the most cost-effective solution but that cost is *very* important, then any departure from the weighted parameter of cost would see a strong variation from the average value with the shape of the exponential value function. In this derivation positive values of the shaping parameter ($\alpha > 0$) represent those parameters that the decision maker is prepared to allow as gradual departures from the desired goal, and negative values of the shaping parameter ($\alpha < 0$) would represent those parameters that the decision maker is not inclined to accept as departures from the stated limits.

In Fig. 8.13 it can be seen that for the value function ($\alpha < 0$) the function rapidly increases initially away from the desired goal x_d. If this function were applied to cost, for example, it would mean that the decision maker would not favor departing from the weighted value of cost. On the other hand, for the value function ($\alpha > 0$) the function only increases gradually away from the desired goal x_d. If this curve was applied to performance (expressed as % leakage), it says that the decision maker is allowing some departure from the *desired goal* in the selection of candidates, but not too far, as shown by the eventual rapid increase in the value function as the (leakage) increases above the desired value. If the desired value of leakage was 0.81% (as used in the preceding example), then this value function would eventually reach undesirable values as the parameter increased. In this case, if the *maximum acceptable level* x_m is say 10% one can see from Fig. 8.13 how this effect is treated.

Purely for the sake of illustration, the horizontal axis has been shown as varying from a low value of zero to a high value of 10. In actual calculations this range of 0–10 would be replaced by a range from a *desired goal* x_d to the *maximum acceptable limit* x_m. Mathematically, it is of interest to note that if the desired goal and maximum acceptable limit are used as the lowest and highest values in the candidate system sets and the shape parameter approaches infinity (so that the value function becomes a straight line) then the results from using Eq. (8.8) are *identical* to those that would be achieved by using the preceding set of equations [see Eq. (8.7)] with unvarying weighting constants. Figure 8.14 shows these value functions for a specific set of shaping parameters α. While theoretically the exponential functions reduce to a straight line for $\alpha = \infty$, in a practical sense this occurs at much lower values, say, $\alpha \rightarrow 100$. For the examples shown later, the particular values of $\alpha = 2$ and -2 are used.

As set up here, the aim is to *minimize* the function $f(p, c, s, r)$; hence, any move away from the *desired goal* (or the origin in Fig. 8.14) for any par-

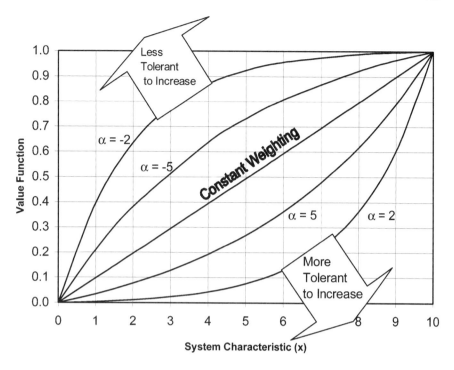

Fig. 8.14   General set of value functions.

ticular candidate will *add* to its score and thus cause it to move away from the minimum value and from being the selection most likely to meet the aims of the decision maker. It is best to show how this approach can be used to advantage in the selection process with an example.

EXAMPLE PROBLEM 8.5—CONDITIONAL DECISION MAKING

The decision maker has reviewed the results of the preceding treatments in selecting the candidate system but is uncomfortable with how the selection fits within some general boundary conditions that have been given him by his staff. Redo the calculations (ignoring the political constraints as used in example problem 8.4), and show how the various candidates fare in relation to the following limits, goals, and constraints. For this exercise performance, cost, schedule, and risk are to be evaluated at the same level, but with the same weightings as before. The basic features are as shown in the tables embodied in example problem 8.1.

For this evaluation the decision maker is willing to let the performance "creep" up to higher values (greater than 0.81%) but is not willing to see the cost, schedule, and risk get too far out of hand. In other words, he wants the best system that he can get but wants to get on with it and not have problems

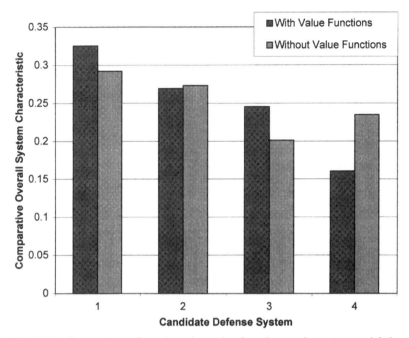

Fig. 8.15   Comparison of results using value functions and constant weighting.

of cost, schedule, and risk. What would be the recommended system candidate?

After some discussion it is agreed with the decision maker that the following bounds the problem including the reluctant departures from preferred values:

1) Leakage *has a desired goal of 0.10% but willing to accept up to 10%.*
2) Cost *has a desired goal of $2 billion that is acceptable but reluctantly will accept up to $5 billion.*
3) Schedule *has a desired goal of 4 years and reluctantly will accept up to 12 years.*
4) Risk *has a desired goal of $200 million years but reluctantly will accept up to $5 billion years.*

Accordingly, the analyst then assumes a value function with $\alpha = 2$ for performance and $\alpha = -2$ for the other factors of cost, schedule, and risk.

The overall results using these value functions are displayed in Fig. 8.15, together with the earlier results obtained by the more direct method where no such constraints were imposed (in example problem 8.1).

It can be seen from Fig. 8.15 that there has been a shift in the selection as a result of this more refined selection process that has taken into consideration how close the system characteristics are to some predetermined set of limits. System 3 (the originally preferred system) has been replaced by system 4 in

Table 8.13 Comparison of results by constant weighting and by value functions

System	By constant weighting		By value functions with constraints	
	Ranking	Relative ranking, %	Ranking	Relative ranking, %
System 1	4	69	4	49
System 2	3	74	3	60
System 3	1	100	2	65
System 4	2	86	1	100

this example, when the concerns about cost, schedule, and risk have been incorporated into the selection process. The other candidate systems have held the same relative positions in the selection process. Numerically, the comparisons are shown in Table 8.13.

Table 8.13 reflects the decisions that cost, schedule, and risk are important in the selection process and that this has now been treated a little more rigorously by setting constraints on the problem rather than a simple comparison of which is the best. The problem has now been treated as which is best, given certain definable constraints. These results dramatically show how the system candidates fare at the level of performance, cost, schedule, and risk when expressed in relation to the issues of concern to the decision maker. Now it is seen that despite the superior performance (approximately 10:1 less leakage) the very real concerns of cost, schedule, and risk have overcome this advantage in a very real sense. Consider how each of the systems have fared given the decision-maker's willingness to either accept or not accept departures from his goals.

Performance: *Figure 8.16 shows the relative value of leakage for each of the system candidates and how close they come to the desired value of minimum leakage. In Fig. 8.16 can be seen the mathematical representation of the decision-maker's willingness to accept higher leakage. This appears as the value function keeping the performance (leakage) acceptably low for reasonable departures from his goal of 0.10%. In this evaluation both System 3 and System 4 are deemed to have a "comfortably acceptable" leakage in the decision maker's eyes and are thus not heavily penalized by the value function. On the other hand, even though the leakage of Systems 1 and 2 are technically "within spec" (ie 9% leakage is less than 10% leakage), the decision maker has made them less desirable by the use of a value function that increases their "weight" (and therefore undesirability) in the evaluation. It is seen that the willingness to accept larger leakage values in the selection process has kept all systems within reasonable limits, and no serious penalty was experienced by any of the candidates.*

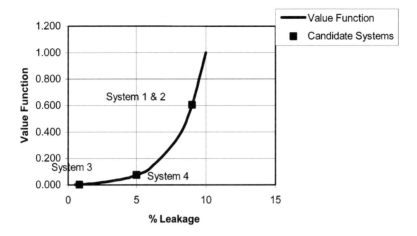

Fig. 8.16   Value function for performance (leakage).

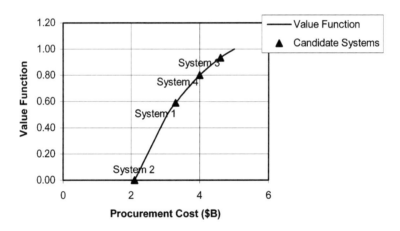

Fig. 8.17   Value function for (procurement) cost.

Cost: *Figure 8.17 shows the relative value of procurement cost for each of the system candidates and how close they come to the desired value of minimum procurement cost. Now by using the less desirable preferences (α < 0), the analysts have captured the reluctance of the decision maker to go much beyond the desired goal of minimum cost of $2 billion for any system. This appears in Fig. 8.17 by the rapid increase in the value function (less desirable) as it moves away from zero. This shows that systems 1, 4, and 3 are more heavily penalized than the cost-attractive system 2. By comparing the relative shape of the value function curve in Fig. 8.16, where slow departures from the desired goal were deemed acceptable, and the value function curve in Fig. 8.17, where departures from the goal were deemed*

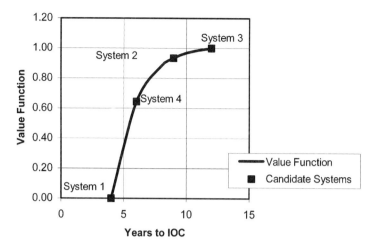

Fig. 8.18 Value functions for schedule (years to IOC).

only reluctantly acceptable, the ability to represent less-precise or conditional decision making in a reproducible and repeatable form has been retained.

Schedule: *Figure 8.18 shows the relative value of schedule (years to IOC) for each of the system candidates and how close they come to the desired value of four years to IOC. Clearly the 12-year schedule for system 3 has had an adverse effect on its otherwise attractiveness in performance (minimum leakage).*

In this case the less desirable value of the schedule of system 3 (12 years to IOC) has pushed it to the boundary (value function of unity) in the comparison. System 2 with nine years to IOC and System 4 with six years have clearly lost out to system 1 with its short four years to IOC. The rapid rise in the value function away from zero (taking the system away from the desired goal) is shown by the shape of the curve in Fig. 8.18.

Risk: *Finally, Fig. 8.19 shows the relative value of risk (billion dollar years of R&D) for each of the system candidates and how close they come to the desired value of minimum risk, represented by minimum expenditure of R&D funds and minimum schedule to milestone B.*

In a similar vein to the preceding comparisons of value functions, Fig. 8.19 displays the reluctance to accept high risk in order to achieve the desired defense system. The very high (relatively), risk associated with system 3 and to a lesser extent systems 2 and 4 is shown by Fig. 8.19.

It is seen from the preceding Example how the issues of imprecision in the decision making can be handled in a consistent, reproducible, and understandable manner. If the decision maker has either 1) a precise set of limits on cost, schedule, etc., or 2) only a general idea of the limits or boundaries, it is seen how both scenarios can be handled by this method. The

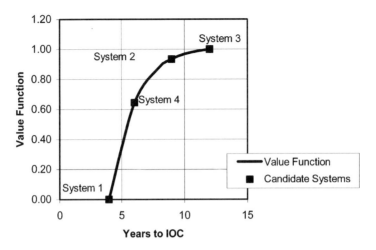

Fig. 8.19 Value function of risk (billion dollar years of R&D).

*graphical display in Figs. 8.16–8.19 provide a picture of such decisions
and suggest how any particular defense system candidate could/should be
modified to meet the requirements laid down by the decision maker.*

ALTERNATIVE MEANS OF DISPLAY

The results from the preceding analysis have been displayed mainly in
tabular and column format, which is a satisfactory means for most analyses.
For some instances, however, it might be beneficial to use a more graphical
form of display. This is especially desirable when then there are more
systems to compare or, as will be shown, more components of the four
measures of performance, cost, schedule, and risk are to be considered. The
form of the analysis has been to seek the *minimum* as the best for
mathematical reasons to avoid anomalies in the matrix equations. One could
imagine that the aim of the comparison all of the time is to seek the bull's-eye
of the dartboard of candidates (Fig. 8.20).

If each of the components of the performance, cost, schedule, and risk (or
whatever set of parameters are agreed upon to evaluate at the outset of the
analysis) are seen as spokes on the dartboard, then the aim is to hit as close to
the center as possible.

This has been done in Fig. 8.21, where just two of the defense system
candidates (systems 1 and 2) have been displayed for clarity. The performance (% leakage), cost, schedule, and risk have been displayed in each of
the four quadrants. Selected spokes have been used in each quadrant to
clearly show the results. In a similar vein to radar charts, it will be recalled
that the object in the analysis is to *minimize* the size of the spokes in all four

Fig. 8.20   Seeking the best solution.

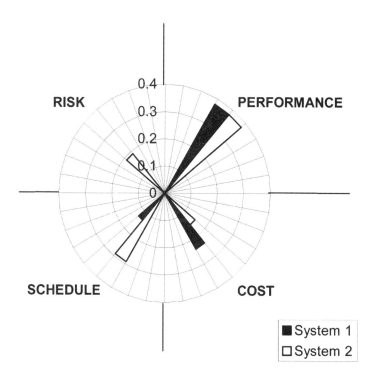

Fig. 8.21   Example problem results for four main factors.

factors or system characteristics for best system. Aiming for the bull's-eye provides the best solution. These results and display can be compared to the tabular values already provided in Fig. 8.7 or Eq. (8.4). Here in the Example Problem, systems 1 and 2 have the same performance (leakage) as seen in the upper right-hand quadrant. The lower right-hand quadrant shows that system 2 costs less than system 1. The schedule and risk quadrants show that system 2 is not as good as system 1. The small value of risk for system 2 is close to the bull's-eye and not visible on this scale.

Such a display allows for an immediate appreciation of all factors. In this case the factors are performance, cost, schedule, and risk, but they could just as easily have been cost-effectiveness, operating and support costs, or some other attributes of concern to the decision maker.

If this concept is expanded further to where many other subcomponents of each system are to be evaluated, this form of dartboard or polar plot is particularly useful. Consider a hypothetical example where the analysis has been expanded to the subcomponent level where the various contributors to each of the main factors have been analyzed in a similar manner to the main factors. Say, for example, the cost factors have been broken down to the costs of each of the main elements of cost, such as interceptor costs, radar costs, etc; these can be displayed in the cost quadrant of Fig. 8.21 so that a main cost driver can be immediately displayed. Similarly, for schedule, which items are contributing to the schedule can also be displayed. If there is a "long pole in the tent" that is contributing to the system not having the desired value, this would be immediately identifiable. Figure 8.22 shows

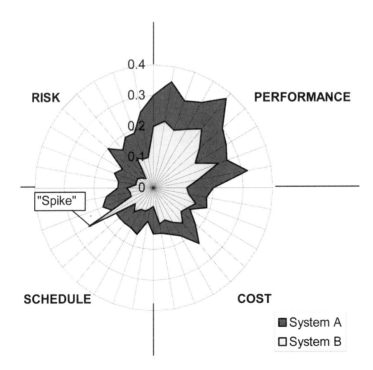

Fig. 8.22   Hypothetical example of displayed subcomponent contributors.

such a hypothetical set of solutions where the linear equations (complete with any value functions included) have been solved as before and the net results shown.

In this hypothetical example systems A and B have been compared. In this radar-plot analogy system B (lighter-shaded diagram) has a lesser area under the curve than system A (darker-shaded area), such that its overall set of system subcomponents is closer to the desired goals or specification values and thus would be the selected system.

In this constructed example one element of system B however, seen in the lower left-hand quadrant on schedule and labeled "Spike," is greater than the schedule shown for system A. If such a schedule contributor was greater than some specification value of schedule, it might nullify every other attribute shown, or it might influence the decision maker to reconsider the options or criteria. [One could include the specification values for performance, cost, schedule, and risk in Fig. 8.22 as circular rings so that any spikes would be immediately obvious (as to not meeting spec) and allow ease of discussion during any management review of the results of the comparison of systems.] Such a display would thus serve to highlight this long pole in the comparison against requirements and allow management focus for any desired remedial action. Such displays can help in the decision making.

SUMMARY

The aim throughout this chapter has been to show how the many complex issues of missile defense can be tackled at a top level first and then to successively delve deeper into each lower level and more detailed treatment as the information becomes available. It is hoped that the main message that *computers do not make decisions, people do* has been recognized. The main purpose of computers here is to take the myriad pieces of data and, hopefully, information and to arrange them in logical packages suitable for digestion and understanding by the various decision makers.

To avoid cluttering the pages with many examples and variations on a theme, only top-level simple examples have been used, more to introduce the method than to solve all of the complex problems in any particular set of defense system candidates to evaluate. The method developed here is applicable to many other management fields in other than the defense field, but the concentration on missile defense serves to focus the approach without too many generalities for a more wide audience.

To emphasize that the purpose here is to present a method that serves more to help the decision maker in his choices rather than force any final solution, the method here might be called the *Management Aids Needed to Logically Evaluate* Method, or perhaps M.A.N.T.L.E. Method for short.

The method has the following characteristics:
1) System characteristics are compared in a systematic and reproducible way.
2) Mathematical singularities have been eliminated.
3) The method allows for less-precise decision making to be included.
4) Specification boundaries are included in the comparisons.
5) The decision maker (at all levels) is kept in the loop as results
   appear.
6) The method can be expanded into lower-level subsystems equally
   well.

EXAMPLE PROBLEMS

8.1 In a similar set of systems characteristics as given in example problem
     8.1, a new set of systems is to be considered. They have the following
     values for performance, cost, schedule, and risk:

System	Performance (leakage), %	Cost (procurement), $ billion	Schedule, (Years to IOC)	Risk, ($ billion × years)
System 1	9	3.3	4	0.20
System 2	9	2.1	9	1.00
System 3	2.5	2.0	3	0.30
System 4	5	4	6	0.60

Given this set of characteristics, is there a unique solution of "best"?

8.2 Suppose that in example problem 8.2, the cost value is changed from
     the stated procurement cost to the operating and support (O&S) cost.
     Assume that the operating lifetime is such that the O&S cost is the same
     numerical value as the procurement cost.

8.3 Using the same system characteristics as in Exercise 8.2, recalculate the
     best solution assuming that the O&S costs are 50% more than the
     procurement costs.

8.4 Repeat the less-precise decision-making example problem 8.5, but
     change the shaping factor in the value functions to $\alpha = \pm 1.5$ for the
     more-tolerant and less-tolerant factors. Plot out the results. Given the
     shape of the value functions, how would you word the level of
     reluctance by the decision maker in the same manner as in example
     problem 8.5?

8.5 Repeat example problem 8.5, but replace the shaping factor with
     $\alpha = \infty$. Did you get the same answer as for the constant-weighting
     method as given in example problem 8.3?

Some Ballistic Missiles Used by Different Nations

Cautionary note: The data collected in Table A1 have been collected over time from many different sources. In each line entry one parameter might be from one source, and another parameter might be from another source, which can give rise to some inconsistencies. Also, the reader will notice, as in other publications, that most values have been rounded to the nearest 10 or 100 kg or 100 km. It is unlikely that in each case all values are precisely the even values so often quoted. It is suggested, however, that statistically the large sampling will provide the necessary trends, whereas specific cases might not be precise.

Table A1 Some ballistic missiles used by different nations

Missile	Fuel	Maximum range, km	Launch weight, kg	Warhead (payload), kg[a]	Accuracy (Circle Error of Probability), m	Country of use[b]	Year in service
Condor 2	—	900	—	500	—	Argentina	Terminated
SS-1 (Scud B)	Liquid 1 stage	300	3,100	985	450	Algeria	—
MB/EE-150	—	150	—	500	—	Brazil	Terminated
SS-300	—	300	—	450	—	Brazil	Terminated
SS-600	—	600	—	500	—	Brazil	Terminated
SS-1000	—	1,200	—	—	—	Brazil	Terminated
VLS	—	5,000	—	500	—	Brazil	In development
CSS[c]2/DF-3[d]	Liquid 2 stage	2,800	64,000	2,150	1,000	China	1971–present
CSS-3/DF-4	Liquid 2 stage	4,750	82,000	2,200	1,500	China	1981–present
CSS-4/DF-5	Liquid 2 stage	13,000	183,000	3,200[e]	500	China	1981–present
CSS-6/DF-21	—	1,800	—	600	—	China	—
CSS-N-3/JL-1[f]	—	1,700	14,700	600	—	China	1983–present
DF-11/M-11	—	290	—	800	—	China	—
DF-15/M-9	—	600	—	950	—	China	—
SS-1 (Scud B)	—	300	3,100	985	450	Egypt	—
Vector	—	900	—	500	—	Egypt	Terminated
SS-21 (Scarab)	—	120	—	480	—	Czech Rep.	—
SS-1 (Scud B)	—	300	—	985	450	Czech Rep.	—
SSBS[g] S2	—	3,300	31,900	—	—	France	1971–1983
S3D	—	3,500	25,800	1,800	—	France	1980–present
SSBS S30	—	3,500	25,800	1,800	—	France	1980–present
MSBS[h] M1	Solid 2 stage	2,500	18,000	1 × 500 kT	1,000	France	1971–1974
M2	Solid 2 stage	3,000	20,000	1 × 500 kT	1,000	France	1974–1977
M20	Solid 3 stage	3,000	20,000	2,200	—	France	1977–1991

M4A	Solid 3 stage	4,000	35,000	1,200	—	France	1985–present
M4B	Solid 3 stage	5,000	35,000	1,200	—	France	1987–present
M5	—	6–11,000	—	—	—	France	In development
A-4 (V-2)	Liquid 1 stage	360	12,853	1,000	—	Germany	1944–1945
SS-21 Scarab	—	120	—	480	—	Hungary	—
SS-1 (Scud B)	Liquid 1 stage	300	—	985	450	Hungary	—
Prithvi 1	—	150	—	1,000	—	India	—
Prithvi 2	—	250	—	500	—	India	—
Prithvi 3	—	350	—	500	—	India	—
Agni II[i]	—	2,000	—	1,000	—	India	2002
Mushak 120	—	120	—	190	—	Iran	—
Mushak 160	—	160	—	190	—	Iran	—
Mushak 200	—	200	—	500	—	Iran	—
M-11	—	290	—	800	—	Iran	—
SS-1 (Scud B)	—	300	—	985	450	Iran	—
Scud C	—	550	—	500	—	Iran	—
TONDAR 68	—	1,000	—	400	—	Iran	In development
SS-1 (Scud B)[j]	—	300	—	985	450	Iraq	—
A1 Husayn[j]	—	650	—	500	1000	Iraq	—
A1 Abbas[j]	—	900	—	350	—	Iraq	—
BADR 2000[j]	—	900	—	500	—	Iraq	—
Tammuz 1[j]	—	2,000	—	750	—	Iraq	—
Lance MGM 52	—	130	—	450	—	Israel	—
Jericho 1/YA-1	—	500	—	500	—	Israel	—
Jericho 2/YA-2	—	1,500	—	1,000	—	Israel	—
Shavit	—	4,500	—	1,100	—	Israel	—
SS-1 (Scud B)	—	300	—	985	450	North Korea	—
Scud C	—	550	—	500	—	North Korea	—
NoDong 1[k]	—	1,000	—	1,000	—	North Korea	—

(Continued)

Table A1 *Continued*

Missile	Fuel	Maximum range, km	Launch weight, kg	Warhead (payload), kg[a]	Accuracy (Circle Error of Probability), m	Country of use[b]	Year in service
NoDong 2	—	1,500	—	1,000	—	North Korea	—
Taepo Dong 1[l]	—	2,000	—	1,000	—	North Korea	—
Taepo Dong 2[m]	—	3,500	—	1,000	—	North Korea	—
Hyon Mu	—	250	—	300	—	South Korea	—
SS-1 (Scud B)	—	300	—	985	450	Libya	—
SS-21 (Scarab)	—	120	—	480	—	Libya	—
Scud C	—	550	—	500	—	Libya	—
Al Fatah	—	950	—	500	—	Libya	—
HATF 1	—	100	—	500	—	Pakistan	—
HATF 2	—	300	—	500	—	Pakistan	—
HATF 3	—	600	—	500	—	Pakistan	—
M-11	—	290	—	800	—	Pakistan	—
Ghauri II[n]	—	—	—	—	—	Pakistan	2001
SS-1 (Scud B)	Liquid 1 stage	300	6,370	985	450[o]	FSU[p]	1955
SS-2 (Sibling)	Liquid 1 stage	600	20,400	—	—	FSU	1952–1960
SS-3 (Shyster)[q]	Liquid 1 stage	1,200	28,600	—	—	FSU	1956–1970
SS-4 (Sandal)[r]	Liquid 1 stage	2,000	42,000	—	2,400	FSU	1957–1991
SS-5 (Skean)	Liquid 1 stage	4,100	35,000	—	1,000	FSU	1961–1984
SS-6 (Sapwood)	Liquid 1 stage	6,200	300,000	—	8,000	FSU[p]	1959–1968
SS-7 (Saddler)	Liquid 2 stage	11,500	140,900	20–25 MT	2,800	FSU	1961–1979
SS-8 (Sasin)	Liquid 2 stage	12,500	80,000	—	2,800	FSU	1965–1977
SS-9 (Scarp)[s]	Liquid 3 stage	12,000	190,000	—	1,800	FSU	1965–1979
SS-11M-1 (Sego)	Liquid 2 stage	10,000	50,100	950 kT	1,800	FSU	1966–1988
SS-11M-2 (Sego)	Liquid 2 stage	13,000	50,100	—	1,100	FSU	1972–1994
SS-11M-3 (Sego)	Liquid 2 stage	10,600	50,100	—	1,100	FSU	1972–1989

SS-13 (Savage)	Liquid 3 stage	9,400	51,000	750 kT	1,800	FSU	1972–1996
SS-16 (Sinner)	Liquid 3 stage	9,000	44,000	—	400	FSU	1976–present
SS-17 (Spanker)	Liquid 2 stage	10,500	71,000	6 MT	400	FSU	1975–1990
SS-18 (Sickle)	Liquid 2 stage	11,000	211,100	8,800	350	FSU	1975–1990
SS-19 (Stiletto)	Liquid 2 stage	10,000	105,600	6 × 500 kT	400	FSU	1975–1985
SS-20 (Saber)	Solid 2 stage	5,000	36,000	—	400	FSU	1975–present
SS-21	—	120	480	—	—	FSU	—
SS-24 (Scalpel)	—	10,000	104,500	3,200[t]	200	FSU	1987–present
SS-25 (Sickle)	—	10,500	45,100	1,200[u]	200	FSU	1985–present
SS-N-4	—	650	13,750	1,598	463	FSU	1961–1973
SS-N-5 (Stark)[v]	Solid 2 stage	1,400	19,650	1,179	3,000	FSU	1961–1991
SS-N-6 (Serb)	Solid 2 stage	2,500	14,200	650	1,300	FSU	1975–1990
SS-N-8 (Sawfly)	Liquid 2 stage	7,800	33,300	1,100	2,400	FSU	1967–present
SS-N-8	Liquid 2 stage	9,100	33,300	1,100	1,600	FSU	1971–present
SS-N-17 (Snipe)	Solid 2 stage	3,900	26,900	450	1,400	FSU	1977–1991
SS-N-18 (Stingray)	Liquid	6,500	35,300	1,650	1,400	FSU	1979–present
SS-N-20 (Sturgeon)	—	8,300	84,000	2,550	500	FSU	1981–present
SS-N-23 (Skiff)	Liquid	8,300	40,300	2,800	500	FSU	1985–present
CSS-2	—	2,650	64,000	2,150	1,000	Saudi Arabia	—
SS-21 (Scarab)	—	120	—	480	—	Syria	—
SS-1 (Scud B)	—	300	—	985	—	Syria	—
Scud C	—	550	—	500	—	Syria	—
M-11	—	290	—	800	—	Syria	—
M-9	—	600	—	950	—	Syria	—
Green Bee	—	130	—	400	—	Taiwan	1975–present
Sky Horse	—	950	—	500	—	Taiwan	1980–present
SS-21 (Scarab)	—	120	—	480	—	Ukraine	—
SS-1 (Scud B)	—	300	—	985	450	Ukraine	—

(*Continued*)

Table A1 *Continued*

Missile	Fuel	Maximum range, km	Launch weight, kg	Warhead (payload), kg[a]	Accuracy (Circle Error of Probability), m	Country of use[b]	Year in service
Scalpel	—	10,000	—	3,200	—	Ukraine	—
Stiletto	—	10,000	—	3,200	—	Ukraine	—
SS-1 (Scud B)	—	300	—	985	—	UAE	—
Polaris A-3TK	Solid 2 stage	4,630	13,600	500	850	United Kingdom	1964–1992
Trident D5	—	12,000	59,090	2,800	90	United Kingdom	1988–present
NIKE Hercules	Solid 2 stage	160	4,500	—	—	USA	—
Minuteman 2	Solid 3 stage	11,500	31,818	2 MT nuc	560	USA	1962–1969
Minuteman 3	Solid 3 stage	14,500	34,500	3 × 200 kT	220	USA	1965–1995
Polaris A-1	—	2,600	12,700	408	1,830	USA	1960–1965
Polaris A-2	—	2,800	13,608	500	1,200	USA	1961–1974
Polaris A-3	—	4,630	13,600	500	850	USA	1964–1981
Poseidon (C-3)	Solid 2 stage	4,600	29,500	1,497[w]	460	USA	1971–1994
Trident I (C-4)	Solid 3 stage	7,400	32,850	1,500	463	USA	1979–present
Trident II (D-5)	Solid 3 stage	12,000	59,090	2,800	90	USA	1988–present
Patriot (MIM-104)	Solid 1 stage	80	700	90	—	USA	1985–present
ATACMS	Solid 1 stage	150	—	—[x]	—	USA	1990–present
HAWK (MIM-23)	—	40	627	54	—	USA	1960–present
SM-2 (RIM-67B)	2 stage	167	1,324	—	—	USA	1981–present
THAAD	—	—	—	—	—	USA	—
SM-3	—	—	—	—	—	USA	—
PAC-3	—	—	—	—	—	USA	—

[a]In some sources, this is listed as throw weight, which according to the SALT II definition includes the *warhead (reentry vehicle)*, the *postboost vehicle*, the *penetration aids*, and any *release mechanisms*. Care should be taken in using these values as exact values of warhead weight.
[b]Might not be the developer. Many missiles have been sold or transferred from the originating nation.
[c]CSS = U.S. designation for Chinese surface-to-surface; N = naval.

[d]DF = *dong feng* (east wind).

[e]Or a single 5 MT nuclear warhead.

[f]JL = *ju lang* (great wave).

[g]SSBS = *sol-sol balistique stratégique* (surface-to-surface ballistic strategic), (land based). Note that although named "strategic" by France, these missiles were deemed "tactical" intermediate-range ballistic missiles (IRBM) in the SALT documents. It depends on the mission.

[h]MSBS = *mer-sol balistique stratégique* (sea-to-ground ballistic strategic); (also see footnote g).

[i]Agni = *fire*.

[j]These have been listed even though banned by UN Security Resolution 687.

[k]Also known as *Rodong* 1.

[l]Also known as *Nodong* 3.

[m]Aso known as *Nodong* 4.

[n]Ghauri = *earth*.

[o]1959 naval version (SS-N-1) had a CEP of 8000 m.

[p]FSU = former soviet union. This includes missiles in Russia, Belarus, Bulgaria, Georgia, and Kazakhstan. Ukraine listed separately.

[q]First to carry nuclear warhead.

[r]Missile in the "blockade" of Cuba on 22 October 1962.

[s]SS-10 (Scrag) was experimental only (1968).

[t]Or 10 × 550 kT nuclear warhead.

[u]Or 1 × 550 kT nuclear warhead.

[v]First Soviet sea-launched ballistic missile to be launched from submarine while submerged.

[w]Or 10–14 × 50 kT nuclear warhead.

[x]950 M42 bomblets.

Some Cruise Missiles Used by Different Nations

Cautionary note: A similar caution is made to that listed for Appendix A on ballistic missiles in that the cruise missile data shown in Table B1 are rounded and collected from various sources with possible inconsistencies. In addition, some interpretation has been made as to the speed values quoted. In some source material the speed is quoted as high or low subsonic or supersonic or hypersonic. By comparison with similar cruise missiles where the speeds *are* quoted, it is possible to group those missiles into specific speed categories, and these are the values shown. It was also noted when plotting the cruise missile data that certain values appeared to be inconsistent with those from more reliable sources. The values quoted are as given but have not been used in setting the trends shown in the main text. In some cases prototype data published by the country of origin have been included even though the missile is not operational, and again this has not been allowed to affect the trend data. [For example, the data on the Indian cruise missile *Lakshya* have been listed as quoted by the source even though at the time of going to press it is known that the engine has not performed as expected (see Ref. 91).] As before, the aim is to collect a large enough database to provide a reasonable statistical sampling of the state of the art of cruise missile design. It has not been the intent to compile *all* cruise missiles (many other publications have already accomplished this quite well), but the intent is to seek trends in the various types to establish a state of the art.

Table B1   Some cruise missiles used by different nations

Cruise missile	Mission	Propulsion	Launch method	Range, km	Launch weight, kg	Payload, kg	Speed, km/s	Country of first use	Year in service
MQ-2 Bigua	LA[a]	TJet[b]	A/G[c,d]	900	—	70	—	Argentina	—
SM-70 Barracuda	AS[e]	SR[f]	A/G/S[c,d,g]	70	—	10	—	Brazil	—
Sy-1/HY-1	AS	LR[h]	G/S	50	2,300	513	0.24	China	1974
HY-2 (Silkworm)	AS	LR	G/S	95	—	513	0.27	China	1978
HY-3/C-301	AS	RJet[i]	A/G/S	100	—	500	0.27	China	1995
HY-4/C-201 (Sadsack)	AS	TJet	A/G/S	150	1,750	500	0.27	China	1985
FL-1	AS	LR	G/S	40	—	513	0.27	China	1980
FL-2/SY-2	AS	LR	G/S	50	—	365	—	China	1983
C-101	AS	RJet	A/G/S	50	—	400	—	China	—
C-601	AS	SR	A	95	—	500	0.27	China	1985
YJ-1/C-801	AS	SR	A/G/S	40	—	165	0.27	China	—
YJ-2/C-802	AS	TFan[j]	A/G/S	120	—	165	0.27	China	—
Exocet MM-38	AS	SR	G/S	42	725	165	0.32	France	—
Exocet MM-40	AS	SR	G/S	70	—	165	—	France	—
Armat	AS	SR	A	90	—	160	—	France	—
Apache	LA	TJet	A	150	1,300	780	—	France	1996–present
ANS	AS	RJet	S	180	—	180	—	France	—
ANL	AS	RJet	A/S	50	—	180	—	France	—
ASMP	ST[k]	RJet	A	300	—	300.kT	0.85	France	—
Apache C	LA	TJet	A	400	—	400	—	France	2001–present
V-1	LA	LR	G	250	2,279	952	0.16	Germany	1944
Kormoran 1 (AS 34)	AS	SR	A	35	600	165	—	Germany	1977
Kormoran 2 (AS 34)	AS	SR	A	60	600	220	—	Germany	1991
Lakshya	LA	TJet	G	500	—	200	—	India	—
HY-2 (Silkworm)	AS	LR	G/S	95	—	513	—	Iran	—

HY-2 Mod	AS	TJet	G/S	450	—	500	—	Iran	—
FAW 70	AS	LR	G/S	70	2,600	500	—	Iraq	—
FAW 150	AS	LR	G/S	150	2,750	500	—	Iraq	—
FAW 200	AS	LR	G/S	200	3,500	500	—	Iraq	—
Ababil	LA	TJet	A	450	1,000	500	—	Iraq	—
Gabriel II	AS	SR	S	36	—	100	—	Israel	—
Gabriel III	AS	SR	A/S	36	—	150	—	Israel	—
Gabriel IV	AS	TJet	A/S	200	—	240	—	Israel	—
Popeye	LA	SR	A	100	1,500	395	—	Israel	—
Otomat Mk 1	AS	TJet	G/S	80	800	210	—	Italy/Fr	—
Otomat Mk 2	AS	TJet	G/S	180	800	210	—	Italy/Fr	—
ASM-1	AS	SR	A	50	550	150	—	Japan	—
SSM-1	AS	TJet	G/S/Sub[d,g,l]	150	680	250	—	Japan	—
SS-N-2a (Styx)	AS	LR	S	43	—	500	—	North Korea	—
HY-1 (Silkworm)	AS	LR	S	95	—	513	—	North Korea	—
Penguin	AS	SR	A	50	—	120	—	Norway	—
Sepal	AS	TJet	G/S	450	—	1,000	—	FSU[m]	—
Starbright	AS	SR	Sub	65	—	500	—	FSU	—
Siren	AS	SR	S/Sub	110	—	500	—	FSU	—
Sandbox	AS	TJet	S/Sub	550	—	1,100	—	FSU	—
Shipwreck	AS	TFan	S/Sub	550	—	750	—	FSU	—
Sampson	ST	TFan	Sub	3,000	—	200 kT	—	FSU	—
Sunburn	AS	SR	S/Sub	110	—	500	—	FSU	—
AS-1 (Kennel)	AS	TJet	A	100	—	1,000	—	FSU	—
AS-2 (Kipper)	AS	TJet	A	120	6,000	1,000	—	FSU	—
AS-3 (Kangaroo)[n]	LA	TJet	A	650	11,000	800 kT	0.68	FSU	1960+
AS-4 (Kitchen)	LA/AS	LR	A	400	6,805	1,000	1.08	FSU	1960+
AS-5 (Kelt)	LA/AS	LR	A	180	—	1,000	—	FSU	1965

(Continued)

Table B1 *Continued*

Cruise missile	Mission	Propulsion	Launch method	Range, km	Launch weight, kg	Payload, kg	Speed, km/s	Country of first use	Year in service
AS-6 (Kingfish)	LA/AS	SR	A	400	4,500	1,000	—	FSU	1970
AS-9 (Kyle)	LA/AS	LR	A	90	—	200	—	FSU	—
AS-11 (Kilter)	LA/AS	SR	A	50	—	130	—	FSU	—
AS-15 (Kent)°	ST	TFan	A	3,000	1,500	200 kT N	0.20	FSU	1992
AS-16 (Kickback)	LA/ST	SR	A	200	—	250	1.5	FSU	—
ASM-MMS	AS	RJet	A	250	—	320	—	FSU	—
SS-N-2a (Styx)	AS	LR	S	43	—	513	—	FSU	—
SS-N-2b (Styx)	AS	LR	S	50	—	513	—	FSU	—
SS-N-2c (Styx)	AS	LR	S	85	—	513	—	FSU	—
SS-N-2d (Styx)	AS	LR	S	100	—	513	—	FSU	—
SS-N-19	AS	TJet	G/S	580	7,000	—	0.75	FSU	—
Shaddock	AS	TJet	Sub	450	—	1,000	—	FSU	—
RB 08A	AS	TJet	G/S	250	—	250	—	Sweden	—
RBS 15	AS	TFan	A/G/S	90	—	250	—	Sweden	—
Hsiung-Feng I	AS	SR	S	36	—	100	—	Taiwan	—
Hsiung-Feng II	AS	TJet	S	170	—	75	—	Taiwan	—
Sea Eagle	AS/LA	TJet	A/S	110	—	230	—	United Kingdom	—
CGM-13 (MACE)	LA	TJet	G	1,947	8,181	1,363	0.29	USA	1955–69
AGM-69 SRAM	ST	TFan	A	200	—	170 kT N	—	USA	—
Scarab-1	—	—	—	1,800	1,100	200	—	USA	—
Scarab-2	—	—	—	600	1,150	500	—	USA	—
Scarab UAV	Recce	TJet	Para	—	1,100	—	0.24	USA	—
AGM-84A Harpoon	AS	TFan	A	120	695	220	0.29	USA	—
RGM 84A Harpoon	AS	TFan	S	120	—	220	0.30	USA	1977
Harpoon Blk 1D	AS	TFan	A/S	277	—	220	—	USA	—

UGM-84A Harpoon	AS	TFan	Sub	120	—	220	—	USA	—
AGM-84E SLAM	LA	TJet	A	110	628	220	0.26	USA	1990
AGM-86B ALCM	ST	TFan	A	2,500	1,270	200 kT N	0.25	USA	1982
AGM-86B CALCM	LA	TFan	A	2,500	—	unknown	—	USA	—
AGM-129A ACM	LA/ST	SR	A	3,000	—	450	—	USA	—
AGM-131 SRAM-2	LA/ST	SR	A	400	—	289	—	USA	—
AGM-65 Maverick	AS/LA	SR	A	22	289	136	0.45	USA	1972
AGM-88 (HARM)	AS/LA	SR	A	100	366	66	0.60+	USA	1984
BGM-109A	LA/AS	SR/TF	S	2,500	1,200[p]	454[q]	0.2	USA	1982
BGM-109B	AS	SR/TF	S	464	—	—	0.25	USA	—
BGM-109C	LA	SR/TF	S	1,300	—	—	0.25	USA	—
AIM-54 (Phoenix)	AA[r]	—	A	204	450	61	1.5	USA	1974
AIM-120 AMRAAM	AA	—	A	74	157	20	1.2	USA	1991
AIM-7 (Sparrow)	AA	SR	A	56	231	39	1.5	USA	1958
AIM-9 (Sidewinder)	AA	SR	A	19	87	10	0.70	USA	1956
AGM-45 (Shrike)	AA	SR	A	40	177	66	0.68	USA	1963
Hound Dog	—	—	—	1,140	—	1 MT	—	USA	1961

[a] LA = land attack.
[b] TJet = turbojet.
[c] A = air launched.
[d] G = ground launched.
[e] AS = antiship.
[f] SR = solid fuel rocket.
[g] S = ship launched.
[h] LR = liquid fuel rocket.
[i] RJet = ramjet.
[j] TFan = turbofan.
[k] ST = strategic.
[l] Sub = submarine launched.
[m] Former Soviet Union.
[n] Largest cruise missile and only one that can be carried by aircraft Tu-95 (Bear-B).
[o] Similar to *Tomahawk*, and 16 missiles can be carried by Tu-95 (Bear-B) or 12 by Tu-160 (Blackjack).
[p] Missile alone. Additional 249 kg for booster or 454 kg capsule for submarine launch.
[q] 95–150 kT (approx 123 kg) nuclear warhead in TLAM-N version.
[r] AA = antiair.

CHARACTERISTICS OF BALLISTIC-MISSILE TRAJECTORIES

The trajectories of ballistic missiles have three distinctive phases, which can be described as 1) a powered or *boost phase*, which begins at launch and ends at thrust cutoff or burnout; 2) a *ballistic trajectory* or orbit phase that constitutes the major part of the trajectory, where the missile is moving under its own momentum (generated by the burnout velocity) and acted upon by gravity alone; and 3) a *reentry phase*, where the missile reenters the atmosphere and is subjected to aerodynamic forces until impact.

In the literature additional phases are often described to highlight some additional feature of the missile. Such additional phases include, for example, *ascent phase*, *midcourse phase*, *terminal phase*, and other designations. Such designations, although useful to illustrate some key feature of a particular missile's trajectory, are still encompassed in one or more of the preceding three main phases. These phases of the total trajectory of the missile are shown in Fig. C1. The relative size of the trajectory to the dimensions of Earth has been greatly exaggerated for illustrative purposes.

The key defining parameters are the burnout altitude h_{bo}, the burnout velocity V_{bo}, and the angle at burnout and start of the ballistic phase θ_{bo}. Additionally, h_{re} is the altitude of reentry, and r_{bo} and ν_{bo} are the polar coordinates of the burnout point. The range of the ballistic phase expressed in degrees is Ψ. The launch point is usually a given, and the impact point is determined by the other parameters. The equations for each part of the trajectory can be quite complex if all factors are rigorously included. Such factors include the nonspherical nature of Earth, the effects of Earth's rotation, the variations in the density of the atmosphere, the effects of the variations in the gravitational forces throughout the trajectory, and other complicating effects of the missile itself. Fortunately, as will be shown, several simplifying assumptions can be made (without jeopardizing the fidelity of the results) that make the analysis and understanding of the driving factors in the characterization of ballistic missile trajectories readily understandable. Each phase, *boost*, *ballistic trajectory*, and *reentry*, will be discussed in turn.

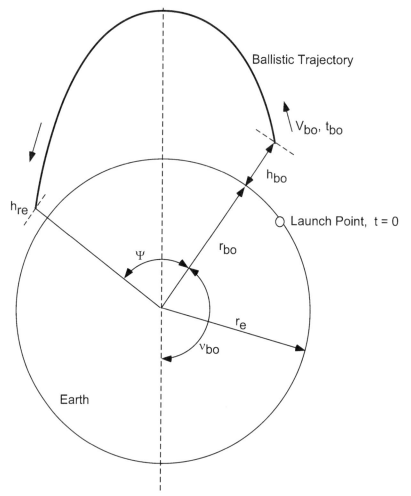

Fig. C1   Basic geometry of a ballistic-missile trajectory.

BOOST PHASE

The trajectory of the missile during the boost phase is influenced by two main factors: the acceleration imparted, measured say in g, and the angle θ of the trajectory as the missile leaves the launcher. Typically, the ballistic missile is launched at 90 deg (straight up) as it clears the launch area and then by preplanned steering guided to the angle θ_{bo} at burnout set for the desired range. [In certain battlefield or shipborne operations the missile can also be launched directly at the desired θ_{bo}. Two examples from today's systems are the ASTER missile, which is typically launched vertically ($\theta_{bo} = 90$ deg), and the Patriot, which is typically launched at the angle for best range.] How this burnout angle θ_{bo} is selected is described in the next section on the discussion of the ballistic trajectory.

The "shaping" of the trajectory angle is a matter of design, and different techniques have been used by different designers and different battlefield (or ship) needs. Figure C2 shows a typical shaping of the trajectory smoothly transitioning from 90 deg to the desired ballistic trajectory launch angle θ_{bo}.

The acceleration of the missile is determined by the propulsive thrust, the aerodynamic resistance, and the mass of the missile (which will decrease as the propellant is used). The thrust will depend on the type of propellant (solid or liquid), which varies widely among designers and users in the various nations of the world. Using the specific impulse I_{sp} as a measure of the thrust of rocket engines, then for solid propellant typical values of I_{sp} range from 165–300 s. [See standard propulsion texts for definitions of specific impulse (I_{sp}), which is the ratio of a kilogram of thrust to the kilogram-per-second propellant flow and thus quoted in seconds. Appendix E provides more details on these features of missile propulsion.] For liquid propellants I_{sp} ranges from 248 s for liquid-oxygen kerosene (With an oxygen–alcohol mixture the V-2 had a specific impulse of 218 s.) to approximately 300 s for the more difficult-to-handle fluorine–hydrazine fuel mixtures. (A good rule of thumb when using I_{sp} is to note that in SI units, where the acceleration caused by gravity at sea level is very nearly 10 m/s^2, then with an I_{sp} of 250 s, this equates directly to 2.50 km/s exhaust velocity and a thrust of

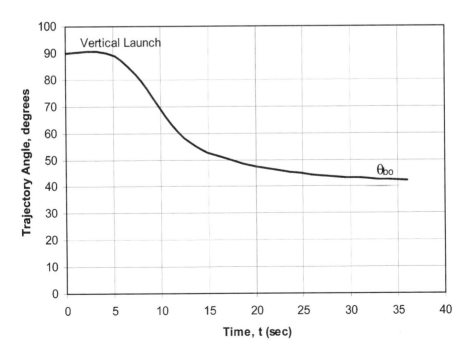

Fig. C2 Typical shaping of boost-phase trajectory.

2500 Ns/kg.) Such thrusts, coupled with typical aerodynamic drag characteristics of ballistic missiles (see later), can produce boost-phase accelerations that can vary over a wide range. Values of accelerations from $n = 2$ to 20 are common. Because there is a wide range of possibilities, it is more useful to examine typical values as determined from simple application of Newton's laws of motion to the quoted boost times and speeds of those missiles that are in use or in planned use today. Figure C3 shows the expected boost-phase burn times for a range of average accelerations. As the propellant weight is a significant part of the missile weight, the acceleration will increase as the missile accelerates and the propellant is used up. The acceleration can easily double or triple as the boost phase continues. The curves in Fig. C3 are for constant or average (over time of boost) acceleration values.

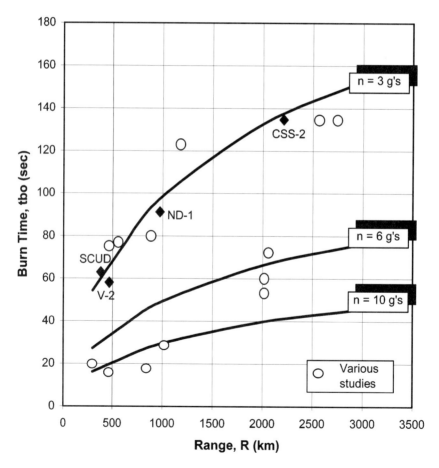

Fig. C3   Boost-phase times for various accelerations.

Superimposed on Figure C3 are the collected data points from a variety of sources on boost-phase times. It can be seen from Fig. C3 that the older designs tend to have low accelerations (around an average value through boost of $n = 3$) and the newer designs that have average accelerations of about $n = 10$. The NIAG studies used the mean value curve (corresponding to an average acceleration of $n = 6$ g) for most of the work, with excursions to slower or faster burns to determine the impact on the defensive systems. [These are a series of studies done by industry for NATO through the auspices of the NATO Industrial Advisory group. These studies included a wide range of international modeling techniques that provide a broad base for comparison. The particular group charged with the analyses of *ballistic-* and *cruise-missile* defense was subgroup NIAG SG-37 (1990–2000).] Based on these evaluations, it is seen that the boost-phase times for tactical ballistic missiles in the present and near future are most likely to be within the values shown in Table C1 (taken from Fig. C3) for a range of likely average accelerations.

The trajectories through boost in terms of the downrange x and altitude y using these typical accelerations will now depend on the shaping of the trajectory using the various steering laws already discussed. These might be a typical exponential steering where the variation of the trajectory angle θ exponentially approaches the required θ_{bo} for the start of the trajectory. The particular steering law used in Fig. C2 is a cubic steering (meaning that the trajectory-angle variation is a cubic function in time from $t = 0$ to t_{bo}), which allows the angle to remain close to vertical for initial climb out before turning over toward the final θ_{bo} value. Using such steering laws and the various acceleration values from Fig. C3 and Table C1, it is possible to prepare typical values of downrange distances and altitudes for the boost-phase portion of the ballistic-missile trajectory.

Table C1   Approximate boost-phase burn times

Missile range, km	Boost-phase times for values of average acceleration n		
	$n = 3$	$n = 6$	$n = 10$
500	60	30	20
1000	100	50	30
1500	120	60	35
2000	130	65	40
2500	145	70	42
3000	155	75	45

To indicate the variations of trajectories possible for the *boost phase*, consider the equations of motion of a ballistic missile accelerating from ignition up to burnout. The x, y components of the trajectory can be expressed by

$$x = \frac{1}{2} n g t^2 \cos \theta \qquad (C.1)$$

$$y = \frac{1}{2} g (n \sin \theta - 1) t^2 \qquad (C.2)$$

where the acceleration n is dependent upon the specific impulse I_{sp} of the missile (see Appendix E), which will normally give a variation of acceleration similar to

$$n = n_0 + n_1 \sqrt{t} \qquad (C.3)$$

where the initial acceleration n_0 and the rate of propellant burning n_1 are dependent upon the specifics of a particular design and choice of propellant. The trajectory angle θ can be varied as discussed according to the steering laws built in by the designer. It can be the more typical exponential steering.

$$\theta = \theta_0 \exp (Kt) \qquad (C.4)$$

where the shape parameter K is given by

$$K = \frac{1}{t_{bo}} \ln \left(\frac{\theta_{bo}}{\theta_0} \right) \qquad (C.5)$$

or by some other steering law such as the cubic steering shown earlier.

These variations give rise to different shapes of the (x, y) trajectory during climb out. Figure C4 shows typical endpoints of boost-phase trajectories, that is, the (x, y) coordinates of the point of burnout (or *brennschluss*) at $y = h_{bo}$. These calculations are for different, but typical, initial accelerations, burn rates, steering laws, and values of V_{bo}.

Although there is a general linear trend, it can be seen from Fig. C4 that as the burnout velocity V_{bo} increases the actual burnout point becomes more sensitive to the specific accelerations imparted to the missile and the steering law used. The particular calculations in Fig. C4 were done for boost-phase accelerations from $n = 2$ to $20\,g$. The steering laws were varied from launching at initially at 90 deg, then rotating toward the desired angle at burnout required for maximum range. It will be noticed that despite these

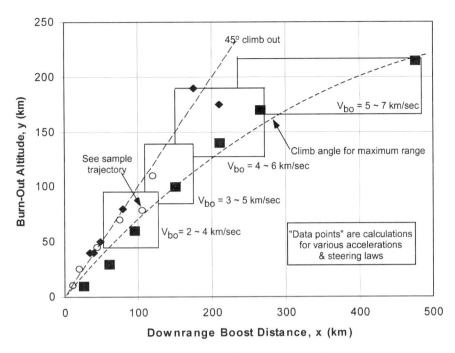

Fig. C4   Typical end of boost-phase coordinates.

variations the (x, y) coordinates of the burnout point correspond roughly as if the missile had been launched at the final burnout angle θ_{bo}. This is because even with exponential or other steering laws the rotation is usually done rapidly in the first few seconds once the missile has cleared the area. A sample boost-phase trajectory, corresponding to the sample trajectory point shown in Fig. C4, is shown in Fig. C5. This particular example is for a ballistic missile launched with an initial acceleration of $n = 2.55\,g$ and accelerated up to $n = 6.74\,g$ at a burnout velocity of 3.91 km/s and a burnout angle of 40.5 deg. As will be shown in the next section (and in Chapter 2, page 98) this corresponds to a ballistic missile launched for a maximum range of just under 2000 km. The burnout altitude (at end of boost) becomes $h_{bo} = 70$ km at a downrange point of $x_{bo} = 106$ km. The burnout time $t_{bo} = 65$ s.

The wide range of design choices of steering laws, propellant burn rates, accelerations, and launch angles makes it difficult to characterize the end of boost-phase coordinates in any precise manner. Table C2 suggests typical values based on the preceding analysis to provide an indication of "where in the sky" one can expect the boost phase to end.

Generally, depending on the steering laws, the *higher* the boost-phase acceleration, the *lower* the burnout altitude is but at a further distance

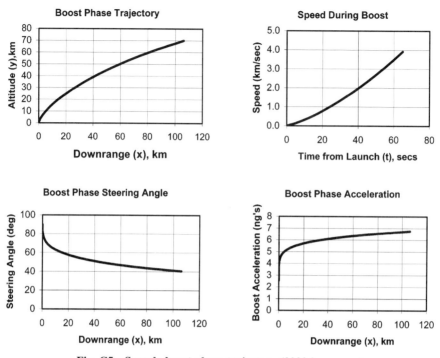

Fig. C5 Sample boost phase trajectory (2000-km range).

downrange. In the preceding compilation the lower values of burnout altitude h_{bo} correspond to average boost-phase accelerations of $n = 4-7\,g$, and the higher values correspond to average boost phase accelerations of $n = 3\,g$. The reverse is true in the case of the listed values of the burnout range, where the lower values of burnout range (x_{bo}) correspond to lower values of boost-phase accelerations. These values of the point in space at which the ballistic trajectory begins are an important parameter to establish when analyzing the ballistic phase of the trajectory.

Table C2   Approximate coordinates of end of boost phase

Missile range, km	Burnout altitude h_{bo}, km	Burnout range x_{bo}, km
500	20–40	25–75
1000	45–70	40–100
2000	70–130	75–150
3000	100–170	125–250
.	.	.
.	.	.
.	.	.
10,000	175–220	425–475

BALLISTIC-PHASE TRAJECTORIES

The second phase of ballistic-missile trajectories is the ballistic phase itself. The various parameters of interest of ballistic trajectories can be determined from at least three different approaches or viewpoints. The first approach comes from knowledge of astrodynamics and orbital mechanics using polar coordinates, the second approach comes from a similar knowledge of motion in a vertical plane over a nonrotating Earth but using Cartesian coordinates, and the third approach comes from standard treatment of the dynamics of a projectile moving over a flat Earth.

Each of these approaches yields useful insights, and, further, provides an aid in the understanding of the difference between "strategic" and "tactical or theater" ballistic missiles.

ORBITAL-MECHANICS APPROACH

The equations of motion for an object launched from the Earth—the so-called "two-body problem"—are generally used in any complete analysis of ballistic-missile trajectories. Such treatments properly take into account the effects of Earth curvature, Earth rotation, variations in the gravitational field (toward the center of the Earth), and other important factors. For those ballistic trajectories where the surface distance traveled is a significant fraction of the Earth's circumference, such factors are important. For shorter distances traveled, such "Earth factors" are less important, and the simpler equations derived from classical projectile dynamics over a flat Earth are quite accurate. (For typical ballistic-missile ranges of 3000 km or less, the distance traveled is less than 7.5% of the circumference of the Earth or less than 27 deg in polar-angle terms.)

It is known that from the basic assumption of a two-body problem the ballistic missile (the small body) must follow a trajectory that is a conic section above the Earth (the big body). The equation of a conic section, in polar coordinates, is

$$r = \frac{p}{1 + e \cos v} \tag{C.6}$$

where r is the distance from the center of the Earth (the focus of the conic section), v is the polar angle, p is the semilatus rectum, and e is the eccentricity, the value of which determines whether the trajectory is a circle ($e = 0$), an ellipse ($e > 1$), a parabola ($e = 1$), or an hyperbola ($e < 1$).

Figure C1 shows the relationship of (r, v) to the ballistic trajectory. In this analysis the semilatus rectum p of the conic section is an artifice and is not used, although it is equal to one-half the width of the conic-section trajectory taken at the focus (center of the Earth in this instance). Even without consideration of the Earth's rotation, the equations of motion that can be derived from such considerations are quite complex and are treated in many texts on astrodynamics. For this analysis the work by Bate et al.[92] will be used.

The solution to the orbital equations as applied to the case of the ballistic-missile trajectory (for the ballistic phase only) can be expressed as a *ballistic-range equation*, as follows:

$$\cos\left(\frac{\psi}{2}\right) = \frac{1 - Q_{bo}\cos^2\theta_{bo}}{\sqrt{1 + Q_{bo}(Q_{bo} - 2)\cos^2\theta_{bo}}} \tag{C.7}$$

The corollary equation to Eq. (C7) for the *trajectory launch angle* θ_{bo} is

$$\sin\left(2\theta_{bo} + \frac{\psi}{2}\right) = \frac{2 - Q_{bo}}{Q_{bo}}\left(\sin\frac{\psi}{2}\right) \tag{C.8}$$

Equations (C.7) and (C.8) provide very useful results in describing the *ballistic phase* of ballistic missiles. In the preceding equations

$$\theta_{bo} = \text{launch angle at end of boost} \tag{C.9}$$

$$Q_{bo} = \text{nondimensional parameter } \frac{V_{bo}^2 r}{\mu} \tag{C.10}$$

$$\Psi = \text{ballistic range expressed in degrees} \tag{C.11}$$

$$r = \text{radial distance to start of ballistic trajectory} = r_e + h_{bo} \tag{C.12}$$

The nondimensional parameter $Q_{bo} = V_{bo}^2 r/\mu$ is a useful parameter that relates the speed of the ballistic missile to the speed that would result in a circular orbit around the Earth [i.e., $Q_{bo} = (V_{bo}/V_{cs}^2)$], where V_{cs} is the ballistic missile speed that would result in a circular orbit around the Earth ($V_{cs} = 7.905$ km/s). (The Earth constants are mean equatorial Earth radius $r_e = 6378.145$ km and the gravitational parameter $\mu = 3.9860 \times 10^5$ km³/s².)

The range R (expressed in kilometers) of the ballistic missile can be determined from the relationship

$$R = K\Psi \tag{C.13}$$

where the range Ψ in degrees is given by Eq. (C.7) and the range R in kilometers is determined according to the orientation on the Earth's surface. The constant K varies around the oblate spheroid of Earth, but, to a good approximation, $K = 111$ km/deg (see Appendix G). Numerically, this corresponds to the range value at latitude $40°$, which is within most expected threat areas of the world.

LOFTED, DEPRESSED, AND MINIMUM-ENERGY TRAJECTORIES

Although not immediately obvious from Eq. (C.7), there are *two* values of θ_{bo} for every value of range Ψ because of the quadratic form of speed and range. This gives rise to the use of the terms "*lofted trajectories*" and "*depressed trajectories*". That is, for the same V_{bo} (or more correctly for the same value of Q_{bo}) it is possible to achieve the same range R with either a *lofted* (high) or a *depressed* (low) trajectory. Also, by launching at some intermediate launch angle it is possible to achieve a greater and *maximum range* value. It can be shown that this corresponds to the *minimum-energy* solution.

Figure C6 shows the unique relationships between the launch angle θ_{bo} (in degrees), the range R (in kilometers), and the nondimensional velocity parameter Q_{bo}, taken from Eq. (C.7), for a set of values of interest to the tactical or theater ballistic-missile designer.

It has become standard practice, in the ballistic-missile community, to consider the *lofted* and *depressed trajectories* to occur at the two-thirds *maximum range* value. This is an arbitrary choice but useful for quick calculations. The adversary might not always make such a convenient choice, of course! If the customary two-thirds maximum-range point is chosen from Fig. C6, one can construct a set of consistent ranges as shown in Table C3.

Although the nondimensional parameter Q is a very convenient parameter in the analysis of the orbit equations, it is not very convenient in the understanding of the speed of ballistic missiles. From Eq. (C.5) however, the direct relationship with speed (and V_{bo} in particular) can be seen from Fig. C7. The relationship depends on the height above the Earth at which the ballistic trajectory starts h_{bo}, but, fortunately, because the radius of the Earth $r_e \gg h_{bo}$ this "refinement", as can be seen from Fig. C7, can be safely dropped.

The effect of the height at which burnout occurs is not influential in the equation, allowing a direct relationship between speed V_{bo} and the parameter Q_{bo}. It will be noticed from Fig. C6 and Table C3 that the launch angle for maximum range decreases with increasing range. This can be seen from examining Eq. (C.3), the trajectory launch-angle equation, where it is seen that maximum range (minimum energy) must occur when

$$\sin\left(2\theta_{bo} + \frac{\Psi}{2}\right) = 1 \qquad (C.14)$$

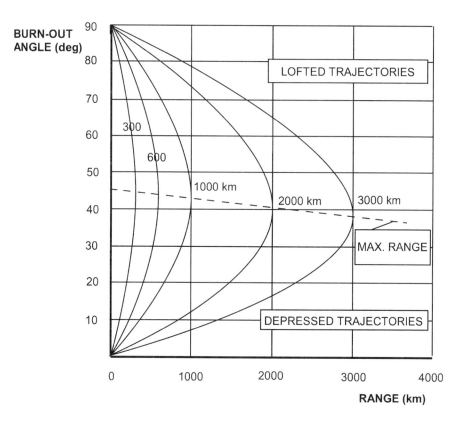

Fig. C6   Lofted, depressed, and maximum range characteristics.

Table C3   Tabulated maximum range, lofted, and depressed trajectories

	Maximum range		Two-thirds maximum range, km	Lofted θ_{bo}, deg	Depressed θ_{bo}, deg
Q_{bo}	R_{max}, km	θ_{bo}, deg			
0.0459	300	44.33	200	68.62	20.49
0.0898	600	43.65	400	68.19	20.02
0.1452	1000	42.76	666	67.60	19.41
0.2701	2000	40.51	1333	66.05	17.97
0.3780	3000	38.26	2000	64.49	16.53

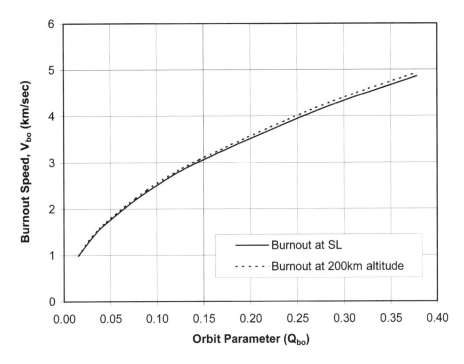

Fig. C7   Relationship of V_{bo} and Q_{bo}.

which would occur when

$$\theta_{bo} = \tfrac{1}{4}(\pi - \Psi) \tag{C.15}$$

Figure C8 shows this relationship expressed in terms of θ_{bo} (degrees) and range R (kilometers). As for the preceding relationships, this is for a spherical nonrotating Earth.

The dotted line in Fig. C8 is the solution for best burnout angle for maximum range taken from Fig. C6. Such a solution of the trajectories worked from the viewpoint of celestial mechanics necessarily provides a solution in a vacuum (no aerodynamic resistance). For simplicity, the effect of the burnout altitude h_{bo} on range and angle has been dropped [see Eq. (C.18)]. The "vacuum solution" also does not take into account the practical range considerations encompassing the boost phase part of the total range (at various accelerations) or the reentry phase contribution to range or the effects of aerodynamic lift on the missile body. To account for these effects, it is "normal" practice to decrease the burnout angle θ_{bo} slightly (by 3–4 deg) to give a better agreement with flight-test data.

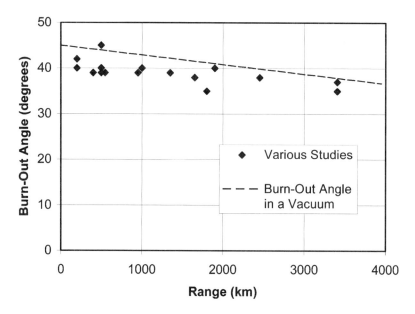

Fig. C8   Launch angle for maximum range.

RELATIONSHIP OF BURNOUT ALTITUDE, SPEED, AND MAXIMUM RANGE

The orbit equations can be further used to show some important relationships between key parameters of interest for understanding ballistic-missile trajectories. As noted earlier, the maximum range condition occurs when

$$\sin\left(\frac{\Psi}{2}\right) = \frac{Q_{bo}}{2 - Q_{bo}} \tag{C.16}$$

Although not shown here, the numerical value of Q_{bo} must always be less than one for ballistic missile trajectories and is frequently much smaller than one. Hence, it becomes possible to express Eq. (C.16) as a series:

$$\sin\left(\frac{\Psi}{2}\right) = \left(\frac{Q_{bo}}{2}\right)\left[1 - \frac{Q_{bo}}{2} + \ldots\right]^{-1} \tag{C.17}$$

Also, for ranges of interest to the tactical ballistic-missile designer Ψ is small (For example, for a missile range of 3000 km or less Ψ is < 27 deg, and sin

$\Psi/2 < 0.23$, which is $\ll 1$), such that to a good approximation

$$\Psi = \frac{V_{bo}{}^2 r_e(1 + (h_{bo}/r_e))}{\mu} \tag{C.18}$$

Rearranging Eq. (C.18) and substituting the appropriate values of the Earth constants, the more familiar form can be seen for the relationship between range R (kilometers) and launch velocity V_{bo} (kilometers/second) and burnout altitude h_{bo}:

$$V_{bo} = \frac{0.099}{\sqrt{1 + (h_{bo}/r_e)}} \cdot \sqrt{R} \tag{C.19}$$

For all practical cases for the ballistic missile (BM), the burnout altitude h_{bo} effect on burnout velocity V_{bo} can be ignored because $h_{bo}/r_e \ll 1$, such that to good approximation

$$V_{bo} = 0.09\sqrt{R} \tag{C.20}$$

Rule of thumb: *The speed of a ballistic-missile (in kilometers/second) is approximately equal to the square root of the range (in kilometers), divided by 10.*

Figure C9 shows the variation of ballistic-missile burnout velocity V_{bo} and ballistic range R. Both the flat-Earth theory and orbital-mechanics results have been included.

The comparison between results of flat-Earth theory (flat Earth can be thought of as $r_e \to \infty$) and those of orbital mechanics, shown in Fig. C9, leads to the assertion that for those BM ranges likely to be encountered in tactical missile defense (i.e., up to 3000+ km range) the flat-Earth theory gives a good representation of BM burnout speeds.

The constant ($=0.09$) in Eq. (C.20) should actually be varied depending on which part of the world is being traversed by the missile because of the nonspherical nature of the globe, but the value quoted is a good average value.

TRAJECTORIES IN A VERTICAL PLANE OVER A NONROTATING EARTH

Whereas in many ways this method is similar to the classical orbital-mechanics approach where polar coordinates were used, this method also keeps the gravitational forces acting toward the center of the Earth but uses Cartesian coordinates. This simple switch of coordinate systems allows several useful factors to be handled in a more direct and simpler manner.

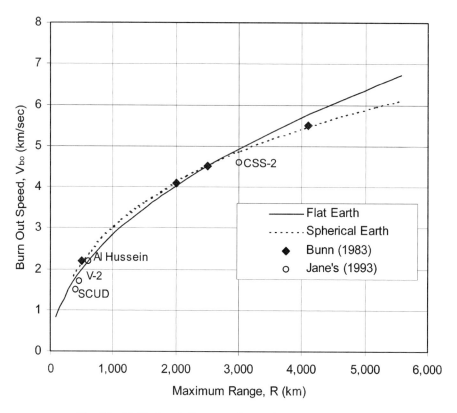

Fig. C9   Ballistic-missile range for various launch velocities.

Figure C10 illustrates the coordinate system, where again the relative size of the ballistic trajectory to the size of the Earth has been greatly exaggerated.

The equations of motion in this vertical (x, y) plane allow for a more direct and familiar representation of the forces acting on the missile. For example, if T and D represent the thrust T and aerodynamic drag D forces on the missile, then

$$\frac{\mathrm{d}V}{\mathrm{d}t} = \frac{T \cos \theta - D - mg \sin \theta}{m} \tag{C.21}$$

and the angle of the trajectory varies with time throughout flight by

$$\frac{\mathrm{d}\theta}{\mathrm{d}t} = \frac{T \sin \theta + L - mg \cos \theta}{mV} + \frac{V \cos \theta}{r_e + h} \tag{C.22}$$

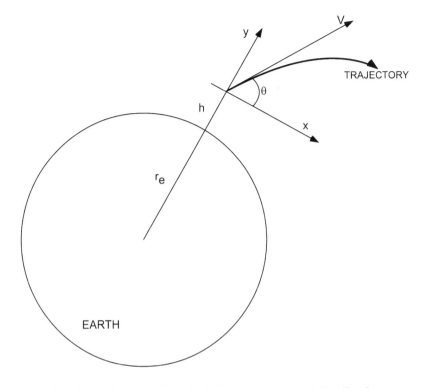

Fig. C10   Ballistic-missile trajectory over a non-rotating Earth.

where the aerodynamic lift on the missile is given by L. Also, with the assumption of a spherical Earth, it is reasonable to represent the variation in gravitational forces with altitude by the relationship

$$g = g_0 \left(\frac{r_e}{r_e + h} \right)^2 \tag{C.23}$$

where g_0 is the gravitational constant at the Earth's surface with a value of 9.806 m/s² (see Appendix G for a collection of applicable constants).

At the point that the *ballistic phase* begins (which corresponds to the end of the *boost phase*), then the thrust $T = 0$. The two remaining terms to be defined are the lift L and drag D of the missile. The classical approach from aerodynamics is to express these forces in the following manner:

$$L = C_L \tfrac{1}{2} \rho V^2 S \tag{C.24}$$

$$D = C_D \tfrac{1}{2} \rho V^2 S \tag{C.25}$$

where C_L and C_D are aerodynamic coefficients based on a reference area (usually the base area in the case of missiles) and vary with the angle of attack α (not shown). The density of the atmosphere ρ varies with altitude in a manner that has been found to be approximately exponential and dependent upon the absolute temperature T and the universal gas constant $\Re$ through the formula

$$\rho = \rho_0 \exp\left(\frac{-g_0 h}{\Re T}\right) \tag{C.26}$$

where the air density at the Earth's surface (on a standard day) is $\rho_0 = 1.226 \text{ kg/m}^3$. This so-called *exponential model atmosphere* can be represented on a standard day with the constant $k = g_0/\Re T = 0.1378$ per km. Appendix D provides more details on the representation of the atmospheric density as it varies both with altitude and with temperature.

The preceding equations of motion for the ballistic phase do not lend themselves well to a closed-form solution, but solutions can be easily performed by numerical integration on a high-speed computer. The short period of time that the ballistic missile spends in the atmosphere during the ascent phase shortly after end of boost means that the lift L and drag D forces (especially if the missile is moving along the trajectory at zero angle of attack) have little affect on the trajectories and act as if conducted in the vacuum of space. The main effect of the lift and drag forces is in the *reentry phase*, which is discussed in the next section.

In what follows, it has been taken that the initial conditions for the ballistic phase are the end of boost-phase coordinates in the two-dimensional space, time as (x_{bo}, h_{bo}, t_{bo}). As was shown in the preceding section on the boost phase, these initial conditions can vary widely, depending upon steering-law assumptions and propellant burn rates and accelerations. Figure C3 shows typical values for state-of-the-art ballistic missiles. For the solution of the ballistic-phase trajectories presented here, the *initial conditions* shown in Table C4 for the start of the ballistic trajectory have been assumed.

Table C4   Initial conditions for ballistic phase

Missile range (nominal), km	Burnout range x_{bo}, km	Burnout altitude h_{bo}, km	Burnout time t_{bo}, s
5000	25	20	33
1000	46	42	45
2000	79	69	60
3000	124	103	80

These values correspond approximately to the mean values discussed in the preceding section where variations occur in boost phase depending upon designer settings. Although it is important to set these values, they do not significantly affect the general characteristics of the ballistic-phase trajectories in range and flight time. The largest effect is in the apogees of the trajectories, and the reader can determine immediately the effect by comparing the trajectories computed and the various altitudes shown in these figures and tables. The intent here is to show the driving parameters in all cases and to show that specific designs for particular purposes follow from these typical results.

Figures 4.6–4.10 in Chapter 4 show the (x, y) coordinates of these typical ballistic-missile trajectories. Although not shown here, these trajectories correspond with those computed from the orbital-mechanics approach shown earlier.

FLAT-EARTH THEORY

It is of interest to include here a simpler treatment of understanding ballistic-missile trajectories that can be found in college textbooks. It provides some useful insight and simple formulas that can be applied to the general understanding of ballistic-missile trajectories. This simpler treatment analyzes the dynamics of a projectile launched from some point in space over a flat Earth. For the ballistic missile the point in space corresponds to the coordinates (x_{bo}, h_{bo}) and the point of thrust cutoff (see Table C4 for typical values). The projectile is now launched at a speed V_{bo} at an angle θ_{bo}. The difference between such a flat-Earth theory and that treated earlier over a spherical Earth, and again using Cartesian coordinates, is that in this case the gravity acts perpendicular to the ground and not toward the center of the Earth. Figure C11 shows the basic features for a ballistic missile launched at the end of boost over a flat Earth.

The equations of motion are (ignoring aerodynamic effects)

$$\frac{dx}{dt} = V_{bo} \cos \theta_{bo} \tag{C.27}$$

$$\frac{dy}{dt} = V_{bo} \sin \theta_{bo} - gt \tag{C.28}$$

Time t can be eliminated from the preceding equations to yield the simple equation for the trajectory after the end of the boost phase:

$$y = \left(\frac{-g}{2V_{bo}^2 \cos^2 \theta_{bo}}\right)x^2 + (\tan \theta_{bo})x + h_{bo} \tag{C.29}$$

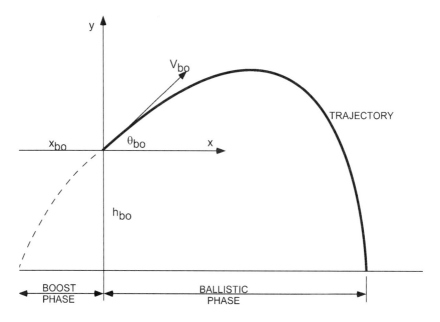

Fig. C11   Ballistic-missile launched over a flat earth.

which shows that the trajectory is a parabola from the end of boost phase (x_{bo}, h_{bo}) and does have a closed-form solution.

The maximum range can be determined from Eq. (C.29) by setting $y = h_{bo}$ such that

$$R = \frac{V_{bo}^2 \sin 2\theta_{bo}}{2g}\left(1 + \sqrt{1 + \frac{4gh_{bo}}{V_{bo}^2 \sin^2 \theta_{bo}}}\right) \qquad (C.30)$$

where for most cases of consideration for tactical ballistic missiles the term containing the effect of h_{bo} can be ignored such that, to a good approximation (It is actually the range at altitude h_{bo}.)

$$R = \frac{V_{bo}^2 \sin 2\theta_{bo}}{g} \qquad (C.31)$$

which provides a maximum range when $\theta_{bo} = 45$ deg independent of the range. It has already been shown for the more exact equations that the burnout angle for maximum range actually decreases with increase in range when proper orientation of the gravity vector is taken into account. However, by adjusting for the burnout angle in the preceding equation the relationship between burnout speed and range becomes

$$V_{bo} = 0.09\sqrt{R} \tag{C.32}$$

which matches the value given before using the more sophisticated approaches provided earlier for ballistic trajectories over a spherical nonrotating Earth.

Two other characteristics of the ballistic trajectories that can be easily calculated using this simple flat-Earth theory are the apogee y_{max} and the time of flight (TOF). These are derived from the preceding equations to give

$$\text{Apogee } (y_{max}) = \frac{(V_{bo} \sin \theta_{bo})^2}{2g} + h_{bo} \tag{C.33}$$

$$\text{TOF} = \frac{V_{bo} \sin \theta_{bo}}{g} \left(1 + \sqrt{1 + \frac{4gh_{bo}}{V_{bo}^2 \sin^2 \theta_{bo}}} \right) \tag{C.34}$$

The potential-energy-ratio term in the square root can be safely dropped for quick calculations (ignoring such terms introduces less than 10% error) such that the TOF can be approximated as

$$\text{TOF} = \frac{f(h_{bo})V_{bo} \sin \theta_{bo}}{g} + t_{bo} \tag{C.35}$$

[This approximation for the total time of (ballistic) flight is actually the value of the flight time to the point along the trajectory that occurs at altitude $y = h_{bo}$, or twice the time to apogee.]

Without the small correction caused by the h_{bo} effect, the function $f(h_{bo})$ is approximately equal to a constant value of two.

It is sometimes more convenient to express these characteristics in terms of the range R of the ballistic missile, which when including the approximation that drops the square-root term gives the following simple formulas:

$$\text{Apogee } (y_{max}) = \frac{R \tan \theta_{bo}}{4} + h_{bo} \tag{C.36}$$

$$\text{TOF} = \sqrt{\frac{f(h_{bo}) \tan \theta_{bo}}{g}} \sqrt{R} + t_{bo} \tag{C.37}$$

which shows that the apogee of the ballistic portion of the trajectory is directly proportional to the range and that the TOF is proportional to the

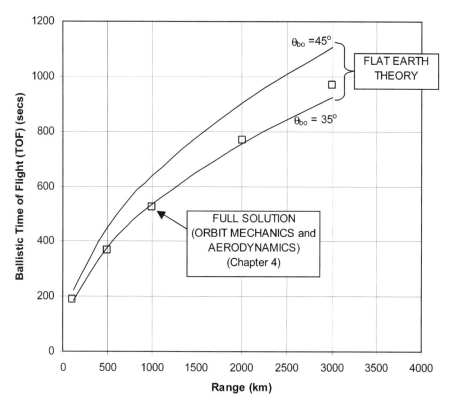

Fig. C12   Time of flight for maximum-range TBM.

square root of the range. For the more correct analyses (using orbital mechanics of trajectories over a spherical, nonrotating Earth) the relationships are a little more cumbersome, but the general trend is the same.

Figure C12 shows the relationship for TOF using the simple flat-Earth theory for two different burnout angles ($\theta_{bo} = 45$ and 35 deg) compared to the results from the more complete analyses (using orbital mechanics) provided earlier.

It is seen that the characteristic square-root relationship is preserved, and that if the lower burnout angles associated with the more correct orbital mechanics approach are used a good representation can be achieved.

For quick calculations the TOF of maximum-range (minimum-energy) ballistic missiles can be represented by

$$\text{TOF (seconds)} = 14\sqrt{R} \qquad\qquad (\text{C.38})$$

Table C5   Ballistic trajectory phase approximate flight times

Maximum range, km	Ballistic flight time, s	Ballistic flight time, min	Boost-phase time, mins (average $n = 6\,g$)	Approximate total TOF, min
120	153	2.6	—	—
300	243	4.0	—	—
500	313	5.2	0.50	5.70
1000	443	7.4	0.83	8.23
1500	542	9.0	1.00	10.00
2000	626	10.4	1.08	11.48
2500	700	11.7	1.17	12.87
3000	767	12.8	1.25	14.05

Rule of thumb: *The TOF of a maximum-range (minimum-energy) ballistic missile (in seconds) is approximately equal to 14 times the square root of the range (in kilometers).*

For typical ranges of ballistic missiles, Table C5 summarizes the flight times. (The equations to predict flight times from an orbital-mechanics approach tend to be cumbersome, and because the numerical values are reasonably predicted by the flat-Earth theory it has been chosen to list just these values for the purposes of decision making sought here.)

It is emphasized that these simplistic results can only be indicative of total flight times because of several factors. First, it is only for the ballistic phase of the trajectory; hence, the time for boost must be added, and a column has been added to Table C5 that shows the boost-phase times taken from Table C1 for a typical case of boost-phase acceleration ($n = 6\,g$). This provides some indication of expected total TOF for a ballistic trajectory, but the reentry phase as treated here assumes no drag ($D = 0$) and thus gives a shorter time of reentry. The actual TOF when the drag effects during reentry are taken into account will be slightly longer (up to a minute or more for the longer range BM). In the next section on the reentry phase, the effect of drag is properly taken into account. The main purpose of Table C5 is to emphasize that the flight times of ballistic missiles are very short, from less than 3 min up to about 15 min for those missiles expected in the near future. Such short flight times will stress any defense against them.

The apogee of ballistic-missile trajectories operating under maximum-range (minimum-energy) conditions using the preceding relationships is shown in Fig. C13, where again a comparison is made between the simple flat-Earth theory and results using the more complete orbital-mechanics approach.

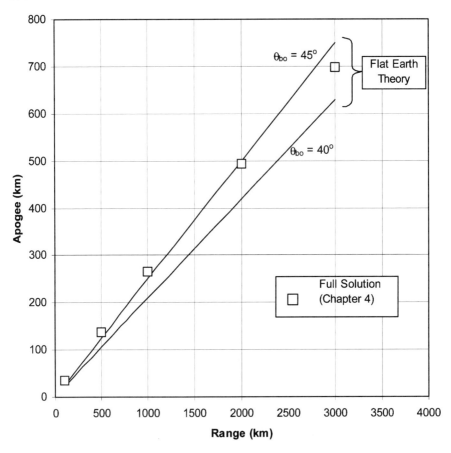

Fig. C13 Apogee of maximum-range ballistic missiles.

It will be noticed from Fig. C13 that to a reasonable approximation the apogee of minimum-energy (maximum-range) ballistic trajectories can be represented by

$$\text{Apogee} = \frac{1}{4}R \qquad (C.39)$$

Rule of thumb: *The apogee of a maximum-range ballistic-missile (in kilometers) is approximately equal to the range (in kilometers), divided by 4.*

To place these values into perspective, typical values of apogee of ballistic missiles can be seen from the following chart (Table C6), where familiar altitudes are provided for comparison.

Table C6 shows that the altitudes associated with ballistic missiles (even those considered to be short range) extend to extreme values when compared

Table C6   Characterization of ballistic missile apogees

Missile maximum range, km	Missile trajectory apogee, km	Atmospheric regions	Other altitudes for comparison
120	24	Troposphere	Mount Everest: approximately 9 km
			Airplane altitude record: 25 km
300	60		
500	100	Stratosphere	
			Aurora Borealis: 160 km
1000	200		V-2 tests at White Sands: 250 km
1500	300		
2000	400	Ionosphere	WAC Corporal tests: 400 km
2500	500		
3000	600	Exosphere	Low-Earth-orbit satellite: 1000 km

to say aerodynamic vehicles known today. The missiles under consideration operate in regions from the troposphere (0 to 25 km) to well into the exosphere ($>$ 500-km altitude).

Using the burnout angles results provided earlier for lofted and depressed trajectories, the typical values for TOF and apogees in Table C7 have been calculated using both flat-Earth theory and orbital-mechanics analyses.

SOME RULES OF THUMB FOR BALLISTIC TRAJECTORIES

Given the preceding simple relationships and the comparison with the results of the more complete analyses done in the earlier sections, it is useful to collect the various relationships for quick appreciation of the order of magnitude of the key parameters in characterization of ballistic-missile trajectories. Table C8 summarizes the various results.

It should be emphasized in Table C8 that these values are only approximations for the purposes of quick calculations to determine the order of magnitude of the key parameters of speed, apogee, and time of flight of ballistic missiles. The values shown take into account the variations of launch angle θ_{bo} and other features that have been given in more detail in the various equations shown earlier and more correct analyses from computer simulations. If the reader is uncomfortable with some of the approximations summarized in Table C8, the complete equations already provided should be used.

Table C7   TOF and apogees for maximum range, lofted and depressed trajectories

Maximum range			Lofted trajectory			Depressed trajectory		
R (km)	TOF (mins)	Apogee (km)	R (km)	TOF (mins)	Apogee (km)	R (km)	TOF (mins)	Apogee (km)
120	2.6	30	80	4.9	52	80	1.9	8
300	4.0	70	200	7.6	128	200	2.9	19
600	5.7	145	400	10.6	250	400	4.1	36
1000	7.4	265	666	13.5	404	666	5.2	59
2000	10.4	495	1333	18.4	750	1333	7.0	108
3000	12.8	695	2000	21.8	1048	2000	8.2	148

Table C8   Some collected rules of thumb for ballistic trajectories

Parameter	Minimum-energy trajectory	Lofted trajectory[a]	Depressed trajectory[b]
Range (km)	R	$2/3\,R$	$2/3\,R$
Burnout speed V_{bo} (km/sec)	$0.09\sqrt{R}$	$0.09\sqrt{R}$	$0.09\sqrt{R}$
Apogee (km)	$1/4\,R$	$1/3\,R$	$1/20\,R$
Time of flight (secs)	$17\sqrt{R}$	$24\sqrt{R}$	$9\sqrt{R}$

[a]The reader should be careful to note that all of the values quoted for *lofted* trajectories use the range value corresponding to its related *minimum energy* range value of R.
[b]The same comment for *lofted* trajectories applies to the *depressed* trajectory values.

REENTRY PHASE

The third and final phase of the ballistic-missile trajectory is the reentry phase. This is most conveniently analyzed using the Cartesian coordinate system of a trajectory over a nonrotating Earth as before, except that now the effects of aerodynamic resistance can be easily incorporated. The equations of motion have already been given in Eqs. (C.7), (C.21), and (C.22) in the discussion of the ballistic phase, except that now the effects of the air density and aerodynamic drag will be included.

As a rule of thumb, numerically, the air density at the (mean) tropopause at $h = 10$ km is about one-fourth of the value at sea level. At the height considered near the limit of aerodynamic sustainable flight (say, 25 km), the air density has fallen to approximately one-thirtieth of that at sea level. These numerical values have a significant impact on the relative importance of the missile characteristics during reentry.

During the *boost phase* and in the *ascending ballistic-trajectory* phase, the aerodynamic effects have little influence on the trajectory. During *reentry*, however, the geometry and angle of attack of the incoming missile (or separated warhead) have a strong influence on the shaping of the trajectory and hence any defense against them.

The drag usually has the greatest effect and is typically treated through the use of a *ballistic coefficient* β, where

$$\beta = \frac{W}{C_D S} \tag{C.40}$$

(The effect of lift on the aerodynamic glide reentry for, say, countermeasure purposes is treated elsewhere in this book.)

The ballistic coefficient is a convenient way to express the resistance effect caused by both weight W and drag $C_D S$ of the incoming missile or warhead. Values of ballistic coefficient β for representative weights and drag areas $C_D S$ are shown in Table C9. [Some writers prefer to quote β in metric units of force that is, in Newtons/square meter. The reader can multiply the quoted values of β given here by g ($=9.806 \text{ m/s}^2$) to obtain numerical values in Newtons/square meter.]

Values of ballistic coefficients, $\beta = 2000–5000 \text{ kg/m}^2$, for *tactical missiles* are common and are used in many simulations that analyze the effects of incoming ballistic missiles (single or multiple stage). *Strategic* missiles, being much larger, typically have ballistic coefficients, $\beta = 5000–15,000 \text{ kg/m}^2$.

For example, for the SCUD B with a diameter of 0.88 cm, a weight of 985 kg and an assumed drag coefficient of $C_D = 0.10$ (see Figure C14), this would give a ballistic coefficient of $\beta = 4048 \text{ kg/m}^2$. The Trident missile, on

Table C9 Ballistic coefficient for typical weights and drag areas

Drag area $C_D S$, m^2	Missile or warhead weight, kg						
	10	20	50	100	200	500	1000
0.01	1000	2000	5000	10,000	20,000	50,000	100,000
0.02	500	1000	2500	5000	10,000	25,000	50,000
0.10	100	200	500	1000	2000	5000	10,000
0.20	50	100	250	500	1000	2500	5000
0.50	20	40	100	200	400	1000	2000
1.00	10	20	50	100	200	500	1000
1.50	7	13	33	67	133	333	667
2.00	5	10	25	50	100	250	500

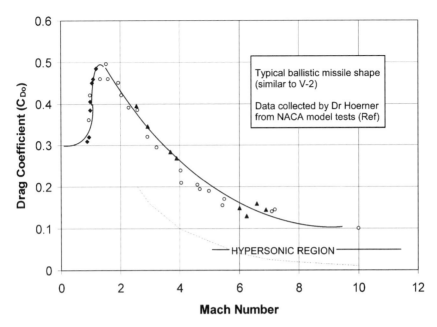

Fig. C14   Variation of drag coefficient with speed for a typical ballistic missile.

the other hand, with a diameter of 2.1 m, a payload weight (second stage) of 2722 kg, and again with an assumed $C_D = 0.10$ would have a ballistic coefficient in the terminal phase of $\beta = 7826 \, \text{kg/m}^2$.

The drag coefficient C_D varies with Mach number, but at the hypersonic speeds associated with ballistic missiles it tends to be a constant. Figure C14 shows a typical variation of C_D with speed for a standard symmetrical missile shape, which tends to support the assumption of $C_D = 0.10$ in the preceding examples.

This particular variation of drag coefficient C_D with Mach number has been reproduced from Hoerner,[93] who reproduced the results of NACA free-flight tests (Fin-stabilized missile configuration flight tests,[94]) of a missile configuration flown at hypersonic speeds (up to Mach 10 or approximately 3 km/s). Today's ballistic missiles have values of drag coefficients not too dissimilar from this characteristic variation with speed. After the transonic drag rise the drag coefficient characteristically falls to a constant value at hypersonic speeds ($M > 5$; $V > 1.5 \, \text{km/s}$).

The deceleration equation of the incoming ballistic missile (after thrust cutoff) can be rewritten to show the effect of the ballistic coefficient directly, such that

$$\frac{dV}{dt} = -g\left(\frac{\rho V^2}{2\beta} + \sin\theta\right) \tag{C.41}$$

which shows that the higher ballistic coefficient (lower drag) provides a slower deceleration. The set of computations for reentry trajectories for a range of ballistic missile trajectories for $100-3000$ km class missiles, using a typical constant value of the ballistic coefficient of $\beta = 5000\,\text{kg/m}^2$, is provided in Chapter 4.

In all of the trajectories considered in this book, it should be emphasized that the case of lifting, reentry bodies (that is, $L \neq 0; D(\alpha) \neq 0$)) has not been considered. It has been assumed that the incoming missile or its reentry vehicle is incoming at a zero angle of attack with no lift or induced drag. The quoted value of the ballistic coefficient β is for a body with a profile (and base) drag only.

STANDARD ATMOSPHERE

GENERAL DESCRIPTION OF THE EARTH'S ATMOSPHERE

Depending on the amount of time that the ballistic missile spends in the atmosphere, both during the boost phase and the terminal phase in particular, the characteristics of the trajectory can be significantly altered by the effects of air density. This effect is quite separate from the effects of the drag through the ballistic coefficient.

As might be expected, the characteristics of the Earth's atmosphere around the globe are far from uniform in both dimensions and in chemical makeup. There have been many changes in the description of an acceptable "standard atmosphere" since the original work by Diehl in 1922 (Ref. 95) and later in 1925 (Ref. 22). In that original work Diehl assumed an atmosphere with a uniform decreasing temperature with increase in altitude. Since that time, with the advances in upper-atmosphere research with improved measurement capabilities, there have been several revisions and new definitions of the characteristics of the atmosphere. After Diehl's assumption of one layer, various official organizations have created tables based on one, four, six, and now seven separate layers, each with different characteristics within the general description of "atmosphere."

In the 1950s the Earth's atmosphere was normally treated as being made up of four layers with the characteristics of each layer changing gradually between the layers and nonuniformly around the globe. The names given to these layers are *troposphere*, *stratosphere*, *ionosphere*, and *exosphere*. The lower layer, closest to the surface of the Earth, is called the *troposphere* and varies in thickness (height) from about 16 + km at the equator to about 8.5 + km at the poles. At the midlatitudes the *tropopause* (the "top" of troposphere or "boundary" between the *troposphere* and the *stratosphere*) is at about 11 km. It is this value that is used in most analyses for ballistic-missile use. The next layer, the *stratosphere*, also varies in height around the globe. The *stratosphere* reaches to heights of about 96 km to greater than 112 km. The next layer, the *ionosphere*, that occurs above about 100+ km is characterized by free electrical charges. The familiar *aurora borealis* (northern hemisphere) and *aurora australis* (southern hemisphere) occur in this layer near the poles and give rise to spectacular displays of light. The *ionosphere* has four sublayers (called the D, E, F, and F_2 layers) with different features that affect, among other things, radio transmission. [D and E layers reflect broadcast waves; F_1 layer reflects short radio waves (during day); and F_2 layer reflects short radio waves (during night)]. The outermost layer of the atmosphere, called the *exosphere*, reaches to altitudes of about 600–1000 km and more, but its makeup is composed of relatively free floating gaseous particles that do not interact with each other and thus have virtually no effect on the characteristics of ballistic missile trajectories. The upper reaches of the *exosphere* contain a high concentration of ozone that absorbs solar ultraviolet radiation, which gives rise to high temperatures that rise to over 2000°C (not read on a standard thermometer scale because of the extremely low particle density at these altitudes).

In 1953 a U.S. Committee on Extension to the Standard Atmosphere (COESA) was established. This led to the various versions that have been used by different groups for various aeronautical purposes including those involved in missile design, test, and operation. These various definitions and tabular data have appeared as the 1958, 1962, 1966, and 1976 U.S. Standard Atmospheres[96] and revised again most recently in 1982 (Ref. 97). Various organizations have participated in the development of the standard atmosphere. They are NASA, the National Oceanic and Atmospheric Administration (NOAA), the U.S. Air Force, and the International Civil Aviation Organization (ICAO). Each organization has published their versions of the data collected by balloons, aircraft, rockets, and satellites. There are slight variations in these various documents of the atmosphere characteristics as a result of the assumptions used in reducing the raw collected data to tabular form. Most of these differences have an insignificant impact on the analysis of ballistic-missile trajectories and are in agreement in the key layers of most interest for ballistic-missile analyses. For example, below 32-km altitude the tabulated data from both the U.S. Government Printing Offices (GPO) issued *1976 Standard Atmosphere* and that of the ICAO are identical. As shown in Chapter 4, it is the region below about 30 km altitude that has the greatest effect on the missile trajectory.

The atmospheric parameter of most interest in trajectory calculations is the air density ρ, and several revisions in the standard atmosphere tables are of interest. In the 1976 U.S. Standard Atmosphere and the 1982 ICAO Standard Atmosphere, the boundary of the second layer (from the Earth's surface) was adjusted from 25 km as published in the 1958 issue to 20 km in the 1976 and 1982 issues. When comparing trajectory analyses that use the various issues of the standard atmosphere, these changes should be recognized. The 1976 (and 1982) versions of the standard atmosphere are used in this book.

Table D1 Layers in the standard atmosphere

Layer	Name	Lower bound, km	Upper bound, km
1	Troposphere	Sea level	11
2	Stratosphere	11	20
3	—	20	32
4	—	32	47
5	—	47	51
6	Mesosphere	51	71
7	—	71	86
—	Ionosphere	Above about 100 km	
—	Exosphere	Upper edges of atmosphere vary from 500–1000-km altitude	

The 1976 Standard Atmosphere processes the collected data by using idealized, steady-state conditions and the assumption of perfect gas theory. In the latest issues, seven distinct layers have been established, as shown in Table D1.

Included in Table D1 are the ionosphere and exosphere as used in earlier publications. It is seen from Table D1 that the refinements in the layers obtained from the collected data are essentially refinements in the description of the *stratosphere* and now include a *mesosphere* as the sixth layer (51–71-km altitude).

The main concern in this book is the calculation of trajectories that are directly influenced by the representation of the air density and aerodynamic resistance. For completeness, however, it should be noted that other characteristics of the standard atmosphere such as electromagnetic wave propagation, scattering, radiation, emission, sky spectral radiance, refractive indices for air, ice, and water, etc. have been published in other documents. One informative source is that published by the National Geophysical Data Center.[98]

IMPACT ON PRESSURE, TEMPERATURE, AND DENSITY

The layers of the atmosphere and their chemical makeup have different characteristics in terms of pressure, temperature, and density. Of these, the pressure of the atmosphere is the most regular in its characteristics, decreasing exponentially from about 101,325 Pa ($10,332 \text{ kg/m}^2$) at sea level on a standard day of $288.15°\text{K}$ ($15°\text{C}$) to about 22,632 Pa (2308 kg/m^2) at the tropopause (11 km). At the outer reaches of the stratosphere (80 km), the air pressure drops to about 1/34,000th that at sea level.

On the other hand, the temperature and density variation throughout the layers of the atmosphere is far from uniform and complicates the formulation of ballistic missile trajectories. From the assumption of a perfect gas and dry air, it is known that the air density is given by

$$\rho = \frac{p}{\Re T} \qquad \text{(D.1)}$$

where p is the air pressure, T is the absolute air temperature, and $\Re$ is the universal gas constant, which for dry air has the value 287 m^2/s^2 per degree Kelvin. This equation is used in the standard atmosphere tables to calculate the air density. Figure D1 shows the values of the temperature measured at the various altitudes. For convenience, the 1976 U.S. Standard Atmosphere temperature has been plotted in degrees Celsius, using the absolute zero as $T_0 = -273.15°C$.

Figure D1 summarizes the temperature variation through the atmosphere up to the outer reaches as provided in the published references. This figure shows clearly the change in temperature (and thus air density) in each of the seven layers. Of particular interest is the set of three layers closest to the Earth's surface that are bounded respectively by the following:

1) Layer 1 (troposphere): Sea level to 11 km (tropopause)
2) Layer 2: >11–20 km
3) Layer 3: >20–32 km

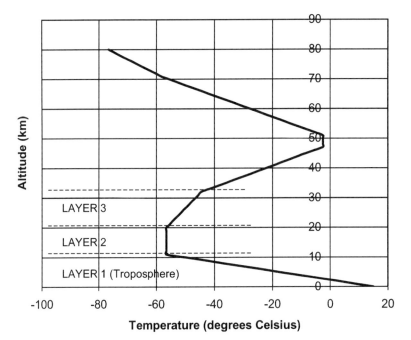

Fig. D1   Temperature variation through the atmosphere.[96]

It is in these layers where the air density has the greatest impact on the trajectories of ballistic missiles, both during the early trajectory phases (boost and ascent) and the terminal phases.

It is seen from Fig. D1 that the temperature varies widely throughout the atmosphere, and only those values up to about 80 km altitude have been shown. These values are for a standard normal day with the temperature at sea level of 15°C. On a standard hot day these values would be different. Because the speed of sound is directly proportional to the square root of the absolute temperature, then the speed of sound will also vary with altitude, and when speeds are quoted in Mach-number terms this causes a variation as discussed in Chapter 2 for the airbreathing threat of cruise missiles (CM). Because there is no appreciable air density above about 30 km altitude, it is convenient to note, using the temperature values shown in Fig. D1, that the speed of sound (on a standard normal day) at sea level is 0.340 km/s and speed of sound at tropopause is given by 0.295 km/s.

From List[99] in layer 1 (below the tropopause) the standard temperature lapse rate is 6.5 deg per 1000 m of altitude.

The variation of air density throughout the atmosphere varies in a manner that is not amenable to a simple formulation for all altitudes if one wishes to match the values of the various quoted texts on the standard atmosphere. However it is possible to use simple exponential forms that give reasonable approximation. Diehl[22] showed that it was possible to express the variation in air density ρ by the formulation

$$\rho = \rho_0 e^{-h/H} \tag{D.2}$$

where ρ_0 is the value at sea level, which on a standard day is 1.225 kg/m³. The air density ρ at altitude h is given by Eq. (D.2). Because the air density does not vary exactly in an exponential manner, the constant H, called a "scale height," is often adjusted for different altitude layers to make the exponential formula fit the published standard atmosphere table values. In SI (Le Système Internationale d'Unités, System of International Standard Units) units, the scale height H varies from about 6–9 km to adjust the results to match the standard atmosphere values. Because the air density has so little effect in the upper layers of the atmosphere, it is not necessary to be precise above the third layer. For the purposes of ballistic-missile trajectory simulation, a good representation of the air density in the first three layers is sufficient.

Table D2 shows the pertinent values of temperature $T(°K)$ and air density ρ taken from the 1976 U.S. Standard Tables together with exponential approximations for those same layers. The exponential atmosphere approximations used in each of the three layers are shown in Table D3.

Table D2   Characteristics of 1976 Standard Atmosphere and exponential approximations

1976 U.S. Standard Atmosphere			Exponential approximations			Ratio of exponential to standard atmosphere
Altitude h, km	Temperature, °K	Density, kg/m^3	Layer 1 (troposphere)	Layer 2	Layer 3	
0	288.15	1.2250	1.2250	—	—	
1	281.65	1.1116	1.0981	—	—	
2	275.15	1.0065	0.9845	—	—	
3	268.65	0.9091	0.8826	—	—	
4	262.15	0.8191	0.7912	—	—	
5	255.65	0.7361	0.7093	—	—	Average 98%
6	249.15	0.6597	0.6357	—	—	
7	242.65	0.5895	0.5700	—	—	
8	236.15	0.5252	0.5110	—	—	
9	229.65	0.4663	0.4581	—	—	
10	223.15	0.4127	0.4107	—	—	
11	216.65	0.3639	0.3682	—	—	
12	216.65	0.3108	—	0.3112	—	
13	216.65	0.2655	—	0.2665	—	
14	216.65	0.2268	—	0.2282	—	
15	216.65	0.1937	—	0.1955	—	
16	216.65	0.1654	—	0.1674	—	Average 101%
17	216.65	0.1413	—	0.1433	—	
18	216.65	0.1207	—	0.1228	—	
19	216.65	0.1031	—	0.1051	—	
20	216.65	0.0880	—	0.0900	—	
21	217.65	0.0749	—	—	0.0770	
22	218.65	0.0637	—	—	0.0652	
23	219.65	0.0543	—	—	0.0552	
24	220.65	0.0463	—	—	0.0467	
25	221.65	0.0395	—	—	0.0395	
26	222.65	0.0337	—	—	0.0335	Average 99%
27	223.65	0.0288	—	—	0.0283	
28	224.65	0.0246	—	—	0.0240	
29	225.65	0.0210	—	—	0.0203	
30	226.65	0.0180	—	—	0.0172	
31	227.65	0.0154	—	—	0.0145	
32	228.65	0.0132	—	—	0.0123	
47	270.65	0.0014	—	—	—	—
51	270.65	0.0009	—	—	—	—
71	214.65	0.0001	—	—	—	—
86	186.95	0.0000	—	—	—	—

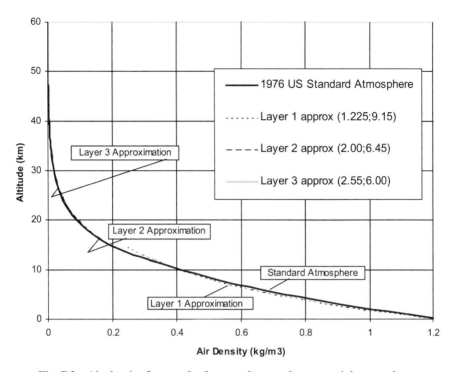

Fig. D2   Air density for standard atmosphere and exponential atmosphere.

In Table D3 the air density ρ is measured in kilograms/cubic meter, and the altitude h is measured in kilometers. As seen in the tabulated values in Table D2, the difference between the values from the 1976 U.S. Standard Atmosphere and the exponential atmosphere approximations used in the trajectory simulations is of the order of $1-2\%$.

Table D3   Exponential atmosphere approximations

Layer	Air density equation
Layer 1 (Troposphere) Decreasing temperature Altitude band (0–11 km)	$\rho = 1.225e^{-h/9.15}$
Layer 2 Constant temperature Altitude band (>11–20 km)	$\rho = 2.000e^{-h/6.45}$
Layer 3 Increasing temperature Altitude band (>20–32 km)	$\rho = 2.550e^{-h/6.00}$

In the trajectory simulations used in Chapter 4, above 32-km altitude the air density is calculated using the exponential approximation for layer 3. The difference between the standard atmosphere and the layer 3 equation for air density increases from the 1–2% difference in the lower critical layers (layers 1–3) to about 20% or more at the upper layers (>32 km), but this has a negligible effect on the trajectories because of the relative magnitude of the terms in the equations of motion that represent the drag and gravity forces acting on the missile. Note that the air density at 70 km, for example, representing the top of the *mesosphere* is approximately 1/12,000th that at sea level.

Figure D2 shows the exponential air density relationships assumed in the trajectory simulation (Chapter 4) compared to the 1976 U.S. Standard Atmosphere values.

For ease of reference, the equations for each of the layers are referenced in Fig. D2 by the appropriate constants in the exponential equations.

NOTES ON MISSILE PROPULSION

In Chapter 4 a set of typical ballistic-missile (BM) trajectories is provided, which shows details for boost phase, midcourse phase, and the terminal phase. In any particular design of missile, of course, the design parameters vary in many ways depending on the final design requirement. The missile can be multistage, liquid, or solid propellant with a variety of fuel types. The burn time could be fast or slow depending on the technology level, and the choices of warheads are also quite different between the missile types. This Appendix summarizes some of the more common features so that the typical trajectories used in Chapter 4 can be viewed in the proper perspective of missile design.

The basic equation of motion for the missile can be expressed as

$$m \frac{dV}{dt} = T - D \tag{E.1}$$

where m is the mass of the missile, V is the speed, T is the thrust, and D is the aerodynamic resistance or drag. For *multistage missiles* the solution to Eq. (E.1) becomes a cumbersome equation as the weight (mass) changes, and speed changes as time progresses through to the final release of the payload that is usually the warhead or kill vehicle. For *single-stage missiles* the equation is easier to express in terms of simple parameters of the missile. The case of the single-stage missile will be treated first.

SINGLE-STAGE MISSILES

Hence, at burnout the speed of a *single-stage missile* is given by

$$V_{\text{bo}} - V_{\text{launch}} = \int_0^{\text{bo}} \frac{(T - D)}{W} \, dt \tag{E.2}$$

If the missile is launched from an aircraft, then the launch speed V_{launch} is the speed of the aircraft. If the missile is launched from a land-based or ship-based system, then $V_{\text{launch}} = 0$. For simplicity, in what follows it will be assumed that $V_{\text{launch}} = 0$.

It is usual in early design to ignore the effect of the aerodynamic drag D during the short time of missile burn, and further to assume that the thrust T is constant, and that the acceleration caused by gravity g is also constant, such that the equation for the missile burnout speed can be expressed by the integral equation:

$$V_{\text{bo}} = \kappa T g \int_0^{bo} \frac{dt}{W} \tag{E.3}$$

A factor κ has been inserted into the equation for the burnout speed V_{bo} to account for the actual reduction in the integral equation for speed to account for the effects of the aerodynamic resistance and for the changes in the acceleration caused by gravity g as a function of altitude. In Chapter 4 and in Appendix D, it is shown that as the missile accelerates during boost phase and altitude is gained the acceleration caused by gravity decrease with increase in altitude. The largest effects are caused by the aerodynamic resistance. It is difficult to give a general expression for these effects, as they depend on the shape and weight of the missile. Further, for multistage missiles the actual value of the factor κ is different for each stage; usually it is the largest for the first stage and has a lesser effect for the second, third, and later stages if used. A good rule of thumb might be to compute the exhaust velocity of missile as per the integral equation (E.3), then apply a factor $\kappa = 0.70$ to obtain the burnout speed (V_{bo}).

The weight of the missile W varies with time t as the propellant is used up, in a manner that can be expressed as

$$W = W_0 - \frac{dW_p}{dt} t \tag{E.4}$$

Hence, the weight of the missile W is equal to the initial weight W_0 in the launcher less the weight of the propellant W_p that is being consumed during the burn at a rate dW_p/dt. The equation for the burnout speed now becomes

$$V_{\text{bo}} = \kappa T g \int_0^{bo} \frac{dt}{W_0 - (dW_p/dt)t} \tag{E.5}$$

Equation (E.5) has the following solution:

$$V_{bo} = \frac{-Tg\kappa}{dW_p/dt} \ln\left[W_0 - \left(\frac{dW_p}{dt}\right) t_{bo} \right] \tag{E.6}$$

The specific impulse of the propellant is given by

$$I_{sp} = \frac{T}{dW_p/dt} \tag{E.7}$$

This convenient expression relates the thrust of the missile T to the burnrate dW_p/dt of the propellant. Hence, at time $t = t_{bo}$ the expression for the burnout speed V_{bo} in terms of the specific impulse and missile design parameters becomes

$$V_{bo} = -I_{sp}g\kappa \ln\frac{W_0}{W_{bo}} \tag{E.8}$$

This simple expression allows for a calculation of the burnout speed V_{bo} in terms of the missile parameters I_{sp} and the start weight W_0 and the weight at end of boost W_{bo} after all of the propellant has been consumed.

Although not derived here, it can be shown that for a *multistage missile* the final speed after all stages have burned out is the sum of all of the individual stages and weights, such that

$$V_{bo} = \sum_{i=1}^{n} \kappa V_{ei} \ln\left(\frac{W_n}{W_{n-1}}\right)_i \tag{E.9}$$

Equation (E.9) is simply a way of saying that the final speed V_{bo} is the sum of the incremental speeds V_{ei} produced by each of the n stages consuming propellant of each of their respective propellant amounts.

The specific impulse I_{sp} is a convenient way of expressing the thrust that can be obtained when a unit weight of propellant is consumed in 1 s. Hence, the specific impulse has the units of time (usually quoted in seconds).

The specific impulse can be calculated from knowledge of the type of propellant and characteristics of the propellant chamber, such that

$$I_{sp} = \text{function}\left(T_c, M, k, \frac{p_e}{p_c} \right) \tag{E.10}$$

The specific impulse I_{sp} is a complex function of the combustion temperature

Table E1   Sample specific impulse values

Propellant	Chamber pressure, $N/m^2 = Pa$	Specific impulse, s	Example missile
Ethyl alcohol water (EtOH + O_2)	120	218	V-2[100]
Liquid oxygen–liquid hydrogen (H_2 + O_2)	113	335	Saturn
Inhibited red fuming nitric acid (RFNA) + unsymmetrical dimethylhydrazine (UDMH)	115	280	SCUD B
Various composite solids	330	165–250	Trident

T_c; the molecular weight of the combustion products M; the specific heat of the combustion products k; and the ratio of the external pressure p_e to the internal chamber pressure p_c. Although a cumbersome function, it is not necessary for the purposes of this book to delve into the intricacies of propulsion system design for I_{sp}. Some typical values of I_{sp}, for typical propellants, are shown in Table E1.

Hence, example missiles of both liquid- and solid-propellant motors exhibit values of specific impulse that vary from anywhere from 160–350 s. Most of the threat missiles that have entered the inventory for tactical use have been in the 260+ s class of specific impulse I_{sp}.

MASS FRACTION

The equation for the burnout speed V_{bo} for a single-stage missile is given in Eq. (E.8). If the takeoff weight is expressed as

$$W = W_{bo} + W_p \qquad (E.11)$$

where W_{bo} is the weight at the end of boost and W_p is the weight of propellant used, then the equation for the burnout speed V_{bo} can be written as

$$V_{bo} = -I_{sp} g \kappa \ln (1 - mf) \qquad (E.12)$$

where the mass fraction mf is given by

$$mf = \frac{W_p}{W_0} \qquad (E.13)$$

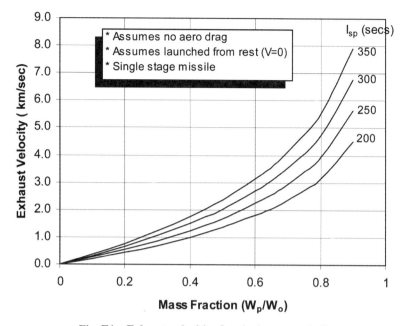

Fig. E1   Exhaust velocities for single-stage missiles.

This result is shown in Fig. E1 for a single-stage missile for a range of values of mass fraction W_p/W_0 and specific impulse I_{sp}.

In Fig. E1 has been shown the calculation of the exhaust velocity V_e for single-stage missiles to emphasize that this velocity must be reduced by the factor κ to account for aerodynamic resistance and gravity terms to achieve the final value of the burnout speed V_{bo}. A general rule of thumb of 70% has been suggested by Ley[100] for quick analyses.

Three single-stage missile examples taken from the literature will be used to indicate how Fig. E1 and the integral equation for the propellant exhaust velocity can be used to estimate the burnout speed V_{bo} for any single-stage missile.

FOR THE V-2 MISSILE

From Eq. (E.12) and the tabulated values in Table E2,

$$V_{bo} = -(0.70)(218)(0.009806)\ln(1 - 0.69)$$
$$= 1.75\,\text{km/s} \tag{E.14}$$

Table E2 Characteristics of selected single-stage missiles

Characteristics	$V\text{-}2$[a]	Al-Hussein	SCUD B
TOW, kg	12,805	6785	6370
Propellant type	$EtOH + O_2$	UDMH propellant RFNA oxidizer	UDMH propellant RFNA oxidizer
Total propellant weight, kg	8796	5000	4000
Specific impulse, s	218	230	230
Mass fraction[b]	0.69	0.74	0.63
Burnout speed, V_{bo}, km/s	1.6	2.2	1.56

[a]Ref. 100, Appendix 2.
[b]These old designs have atypical values of mass function W_p/W_o, and modern missiles tend to have lower values of mass fraction, especially airborne systems.

FOR THE AL-HUSSEIN MISSILE

From Eq. (E.12) and the tabulated values in Table E2,

$$V_{bo} = -(0.70)(230)(0.009806)\ln(1 - 0.74)$$
$$= 2.13\,\text{km/s} \tag{E.15}$$

FOR THE SCUD B MISSILE

From Eq. (E.12) and the tabulated values in Table E2,

$$V_{bo} = -(0.70)(230)(0.009806)\ln(1 - 0.63)$$
$$= 1.57\,\text{km/s} \tag{E.16}$$

These values of V_{bo} for each of the three example single-stage missiles (Fig. E2) can be compared to the computed values of V_{bo} shown in Table E2, which used the *minimum-energy* formula for V_{bo} (see Appendix C).

Note: The factor κ is provided as a first estimate of the complex effect of weight, drag, shape, ballistic coefficient, number of stages, and other factors that convert the missile *exhaust velocity* to the *burnout velocity*. Depending on these factors, the actual correction could be greater than or less than the median value of $\kappa = 0.70$.

These computations thus allow for *single-stage missile* designs to be associated with the generic trajectories for the ballistic missile provided in Chapter 4.

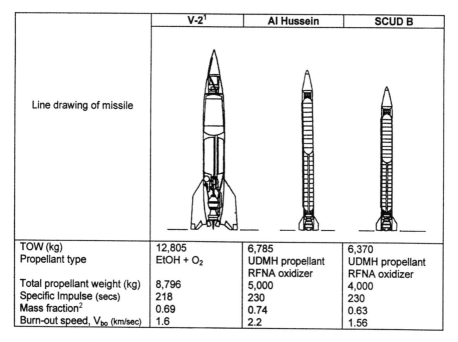

	V-2[1]	Al Hussein	SCUD B
Line drawing of missile			
TOW (kg)	12,805	6,785	6,370
Propellant type	EtOH + O_2	UDMH propellant RFNA oxidizer	UDMH propellant RFNA oxidizer
Total propellant weight (kg)	8,796	5,000	4,000
Specific Impulse (secs)	218	230	230
Mass fraction[2]	0.69	0.74	0.63
Burn-out speed, V_{bo} (km/sec)	1.6	2.2	1.56

Fig. E2   Line drawing of missiles a) V-2, b) Al-Hussein, c) SCUD B.

MULTISTAGE MISSILES

A quick glance at Fig. E1 will show that very high values of mass fraction are required if high exhaust and high burnout velocities are required for either boost-phase intercept (say, ~6 km/s) or to achieve the long ranges of the missiles under discussion (say, > 1000 km). This brings in the concept of multistaging. It can be shown[101] that the weight of the last stage in a multistage missile W_n is related to the payload of that stage W_{PLn} by

$$W_n = W_{PLn} \frac{mf_n \cdot \exp\left(\Delta Ve_n / gI_{sp}\right)}{1 - (1 - mf_n)\exp\left(\Delta Ve_n / gI_{sp}\right)} \qquad (E.17)$$

In the preceding equation the slight change in description of the exhaust velocity to ΔVe_n has been used to emphasize that each stage yields a delta change in speed over the next stage of the missile. This formulation would be the same for the next lower stage $(n-1, n-2, \dots)$ depending on the number of stages. Each stage could have its own mass fraction mf and its own propellant properties and I_{sp}. The payload W_{PLn} of any stage is the total weight of the stages above it in the chain.

EXAMPLE 1

The U.S. Air Force Scientific Board[101] used this formulation together with a knowledge of the state of the art of air-to-air missile design to determine the feasibility of a high-speed interceptor that would be capable of achieving boost-phase intercepts and be capable of being carried on a fighter aircraft. It was found that a missile of the following characteristics could meet that requirement: intercept gross weight of 909 kg, mass fraction of 0.83 (held constant on all three stages), three stages plus kill vehicle (45 kg), and exhaust velocity of 5.7 km/s. Such a missile fits the constraints discussed on aircraft plus interceptors for Boost Phase Intercept (BPI) in Chapter 4.

EXAMPLE 2

Zarchan[102] also analyzes the many possibilities of staging and provides the equations and example curves for two, three and more stage missiles that show the advantage of staging if both reduced weight and high speed are required. One particular case shows that a 6-km/s missile can be achieved to deliver a 4.5-kg kill vehicle for less than 90-kg gross weight if three stages are used, each with a mass fraction of 0.90 and a propellant with an I_{sp} of 250 s. The missile gross weight would increase to greater than 90 kg if two stages are used and would not be possible ($W \to \infty$) if a single stage were considered.

RADAR HORIZON

It is necessary when considering radar range to take into account the curvature of the Earth when the distances become large. The optical horizon can be easily computed from simple geometry. Figure F1 shows a tangent to the Earth's surface and identifies the key parameters. Specifically, R_e is the radius of the Earth; h_r is the height of the radar above the surface; and R_H is the distance from the radar to the geometric horizon, assuming the Earth is a perfect sphere. If the radar is to look beyond the horizon, then Fig. F1 shows how a "target" of height h_r subtends the same tangent to the radar.

The distance from the radar to the horizon $(A \rightarrow B)$ can then be determined from

$$R_H{}^2 = (R_e + h_r)^2 - R_e{}^2 \tag{F.1}$$

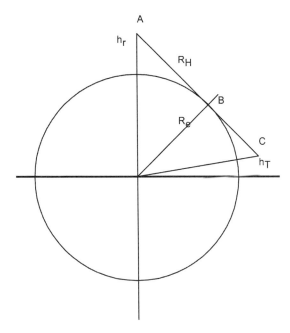

Fig. F1   Line of sight on the Earth.

399

Hence,

$$R_H = \sqrt{(h_r^2 + 2R_e h_r)} \tag{F.2}$$

which, because $h_r \ll R_e$, can be approximated by

$$R_H = \sqrt{2R_e h_r} \tag{F.3}$$

Equation (F.3) is strictly the line-of-sight distance ($A \rightarrow B$). It is found that for radars emitting radar waves that this line of sight bends away from the optical line of sight and that a correction is usually applied called the 4/3rds Earth Rule, such that the radar horizon is more correctly written

$$R_H = \sqrt{\frac{4}{3}} \sqrt{2R_e h_r} \tag{F.4}$$

For a value of Earth's radius $R_e = 6378$ km, the radar horizon can be written as

$$R_H = 4.124\sqrt{h_r} \tag{F.5}$$

where the radar horizon R_H is expressed in units of kilometers and the radar height is given in units of meters.

If now the radar range is to "see" beyond the horizon to detect a target of height h_T above the Earth's surface at a distance ($A \rightarrow C$), then the preceding equations can be modified to give the result, for the radar line of sight to this target, as

$$R_H = 4.124(\sqrt{h_r} + \sqrt{h_T}) \tag{F.6}$$

where, as before, the line of sight R_H is measured in kilometers and the heights of the radar and the target are in meters. For the case of $h_T = 0$, Fig. F2 shows both the line of sight (for optical and infrared sensors) and the radar horizon.

Figure F2 shows the following typical values for line of sight: the distance to horizon for a standard person (1 m tall) is 3.57 km; for an aircraft at 10-km altitude it is 357 km, and for a low-Earth-orbit satellite at 1000 km altitude it is 3570 km.

Figure F2 can also provide information "in reverse" in that it provides the information as to whether a radar (or other sensor) can see targets of certain heights. Consider, for example, a radar at position B on the Earth (on Fig.

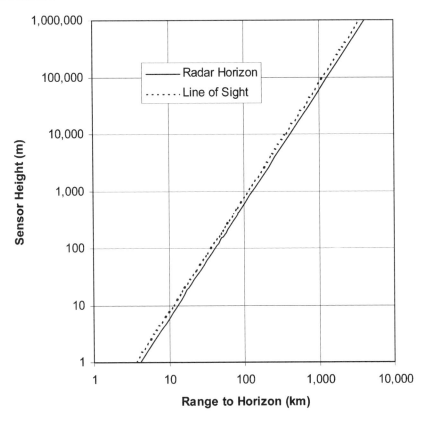

Fig. F2   Line of sight and radar horizon.

F1), and it is looking south towards point C (on Fig. F1). Because of the curvature of the Earth, the ground "falls away" from the horizon such that a target would have to be of height h_T in order for the sensor to see it. (In practice, the target would have to be much higher than h_T because the radar would have to have a lower elevation limit of say 5–7 deg to avoid ground clutter.) Hence, if a radar had a 3570-km detection range the target would have to be *at least* 1000-km high at that distance in order to be seen.

Consider the example of the high-power (approximately 2.5 MW) ballistic-missile early-warning system (BMEWS) radar at RAF Fylingdales in England with a range of greater than 4800 km searching for missiles. (To avoid classification issues, the information on BMEWS has been taken from the Royal Air Force, *Visitors Booklet* for Fylingdales.) At that distance the target must come into view when approximately 1355 km above the surface of the Earth, which is approximately *twice* the apogee of a 3000-km-range ballistic missile. Such considerations lead to architectures of a network of long-range radars to cover large surfaces of the Earth. Figure F3 shows a

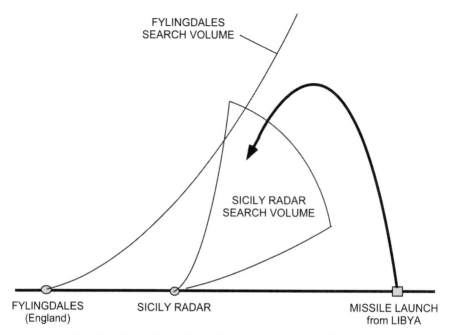

Fig. F3   Radar in southern Europe to augment Fylingdales.

schematic arrangement of a Fylingdales radar located in Northern England looking south, augmented by another long-range radar in southern Europe (say near Raduga on Sicily) to monitor long-range missiles coming from Libya. The shorter-range radar on Sicily "catches" any missiles that underfly the northern located Fylingdales because of the curvature of the Earth.

In Fig. F3 the Earth's surface has been flattened and the radar search lines drawn curved for illustrative purposes. Such a diagram illustrates the effect of the radar horizon on the possible architectures that must be considered.

SOME USEFUL CONSTANTS

Table G1   Conversion factors

1 ft	0.3048 m
1 kg	2.204 lb
1 statute mile	1.6093 km
1 statute mile	5280 ft
1 n mile	1.8520 km
1 n mile	6,076.115 ft
1 radian	57.295779 deg
1 slug/ft^3	515 kg/m^3
1 hp	746 W
°C	Kelvin °K − 273.15
1 short ton	907.18474 kg
1 hectare	100 ares
1 are	100 m^2
1 knot	1.15 statute mph
1 km	3280.8398 ft
1 lb (mass)	0.454 kg
1 lb (force)	4.45 N
1 lb/ft^2 (force)	47.88 N/m^2
1 lb/ft^2 (mass)	4.882 kg/m^2
1 U.S. liquid gal	0.003785 m^3
1 U.S. liquid gal	3.785306 liters
1 ton (long, 2240 lb)	1016.047 kg
1 tonne	1000 kg
1000 ft/s	0.3048 km/s
1 ft^2	0.09290304 m^2
1 m^2	10.763910 ft^2
1 hectare	10,000 m^2
1 knot	1.85 km/h

Table G2   Orbital mechanics

Circular speed at Earth's surface	7.905 km/s
Escape speed from Earth	11.18 km/s
Gravitational acceleration at surface	9.80621 m/s^2 ($=32.1725$ ft/s^2)
Radius of Earth (mean equatorial)	6378.14 km
Radius of Earth (mean polar)	6356.75 km
Kilometers/degree at Equator	111.32 km/deg
Air density at Earth surface	1.225 kg/m^3 ($=0.00238$ slugs/ft^3)

Table G3   Other constants

Speed of sound	
at Earth's surface	0.3404 km/s on a standard day (15°C)
at 10-km altitude	0.3011 km/s on a standard day (15°C)
Speed of light	2.997924 $\times$ 10^8 m/s

Table G4   Speed and Mach number relationship

Speed, km/s	Mach no. (at speed of light)	Mach no. (at 10-km altitude)
1	2.94	3.32
2	5.88	6.64
3	8.81	9.96
4	11.75	13.28
5	14.69	16.61
6	17.63	19.92

Appendix H

CONTENTS OF CD-ROM

This Appendix describes the contents of the companion CD-ROM.

CD-ROM OVERVIEW

The CD-ROM contains a set of files that can be used to perform calculations based on the equations developed and described in the body of the book. The Exercises provided at the end of the Chapters can be solved using spreadsheets provided in the pertinent files on the CD-ROM. Wherever possible, generic tables and charts have been provided that will adjust automatically, as inputs (e.g. data values and chart titles) are provided by the user in the appropriate cells of the spreadsheets.

In most cases, the input sheets contain both locked and unlocked cells. In this manner, the possibility of the user inadvertently erasing or contaminating the basic program is minimized.

The CD-ROM contains the following files:

File 1: Contents of CD-ROM

File 2: Chapter 4: Performance: Footprint Radius

File 3: Chapter 5: Cost of Missile Defense (including Sensitivity Analysis)

File 4: Chapter 8: Evaluation: Solve Linear Equations (with Constant Coefficients)

File 5: Chapter 8: Evaluation: Displaying More than Two Variables in the Evaluation

File 6: Chapter 8: Evaluation: Value Functions

File 7: Chapter 8: Evaluation: Solve Linear Equations (with Value Functions)

Comments on Each File

While each file has been designed to be self-explanatory and stand-alone, the following comments may help the user in setting up particular problems and solutions. For the purposes of this Appendix, those cells where the reader may insert values have been highlighted. All other cells on the spreadsheets in the Files are calculated automatically.

File 1: Contents of CD-ROM

This File is a copy of this Appendix so that the user can refer to it without the need to open the book. A copy of each file complete with example inputs, charts and solutions are also provided in this File so that the user can see immediately the form of the spreadsheets and charts as a guide to using other data inputs that the user may wish to insert.

File 2: Chapter 4: Performance: Footprint Radius

In Chapter 4, various equations are developed for the Footprint Radius of missile defense systems. This File specifically allows the exploration of the Forward Edge Footprint Radius (R_{FE}) for a collocated defense.

First, a Base Case can be constructed using this File. This Base Case allows for variations in the parameters k_1, k_2, k_3, t_d, h_i, θ_{re} and ϕ_1, which form the basic form of a Forward Edge Footprint Radius (R_{FE}). This Base Case includes both radar Face Acquisition and radar Horizon Acquisition. These parameters and terms are defined in Chapter 4.

Second, once the Base Case has been constructed, three major variations are possible as a sensitivity analysis around the Base Case. These variations are:

(1) variation of the interceptor speed (V_{bo})
(2) variation of the detection range (R_D)
(3) variation of reaction time (t_r)

As each variation is explored, the chart titles will automatically adjust to the user inputs.

A sample output from this File is shown here.

FOOTPRINT RADIUS (COLLOCATED DEFENSE)

From Chapter 4, it is seen that there are two basic cases in the determination of the Forward Edge Footprint Radius (R_{FE}). They are:

* **Horizon Acquisiton**
         Where the Detection Range is Greater than the Threat Range
* **Face Acquisition**
         Where the Detection Range is Less than the Threat Range

Footprint Radius (Horizon Acquisition)

From Chapter 4, the Forward Edge Footprint Radius is given by:

$$R_{FE} = V_{ix}(t_F - t_d - t_r) - h_i \cot \theta_{re}$$

where R_{FE} = Forward Edge Footprint Radius
        V_{ix} = Average Interceptor Speed over the ground

t_F = Flight Time of Threat (TOF)
t_d = Time to detect threat
t_r = Time for defense to react, process and launch interceptor
h_i = Desired altitude of Intercept (Keep Out Altitude)
θ_{re} = Reentry Angle of Threat (at intercept altitude)

Footprint Radius (Face Acqusition)

From Chapter 4, the Forward Edge Footprint Radius is given by:

$$R_{FE} = \frac{[R_D \cos \phi_1 - V_{TX} \cdot t_r] \cdot V_{ix}}{V_{ix} + V_{TX}} - h_i \cot \theta_{re}$$

where the new terms are:

$$R_D = \text{Radar (or other sensor) detection range}$$
$$\phi_1 = \text{Minimum search angle}$$
$$V_{TX} = \text{Speed of threat over the ground}$$

all other terms are as for the Case of Horizon Acquisition.

Relationships between speeds, and speed and range are approximated by:

$$V_{ix} = k_1 V_{bo} \qquad\qquad V_T = k_2 \sqrt{R_T} \qquad\qquad V_{TX} = V_T \cos \theta_{re}$$

where k_1 and k_2 are constants and R_T is the threat range.

The time of flight (TOF) of the threat is given by the approximation (see Appendix C):

$$t_F = k_3 \sqrt{R_T}$$

SAMPLE CHARTS
Common inputs (for all three following charts):

$k_1 =$	0.60		$t_d =$	78	secs
$k_2 =$	0.09		$h_i =$	15	km
$k_3 =$	14		$\theta_{re} =$	45	degrees
			$\phi_1 =$	7	degrees

Variable inputs (for three charts):

Three variations on charts of footprint radius are provided as follows:

Effect of Interceptor Speed		Effect of Detection Range		Effect of Reaction Time	
Chart 1		Chart 2		Chart 3	
$R_D =$	200	$V_{bo} =$	1.30	$R_D =$	200
$t_r =$	20	$t_r =$	20	$V_{bo} =$	1.30
$V_{bo} =$	1.00	$R_D =$	200	$t_r =$	20
$V_{bo} =$	1.30	$R_D =$	600	$t_r =$	60
$V_{bo} =$	2.00	$R_D =$	1000	$t_r =$	120

The reader may examine different effects of interceptor speed (V_{bo}); detection range (R_D) and defense system reaction time (t_r) by inserting different numerical values in the cells above that are "No Fill". All other values are calculated automatically and displayed on the charts. Further variations on these charts can be made by changing any of the parameters listed above, i.e. k_1, k_2, k_3, t_d, h_i, θ_{re} and ϕ_1 in the table of common inputs.

Chart 1: Effect of Interceptor Speed

Horizon Acquisition:

		V_{bo}		
		1.00	1.30	2.00
R_T	t_F	R_{FE}	R_{FE}	R_{FE}
100	140.0	10.2	17.8	57.0
200	198.0	45.0	63.0	126.6
300	242.5	71.7	97.7	180.0
400	280.0	94.2	127.0	225.0
500	313.0	114.0	152.7	264.7
600	342.9	132.0	176.0	300.5
700	370.4	148.4	197.5	333.5
800	396.0	163.8	217.4	364.2
900	420.0	178.2	236.2	393.0
1000	442.7	191.8	253.9	420.3
1200	485.0	217.2	286.8	471.0
1400	523.8	240.5	317.1	517.6
1600	560.0	262.2	345.4	561.0
1800	594.0	282.6	371.9	601.8
2000	626.1	301.9	396.9	640.3
2500	700.0	346.2	454.6	729.0
3000	766.8	386.3	506.7	809.2
3500	828.3	423.2	554.6	882.9

Face Acquisition:

V_{bo}		
1.00	1.30	2.00
R_{FE}	R_{FE}	R_{FE}
75.2	87.3	106.4
57.2	68.8	88.1
47.2	58.1	77.0
40.4	50.8	69.0
35.4	45.2	62.8
31.5	40.8	57.8
28.3	37.2	53.6
25.6	34.1	50.0
23.3	31.5	46.9
21.3	29.2	44.1
18.0	25.4	39.4
15.4	22.2	35.6
13.2	19.6	32.3
11.3	17.4	29.5
9.7	15.5	27.0
6.4	11.6	21.9
3.9	8.6	18.0
1.9	6.1	14.8

Chart 1: Effect of Interceptor Speed (V_{bo})

Detection Range, R_D	200	km
Intercept altitude, h_i	15	km
Reaction time, t_r	20	secs

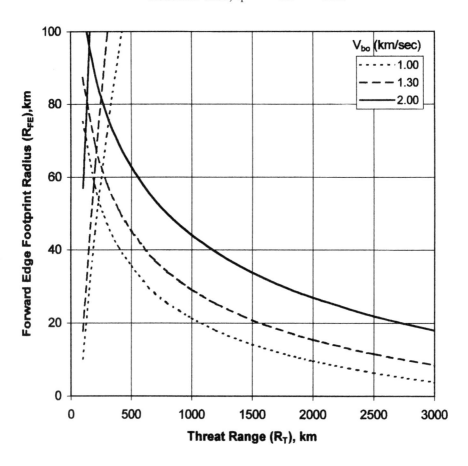

Chart 2: Effect of Detection Range

Horizon Acquisition: *Face Acquisition:*

R_T	t_F	R_D 200 R_{FE}	600 R_{FE}	1000 R_{FE}	R_D 200 R_{FE}	600 R_{FE}	1000 R_{FE}
100	140.0	17.8	17.8	17.8	87.3	305.9	524.6
200	198.0	63.0	63.0	63.0	68.8	253.1	435.7
300	242.5	97.7	97.7	97.7	58.1	222.6	387.2
400	280.0	127.0	127.0	127.0	50.8	201.6	352.5
500	313.0	152.7	152.7	152.7	45.2	185.8	326.3
600	342.9	176.0	176.0	176.0	40.8	173.2	305.6
700	370.4	197.5	197.5	197.5	37.2	162.9	288.6
800	396.0	217.4	217.4	217.4	34.1	154.2	274.2
900	420.0	236.2	236.2	236.2	31.5	146.7	261.8
1000	442.7	253.9	253.9	253.9	29.2	140.1	251.0
1200	485.0	286.8	286.8	286.8	25.4	129.1	232.9
1400	523.8	317.1	317.1	317.1	22.2	120.2	218.2
1600	560.0	345.4	345.4	345.4	19.6	112.7	205.9
1800	594.0	371.9	371.9	371.9	17.4	106.4	195.4
2000	626.1	396.9	396.9	396.9	15.5	100.9	186.3
2500	700.0	454.6	454.6	454.6	11.6	89.7	167.9
3000	766.8	506.7	506.7	506.7	8.6	81.1	153.7
3500	828.3	554.6	554.6	554.6	6.1	74.3	142.4

Chart 2: Effect of Detection Range (R_D)

Interceptor Speed, V_{bo}	1.30	km/sec
Intercept altitude, h_i	15	km
Reaction time, t_r	20	secs

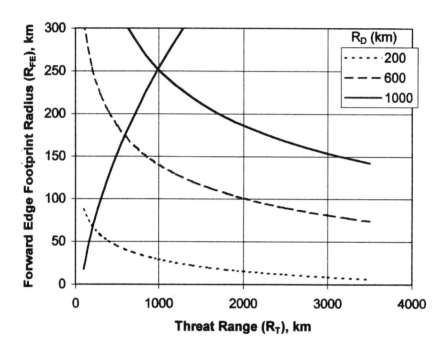

Chart 3: Effect of Reaction Time

Horizon Acquisition:

R_T	t_F	t_r 20 R_{FE}	60 R_{FE}	120 R_{FE}
100	140.0	17.8	− 13.4	− 60.2
200	198.0	63.0	31.8	− 15.0
300	242.5	97.7	66.5	19.7
400	280.0	127.0	95.8	49.0
500	313.0	152.7	121.5	74.7
600	342.9	176.0	144.8	98.0
700	370.4	197.5	166.3	119.5
800	396.0	217.4	186.2	139.4
900	420.0	236.2	205.0	158.2
1000	442.7	253.9	222.7	175.9
1200	485.0	286.8	255.6	208.8
1400	523.8	317.1	285.9	239.1
1600	560.0	345.4	314.2	267.4
1800	594.0	371.9	340.7	293.9
2000	626.1	396.9	365.7	318.9
2500	700.0	454.6	423.4	376.6
3000	766.8	506.7	475.5	428.7
3500	828.3	554.6	523.4	476.6

Face Acquisition:

t_r 20 R_{FE}	60 R_{FE}	120 R_{FE}
87.3	73.3	52.3
68.8	52.1	27.0
58.1	39.9	12.4
50.8	31.4	2.4
45.2	25.1	− 5.2
40.8	20.0	− 11.2
37.2	15.9	− 16.1
34.1	12.4	− 20.3
31.5	9.4	− 23.9
29.2	6.7	− 27.0
25.4	2.3	− 32.3
22.2	− 1.3	− 36.5
19.6	− 4.3	− 40.1
17.4	− 6.8	− 43.1
15.5	− 9.0	− 45.8
11.6	− 13.5	− 51.1
8.6	− 16.9	− 55.2
6.1	− 19.7	− 58.5

Chart 3: Effect of Reaction Time

Dection Range, R_D	200	km
Interceptor Speed, V_{bo}	1.3	km/sec
Intercept altitude, h_i	15	km

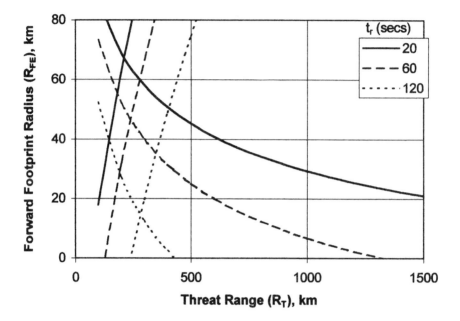

File 3: Chapter 5: Cost of Missile Defense (including Sensitivity Analysis)

This File contains two major sections. The first section constructs the cost of a Base Case of a Two Layer Active Layered Missile Defense System (for defense against both ballistic missiles and cruise missiles). This is a spreadsheet that allows the user to insert values of all key parameters for both Lower Layer and Upper Layer missile defense systems either using own data or the default values from the Chapters of this book.

The second section of this File, variations in the cost of the Base Case can be explored using different values of key parameters. Specifically, the effects of changes in system component unit costs (interceptor, radar, launcher, support system) on the total system cost can be easily explored using the supplied input sheets. Other key parameters, such as share of cruise missile (CM) defense; choice of salvo size; interceptor kill capability, can all be varied and evaluated as to their effect on system cost can be determined.

As the user inserts different values, both the Base Case and the specific variation being studied remain internally consistent. All chart titles automatically adjust to the user inputs as necessary.

A sample output from this File is shown here.

COST OF MISSILE DEFENSE

From Chapter 5, the Procurement Cost of a Two Layer Active Missile Defense can be represented by the sum of the Upper Layer and Lower Layer Missile Defense Systems costs:

Upper Layer Missile Defense System Cost ($UL):

$$\$UL = (\$_{site\,UL}) \times (\text{No. of sites}) + (\$_{int.UL}) \times (Salvo_{UL}) \times (N_{threat})$$

where $\$UL$ = Cost of all Upper Layer sites and interceptors used
$\quad\$_{siteUL}$ = Procurement Cost of Upper Layer "battery" per site
$\quad\$_{int.UL}$ = Upper Layer Interceptor Unit Cost
$\quad Salvo_{UL}$ = Number of UL interceptors launched per incoming attacking missile
$\quad N_{threat}$ = Number of incoming threat missiles launched during conflict

Lower Layer Missile Defense System Cost ($LL):

$$\$LL = (\$_{site\,LL}) \times (\text{No. of sites}) \times (1 + f_{L\&D}) + (\$_{int.LL}) \times (Salvo_{LL})$$
$$\times (\text{BM leakage} + \text{CM share}$$

where $\quad\$LL$ = Cost of all Lower Layer sites and interceptors used
$\quad\$_{siteLL}$ = Procurement Cost of Lower Layer "battery" per site
$\quad\$_{int.LL}$ = Lower Layer Interceptor Unit Cost
$\quad Salvo_{LL}$ = Number of LL interceptors launched per incoming attacking missile
$\quad$ BM Leakage = Number of incoming threat BM not intercepted by UL system
$\quad$ CM share = Share of CM attack to be intercepted (and not allocated to other AD means)
$\quad f_{L\&D}$ = Factor to account for additional LL sites required to provide defense against lofted and depressed trajectories[1]

(1) While not discussed in detail in this book, it can be shown that the center of defended footprints can move and leave Key Assets undefended unless additional LL systems are in place.

IMPORTANT ASSUMPTIONS

This formulation of the Cost of Missile Defense provides a rapid "appreciation" of the costs required to defend against missile attack, but three important assumptions should be kept in mind:

1. The above formulation does not take into account the cost of preparing the sites (e.g. site preparation, cost of transporting equipment to any OOA sites, etc).
2. The above formulation also does not take into account the Operating & Support costs (e.g. fuel, oil, potable water, supplies, personnel, etc). As discussed in Chapter 5, such costs can be of equal magnitude as the procurement costs which are represented in the above.
3. As written, this formulation is for Land Based Systems. For Air Based & Sea Based missile defense systems, separate decisions have to be made as the share of "ship" or "aircraft" platform costs that are to be allocated to missile defense.

SAMPLE CALCULATIONS (BASE CASE):

Each of the component costs are obtained from Chapter 5 or the reader may substitute his or her own values are appropriate.

Number of upper Layer sites required to cover defended Wide Area	5
Number of Key Assets within Wide Area to receive two layer protection	15

Upper Layer Single Site Costs:

Radar Unit Cost	150	$M
Number of radars per site	1	
Radar Support costs per site	50	$M
Launcher Unit Cost	3	$M
Number of Launchers per site	2	
BM/C^3 trucks Unit Cost	5	$M
Number of BM/C^3 trucks per site	4	
Upper Layer System Single Site Cost:	*226*	*$M*

Lower Layer Single Site Costs:

Radar Unit Cost	40	$M
Number of radars per site	1	
Radar Support costs per site	0	$M
Launcher Unit Cost	3	$M
Number of Launchers per site	5	
BM/C^3 trucks Unit Cost	5	$M
Number of BM/C^3 trucks per site	4	
Lower Layer System Single Site Cost:	*75*	*$M*

Interceptor Costs & Performance:

Upper Layer Interceptor Unit Cost:	2.5	$M
Lower Layer Interceptor Unit Cost:	2.5	$M
Upper Layer Interceptor SSPK:	0.70	
Lower Layer Interceptor SSPK:	0.70	

Upper Layer System Salvo   2   Note: A change to
Lower Layer System Salvo   2   Salvo-Look-Salvo
                                         firing doctrine
                                         would require an
                                         increase in the
                                         number of sites
                                         required to provide
                                         same defense cover-
                                         age (see Chapter 4)

Factor ($f_{L\&D}$) for lofted & depressed trajectory attacks: 50   %

Attacking Missiles:

Ballistic Missiles (BM):	500	
Cruise Missiles (CM):	250	
Share of CMD by LL System:	45	%

COST OF TWO LAYER DEFENSE

Upper Layer	3.63 $B	BM Leakers:	4
Lower Layer	2.48 $B	CM Leakers:	10
TOTAL COST:	**6.11 $B**	UL Interceptors req'd:	1000
		LL Interceptors req'd:	315

which shows the total (two layer) defense cost for the specified defense level (i.e. accepted leakage).

SENSITIVITY ANALYSES (AROUND BASE CASE)

The following formulation allows for various key parameters to be varied so that the impact on the total system cost of layered defense can be determined. In each case, all inputs will be as for the Base Case, except for the parameter being varied.

Effect of Upper Layer Interceptor Unit Cost

The values of the Base Case remain the same, except for the Upper Layer Interceptor Unit Costs:

Upper Layer Interceptor Unit Cost ($M):	1.50	2.50	3.00	5.00	7.50
Upper Layer System Cost ($B):	2.63	3.63	4.13	6.13	8.63
Lower Layer System Cost ($B):	2.48	2.48	2.48	2.48	2.48
Total Cost for Two Layer Defense ($B)	**5.11**	**6.11**	**6.61**	**8.61**	**11.11**

System Effectiveness[1]:

BM Leakers:	4	4	4	4	4
CM Leakers:	10	10	10	10	10
UL Interceptors used:	1000	1000	1000	1000	1000
LL Interceptors used:	315	315	315	315	315

(1) To nearest integer

Effect of Upper Layer Interceptor Unit Cost

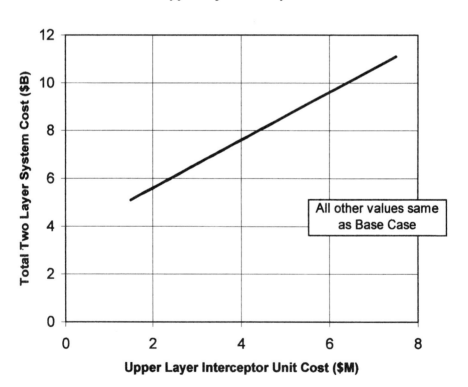

EFFECT OF LOWER LAYER INTERCEPTOR UNIT COST

The values of the Base Case remain the same, except for the Lower Layer Interceptor Unit Costs:

	1.50	2.50	3.00	5.00	7.50
Upper Layer Interceptor Unit Cost ($M):					
Upper Layer System Cost ($B):	3.63	3.63	3.63	3.63	3.63
Lower Layer System Cost ($B):	2.16	2.48	2.63	3.26	4.05
Total Cost for Two Layer Defense ($B):	**5.79**	**6.11**	**6.26**	**6.89**	**7.68**

System Effectiveness:

BM Leakers[1]:	4	4	4	4	4
CM Leakers[1]:	10	10	10	10	10
UL Interceptors used:	1000	1000	1000	1000	1000
LL Interceptors used:	315	315	315	315	315

(1) To nearest integer

Effect of Lower Layer Interceptor Unit Cost

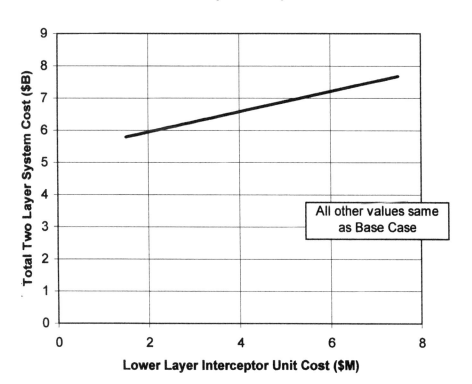

All other values same as Base Case

EFFECT OF UPPER LAYER INTERCEPTOR SSPK

The values of the Base Case remain the same, except for the Upper Layer
Interceptor SSPK:

Upper Layer Interceptor SSPK:	0.40	0.50	0.70	0.80	0.90
Upper Layer System Cost ($B):	3.63	3.63	3.63	3.63	3.63
Lower Layer System Cost ($B):	3.15	2.88	2.48	2.35	2.28
Total Cost for Two Layer Defense					
($B):	**6.78**	**6.51**	**6.11**	**5.98**	**5.91**
System Effectiveness:					
BM Leakers[1]:	16	11	4	2	0
CM Leakers[1]:	10	10	10	10	10
UL Interceptors used:	1000	1000	1000	1000	1000
LL Interceptors used:	585	475	315	265	235

(1) To nearest integer

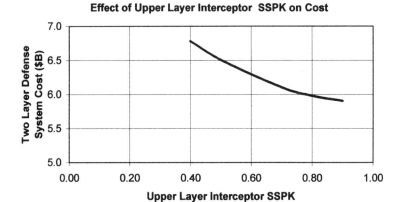

Effect of Upper Layer Interceptor SSPK on Cost

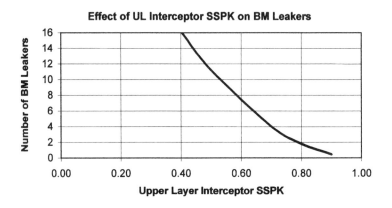

Effect of UL Interceptor SSPK on BM Leakers

EFFECT OF LOWER LAYER INTERCEPTOR SSPK

The values of the Base Case remain the same, except for the Lower Layer
Interceptor SSPK:

Lower Layer Interceptor SSPK:	0.40	0.50	0.70	0.80	0.90
Upper Layer System Cost ($B):	3.63	3.63	3.63	3.63	3.63
Lower Layer System Cost ($B):	2.48	2.48	2.48	2.48	2.48
Total Cost for Two Layer Defense					
($B):	**6.11**	**6.11**	**6.11**	**6.11**	**6.11**

System Effectiveness:

BM Leakers[1]:	16	11	4	2	0
CM Leakers[1]:	41	28	10	5	1
UL Interceptors used:	1000	1000	1000	1000	1000
LL Interceptors used:	315	315	315	315	315

(1) To nearest integer

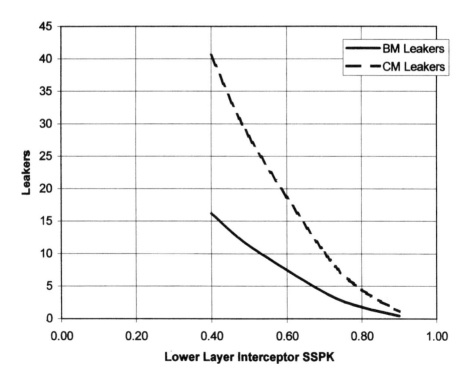

Effect of Lower Layer Interceptor SSPK on Leakage

NOTE: In above chart, the total system cost is constant at a value
of: **6.11 $B**

EFFECT OF UPPER LAYER SALVO SIZE

The values of the Base Case remain the same, except for the Upper Layer
Interceptor Salvo size

Upper Layer Interceptor Salvo size:	1	2	3	4	5
Upper Layer System Cost ($B):	2.38	3.63	4.88	6.13	7.38
Lower Layer System Cost ($B):	3.00	2.48	2.32	2.27	2.26
Total Cost for Two Layer Defense ($B):	**5.38**	**6.11**	**7.20**	**8.40**	**9.64**

System Effectiveness:

BM Leakers[1]:	14	4	1	0	0
CM Leakers[1]:	10	10	10	10	10
UL Interceptors used:	500	1000	1500	2000	2500
LL Interceptors used:	525	315	252	233	227

(1) To nearest integer

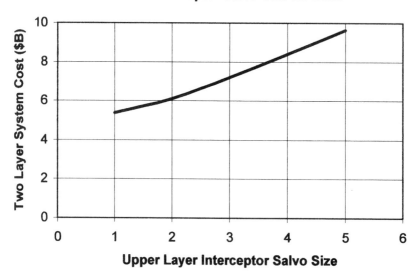

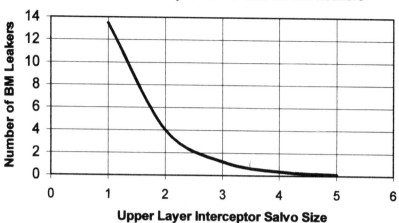

BETTER BULLETS OR MORE SHOTS?

The values of the Base Case remain the same, except for the Upper Layer Interceptor salvo size and Upper Layer Interceptor SSPK.

This combination will allow inspection of the combined effects of changing both SSPK and the salvo firing doctrine. This will show the net result on the cost of the two layer defense system and the resultant leakers as both parameters are varied. The chart will then show from a cost and a leaker viewpoint if it is better to improve the interceptor SSPK or simply launch more lower performance interceptors.

Upper Layer Interceptor SSPK =	0.5				
Upper Layer Interceptor Salvo size:	1	2	3	4	5
Upper Layer System Cost ($B):	2.38	3.63	4.88	6.13	7.38
Lower Layer System Cost ($B):	3.50	2.88	2.56	2.41	2.33
Total Cost for Two Layer Defense ($B):	**5.88**	**6.51**	**7.44**	**8.54**	**9.71**

System Effectiveness:

BM Leakers[1]:	23	11	6	3	1
CM Leakers[1]:	10	10	10	10	10
UL Interceptors used:	500	1000	1500	2000	2500
LL Interceptors used:	725	475	350	288	256

(1) To nearest integer

Upper Layer Interceptor SSPK =	0.6				
Upper Layer Interceptor Salvo size:		1	2	3	4	5
Upper Layer System Cost ($B):		2.38	3.63	4.88	6.13	7.38
Lower Layer System Cost ($B):		3.25	2.65	2.41	2.31	2.28
Total Cost for Two Layer Defense ($B):		**5.63**	**6.28**	**7.29**	**8.44**	**9.66**

System Effectiveness:

BM Leakers[1]:	18	7	3	1	0
CM Leakers[1]:	10	10	10	10	10
UL Interceptors used:	500	1000	1500	2000	2500
LL Interceptors used:	625	385	289	251	235

(1) To nearest integer

Upper Layer Interceptor SSPK =	0.7				
Upper Layer Interceptor Salvo size:		1	2	3	4	5
Upper Layer System Cost ($B):		2.38	3.63	4.88	6.13	7.38
Lower Layer System Cost ($B):		3.00	2.48	2.32	2.27	2.26
Total Cost for Two Layer Defense ($B):		**5.38**	**6.11**	**7.20**	**8.40**	**9.64**

System Effectiveness:

BM Leakers[1]:	14	4	1	0	0
CM Leakers[1]:	10	10	10	10	10
UL Interceptors used:	500	1000	1500	2000	2500
LL Interceptors used:	525	315	252	233	227

(1) To nearest integer

Upper Layer Interceptor SSPK =	0.8				
Upper Layer Interceptor Salvo size:		1	2	3	4	5
Upper Layer System Cost ($B):		2.38	3.63	4.88	6.13	7.38
Lower Layer System Cost ($B):		2.75	2.35	2.27	2.25	2.25
Total Cost for Two Layer Defense ($B):		**5.13**	**5.98**	**7.15**	**8.38**	**9.63**

System Effectiveness:

BM Leakers[1]:	9	2	0	0	0
CM Leakers[1]:	10	10	10	10	10
UL Interceptors used:	500	1000	1500	2000	2500
LL Interceptors used:	425	265	233	227	225

(1) To nearest integer

Effect of UL SSPK and Salvo Size on Cost

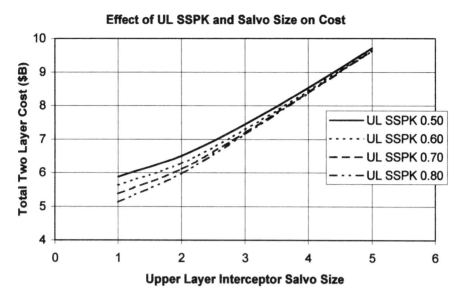

Effect of UL SSPK and Salvo Size on BM Leakers

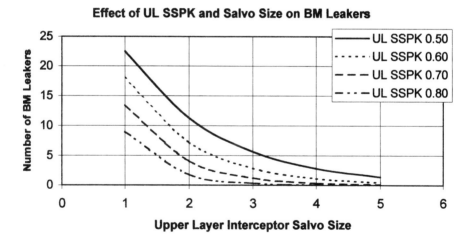

File 4: Chapter 8: Evaluation: Solve Linear Equations (with Constant Coefficients)

This File provides a direct solution to the case of linear equations with constant coefficients. The specific case of "four equations, four unknowns" is presented. For convenience the specific case of performance (leakage); procurement cost; schedule to IOC; and risk (measured in billion dollar years) is used. The user may substitute other combinations of four parameters to determine the results. By clicking on the results column, the results can be rearranged in descending order of importance for ease of reading.

This File also includes the test for consistency. This provides an aid in the determination if any system has its four parameters directly proportional to another systems parameters.

A sample output from this File is shown here.

Solve Linear Equations (with Constant Coefficients)

From Chapter 8, four systems are to be compared as follows:

$$w_1p_1 + w_2c_1 + w_3s_1 + w_4r_1 = k_1$$
$$w_1p_2 + w_2c_2 + w_3s_2 + w_4r_2 = k_2$$
$$w_1p_3 + w_2c_3 + w_3s_3 + w_4r_3 = k_3$$
$$w_1p_4 + w_2c_4 + w_3s_4 + w_4r_4 = k_4$$

where the four parameters in each system are designated by:

$$p_i = \text{performance} \qquad i = 1 \text{ to } 4$$
$$c_i = \text{cost}$$
$$s_i = \text{schedule}$$
$$r_i = \text{risk}$$

The relative weighting (w_j) for each of the four parameters must satisfy:

$$\sum_{j=1}^{4} w_j = 1$$

EXAMPLE PROBLEM

The four systems to be compared are:

System	Perf. (Leakage) (%)	Cost ($B)	Schedule (Years to IOC)	RDT&E Cost ($B/year)	Years to MS III	Risk ($B.years)
1	9	3.3	4	0.2	1	0.2
2	9	2.1	9	0.5	2	1
3	0.81	4.6	12	0.7	5	3.5
4	5	4	6	0.3	2	0.6

The weighting factors for each of the system parameters (performance, cost, ...) are:

$$
\begin{array}{llcl}
\text{Performance } w_1 = & 50 & \% \\
\text{Cost} & w_2 = & 40 & \% \\
\text{Schedule} & w_3 = & 5 & \% \\
\text{Risk} & w_4 = & 5 & \% \\
\text{Sum of weighting factors} = & 100 & \% & \text{(Ensure that sum} = 100\%)
\end{array}
$$

Note: Cells with "No Fill" are for inputs; all other cells are chart automatically calculated.

Solve Linear Equations (with Constant Coefficients) (cont'd)

First, the equations are expressed in matrix form as:

System

$$
\begin{matrix} 1 \\ 2 \\ 3 \\ 4 \end{matrix}
\begin{pmatrix}
0.378 & 0.236 & 0.129 & 0.038 \\
0.378 & 0.150 & 0.290 & 0.189 \\
0.034 & 0.329 & 0.387 & 0.660 \\
0.210 & 0.286 & 0.194 & 0.113
\end{pmatrix}
\begin{pmatrix}
0.500 \\
0.400 \\
0.050 \\
0.050
\end{pmatrix}
=
\begin{pmatrix}
0.292 \\
0.273 \\
0.201 \\
0.235
\end{pmatrix}
$$

Note: Numbers may not add due to rounding (3 decimal places).

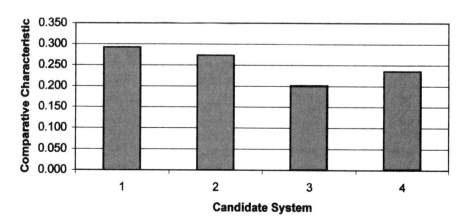

Comparison of Candidate Systems
(Compared on Weighted Performance, Cost, Schedule & Risk)

Relative Comparison:

System	Net Factor	Ratio to "Best"
3	0.201	100%
4	0.235	86%
2	0.273	74%
1	0.292	69%

NOTE: In Net Factor Column, highlight any cell, then click "Data" in Toolbar, then click "Sort", then click "Ascending Order" to align Systems in proper sequence.

This shows which system is "best" (100%) when all factors of performance, cost, schedule and risk are considered with the designated weighting factors applied. The other systems are shown as a percentage of "best".

Test for Uniqueness of Solution

From Cramer's Rule, the solution to the nonhomogeneous equations is unique when the determinant of the coefficients (of p, c, s, r) is non zero.

From the matrix equation above, the determinant has the value: $\boxed{0.005}$

If the determinant has a value greater than zero, the solution is valid.

If the determinant has a value equal to zero, the solution is not unique, and other means must be applied to determine the "best" solution.

Example of Non-Uniqueness

Exercise No 1 in Chapter 8 provided the following set of performance, cost, schedule and risk values for four candidate systems:

System	Leakage (%)	Cost ($B)	Years to IOC	Risk ($B.yrs)
1	9	3.3	4	0.2
2	9	2.1	9	1
3	2.5	2	3	0.3
4	5	4	6	0.6

The determinant of the matrix coefficients becomes, for this set: $\boxed{0}$

Hence, the solution is not unique. By inspection it is seen that Systems 3 and 4 are simply multiples of each other. Some other discriminant must be used to rate one better than the other.

Schedule or Cost, for example, could be used to choose one over the other.

File 5: Chapter 8: Evaluation: Displaying More than Two Variables in the Evaluation

As discussed in Chapter 8, it is frequently desirable to display the results of the Evaluation graphically such that various "boundary conditions" or "limits" can be displayed to guide the evaluation. This File allows the user to examine three basic variations. The first allows the four parameters of performance, cost, schedule and risk to be displayed on a two-dimensional chart. The second allows for the particular combination of three derived parameters of cost-effectiveness, schedule and risk to be displayed on a similar chart. Finally, the third chart allows for any set of three parameters that the user may choose to be displayed on a chart so that any particular limits can be overlayed for review. As the user types in the column headings and inputs the data, the chart will automatically display these user inputs.

A sample output from this File is shown here.

Displaying More than Two Parameters in the Evaluation

In Chapter 8, it is noted that frequently more than two parameters are involved in the selection of systems. "Bubble charts" are a useful means of displaying up to three parameters. This feature coupled with overlaid constraints help in the evaluation of choices.

Basic Set of Parameters:

The following example will use the basic five parameters of:

    Performance (% leakage)
    Cost (of Procurement)
    Schedule (years to Milestone III)
    Schedule (Years to IOC)
    RDT&E Cost

which are summarized in the following Table:

System	Performance (%Leakage)	Cost (Procurement) ($B)	Schedule Years to IOC	RDT&E Cost ($B/year)	Years to Milestone III	Risk ($B.years)
1	9	3.3	4	0.200	1	0.2
2	9	2.1	9	0.500	2	1
3	0.81	4.6	12	0.700	5	3.5
4	5	4	6	0.300	2	0.6

Constructed Parameters from Basic Set:

From the basic set of parameters, various combinations of interest to the evaluator can be constructed. This particular Example combines five basic parameters into a set of three parameters as follows:

System	Cost Effectiveness (% leakage per $B Cost)	Schedule Years to IOC	Risk ($B.years)
1	2.73	4	0.2
2	4.29	9	1
3	0.18	12	3.5
4	1.25	6	0.6

The following two charts display the above information in two combinations.

Chart 1 shows the basic parameters of Performance, Cost, Schedule and Risk.

Chart 2 shows one particular combination of the same parameters.

Each version displays different features of the same basic data.

Different values may be substituted in the above Tables to examine the effects.

NOTE: Cells with "No Fill" are for inputs; all other cells & charts are calculated automatically.

NOTE: Due to programming, data labels may move when other values are substituted.

Performance, Cost, Schedule and Risk

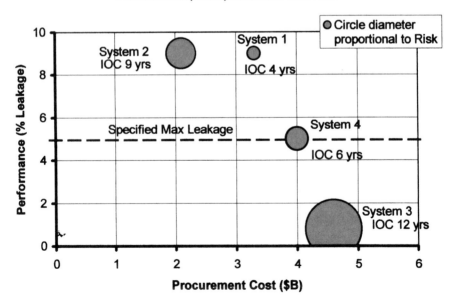

Additional Plots

The reader may wish to plot different combinations of the basic parameters than those shown above. This can be accomplished simply by

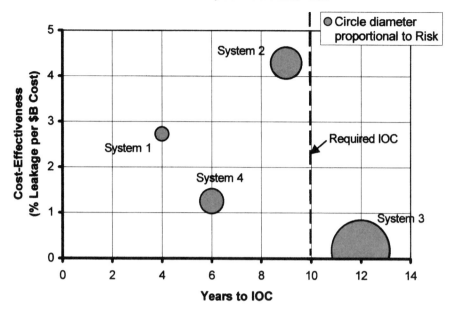

inserting the desired combinations in the following Table using Excel formulas to calculate the parameters. By typing the required axis title in the title cells ($C110-$E110), the graph will automatically display the desired axis titles.

System	X (%.$B)	Y (yrs to IOC)	Z ($B.yrs)
1	29.70	4	0.2
2	18.90	9	1
3	3.73	12	3.5
4	20.00	6	0.6

*The three parameters (X, Y, Z) are as desired by the reader
*Type the title in Cells C110, D110 and E110 as to be displayed on chart
*Perform desired calculation or provide listing of values in cells $C111-$E114
*Chart below will automatically display correct axis titles and values

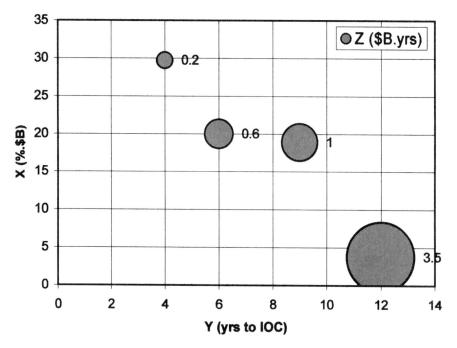

Each system can be identified by its labeled value of "Z".

File 6: Chapter 8: Evaluation: Value Functions

This File allows for any combination of value functions to be calculated and displayed in chart form. This provides a graphic display of the shape of the value functions that the user may wish to examine in the determination of how strongly the departure from some desired or "requirement or specification" value is appropriate. Both the limits ("max" and "min") and the shape of the variation can be explored with these charts.

A sample output from this File is shown here.

VALUE FUNCTIONS

From Chapter 8, the value function was given as:

$$v(x) = \frac{1 - e^{(x-x_d)/\alpha}}{1 - e^{(x_m-x_d)/\alpha}}$$

where x = system characteristic to be minimized (e.g. cost)
x_m = *maximum acceptable limit* of the system characteristic
x_d = *desired goal* of the system characteristic
α = shaping parameter ($\alpha \neq 0$)

With a shaping parameter, $\alpha > 0$ means gradual departure from the desired goal is acceptable and for a shaping parameter, $\alpha < 0$ means any departure from desired goal is unfavorable.

SAMPLE CALCULATION:

For convenience, four system characteristics ($x = p, c, s, r$) are shown. The reader may insert other titles and values in the cells, as appropriate. The tables & charts will display the reader's choices.

Parameters to be shown are:

Leakage (%)	Cost ($B)	ule (Years)	Risk ($B.years)

(reader to substitute own titles, which will read fully in the Formula Bar)

	Leakage (%)	Cost ($B)	ule (Years)	Risk ($B.years)
Maximum acceptable limit:	10	5	12	5
Desired goal:	0.1	2	4	0.2
Shaping parameter[1]:	2	-0.5	-2	-4

(1) See Chapter 8 for typical values of shaping parameter (α).

HINT: Choose input values in cells below that provide a smooth curve between max and min values of each system characteristic in above table.

Value Function for system characteristic   **Leakage (%)**   (see Chart 1)

Leakage (%)	10	8	6	4	2	0.1
Value Function:	1	0.363	0.129	0.043	0.011	0

Value Function for system characteristic   **Cost ($B)**   (see Chart 2)

Cost ($B)	5	4	3.5	3	2.5	2
Value Function:	1	0.984	0.953	0.867	0.634	0

Value Function for system characteristic   **Schedule (Years to IOC)**   (see Chart 3)

Schedule (Years to IOC)	12	10	8	6	5	4
Value Function:	1	0.968	0.881	0.644	0.401	0

Value Function for system characteristic   **Risk ($B.yrs)**   (see Chart 4)

Risk ($B.yrs)	5	4	3	2	1	0.2
Value Function:	1	0.878	0.720	0.519	0.259	0

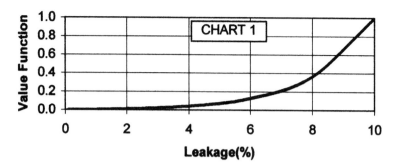

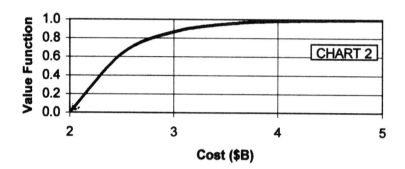

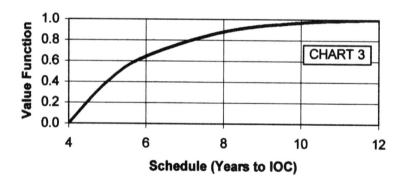

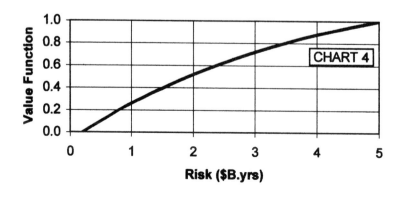

File 7: Chapter 8: Evaluation: Solve Linear Equations (with Value Functions)

This File is similar to that in File 4 in the solution of linear equations, except that in this File each "constant" in the equation is modified by the appropriate value function.

The first section in this File gives the solution as if the coefficients of the equations are constants. This gives the same solution as provided in File 4. The second section in this File then incorporates the "modifiers" of the constants by the Value Functions. A comparison of both approaches is then displayed in the accompanying chart in File 7.

A sample output from this File is shown here.

Solve Linear Equations (with Value Functions)

Use this to test various features such as consistency, check on value function effects, etc.

General equations are:

$$w_1v(p)p_1 + w_2v(c)c_1 + w_3v(s)s_1 + w_4v(r)r_1 = k_1$$
$$w_1v(p)p_2 + w_2v(c)c_2 + w_3v(s)s_2 + w_4v(r)r_2 = k_2$$
$$w_1v(p)p_3 + w_2v(c)c_3 + w_3v(s)s_3 + w_4v(r)r_3 = k_3$$
$$w_1v(p)p_4 + w_2v(c)c_4 + w_3v(s)s_4 + w_4v(r)r_4 = k_4$$

where w_i are weighting factors for p_i, c_i, s_i, and r_i $i = 1-4$

and $v_j(\cdots)$ are the value functions $j = 1-4$

EXAMPLE SET

System	p	c	s	r
1	9	3.3	4	0.2
2	9	2.1	9	1
3	0.81	4.6	12	3.5
4	5	4	6	0.6

Weighting factors:

$$w_1 = 50\ \%$$
$$w_2 = 40\ \%$$
$$w_3 = 5\ \%$$
$$w_4 = 5\ \%$$

$$\sum_{j=1}^{4} w_j = 100\%$$

First, assume all value functions are unity, then equations become:

$$
\begin{array}{c}
\text{System} \\
1 \\
2 \\
3 \\
4
\end{array}
\begin{pmatrix}
\mathbf{p} & \mathbf{c} & \mathbf{s} & \mathbf{r} \\
0.378 & 0.236 & 0.129 & 0.038 \\
0.378 & 0.150 & 0.290 & 0.189 \\
0.034 & 0.329 & 0.387 & 0.660 \\
0.210 & 0.286 & 0.194 & 0.113
\end{pmatrix}
\begin{pmatrix}
\mathbf{w} \\
0.500 \\
0.400 \\
0.050 \\
0.050
\end{pmatrix}
$$

$$
=
\begin{pmatrix}
\mathbf{k} \\
0.292 \\
0.273 \\
0.201 \\
0.235
\end{pmatrix}
$$

Determinant of the coefficients has the value: $\boxed{0.005}$

NOTE: It is not immediately obvious from the resultant "k" array that one system is a multiple of another. One must rely on either inspection or the calculation of the determinant of the coefficients (i.e. if determinant is zero then two systems are indeed a multiple of each other).

IN THE CONTEXT OF THIS ANALYSIS,
IT IS PROBABLY NOT IMPORTANT THAT ONE SYSTEM MIGHT BE A MULTIPLE OF ANOTHER BECAUSE BETTER PERFORMANCE, LOWER COST, SHORTER SCHEDULE, LOWER RISK IS PROBABLY ACCEPTABLE.

Now include the effect of value functions.

$$
v(x) = \frac{1 - e^{(x-x_d)/\alpha}}{1 - e^{(x_m-x_d)/\alpha}}
$$

where x = system characteristic to be minimized
x_m = maximum acceptable limit of the system characteristic
x_d = desired goal of the system characteristic
α = shaping parameter

With a shaping parameter, $\alpha > 0$ means gradual departure from the desired goal is acceptable and for a shaping parameter, $\alpha < 0$ means any departure from desired goal is unfavorable.

Assume the following "limits" and degree of reluctance or acceptance:

	p Leakage (%)	**c** Cost ($B)	**s** Yrs to IOC	**r** Risk ($B.yrs)
Maximum acceptable limit:	10	5	12	5
Desired goal:	0.1	2	4	0.2
Shaping parameter:	2	-2	-2	-2

Value Functions for Example Set above:
System

1	0.604	0.615	0.000	0.000
2	0.604	0.063	0.935	0.363
3	0.003	0.936	1.000	0.889
4	0.076	0.814	0.644	0.199

The table of Value Functions clearly shows how any system value that is close to the desired goal is equal to or close to zero and thus contributes to a low "score" for that system and is more likely to be the "minimum" value overall.

The net result of multiplying the value functions by the values in the basic table above gives the result:

System	**p**	**c**	**s**	**r**
1	5.434	2.030	0.000	0.000
2	5.434	0.132	8.415	0.363
3	0.002	4.307	12.000	3.110
4	0.378	3.255	3.863	0.120

which when normalized, becomes:

System	**p**	**c**	**s**	**r**	**w/value fn** **k**	**no value fn** **k**
1	0.483	0.209	0.000	0.000	0.325	0.292
2	0.483	0.014	0.347	0.101	0.269	0.273
3	0.000	0.443	0.494	0.866	0.245	0.201
4	0.034	0.335	0.159	0.033	0.160	0.235
Check sum:	1.000	1.000	1.000	1.000	1.000	1.000

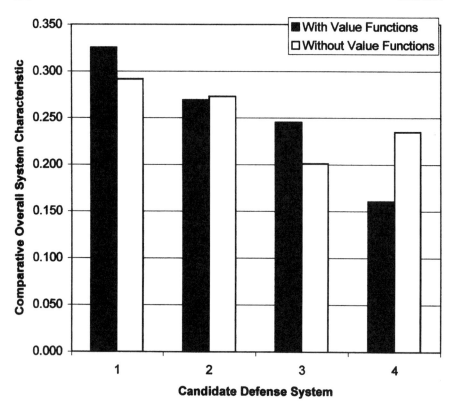

Comparison of Evaluation:

	Constant Weighting		by Value Function	
	Ranking	Relative Ranking %	Ranking	Relative Ranking %
System 1	4	69	4	49
System 2	3	74	3	60
System 3	1	100	2	65
System 4	2	86	1	100

REFERENCES

[1]*Guinness Book of Records.*

[2]Siemienowicz, Casimirus, *Artis Magnae Artillerae, Pars Prima*, 1650.

[3]Congreve, William, *A Concise Account on the Origin and Progress of the Rocket System*, Dublin, 1817.

[4]*American Heritage Dictionary*, New college ed., Houghton Mifflin, 1976.

[5]*The Seattle Times*, 19 April 2001.

[6]Oberth, Hermann, *Die Rakete zu den Planetenräumen*, Munich, 1923.

[7]Lennox, Duncan (ed.), *Jane's Strategic Weapon Systems*, No. 33, 1999.

[8]Arnold, Gen. H. H., *Global Mission*, Harper, New York, 1949.

[9]Hill, Air Chief Marshal Sir Roderic, "Air Operations by Air Defence of Great Britain and Fighter Command in Connection with the German Flying Bomb and Rocket Offensives, 1944–1955," U.K. Secretary of State for Air, London, Oct. 1948.

[10]Hölsken, Dieter, *V-Missiles of the Third Reich: The V-1 and V-2*, Monogram Aviation Publications, Sturbridge, MA, 1994.

[11]*Desert Score*, Carroll Publishing, July 1991.

[12]*Stockholm International Peace Research Institute (SIPRI) Yearbook of World Armaments and Disarmament*, 1969/70.

[13]Glasstone, Samuel (ed.), *The Effects of Nuclear Weapons*, U.S Dept. of Defense and U.S. Atomic Energy Commission, 1964.

[14]Office of Technology Assessment, *The Effects of Nuclear War*, U.S. Congress, Washington, DC, 1979.

[15]Chant, Christopher, and Hogg, Ian, *Nuclear War in the 1980's?* Harper and Row, 1983.

[16]Ziegler, Philip, *The Black Death*, John Day Co., New York, 1969.

[17]The Final Report of the Provost General, 1863–1866.

[18]"Technologies Underlying Weapons of Mass Destruction," Office of Technology Assessment, OTA-ISC-559, Washington, DC, Aug. 1993.

[19]"Report of the Commission to Assess the Ballistic Missile Threat to the United States (The Rumsfeld Report)," 15 July 1998.

[20]*The Military Balance 1986–1987*, International Inst. for Strategic Studies, London, 1986.

[21]*Defense News*, Vol. 16, No. 12, 26 March 2001.

[22]Diehl, Walter, "Standard Atmosphere Tables and Data," NACA Rept. 218, 1925.

[23]Diehl, Walter, "Standard Atmosphere Tables and Data," NACA Rept. 1235, 1955.

[24]Bunn, Matthew, *Ballistic Missile Guidance and Technical Uncertainties of Countersilo Attacks*, MIT Press, Cambridge, MA, 1983.

[25]*Jane's Weapon Systems*, 1993.

[26]Miranda, J., and Mercado, P., *Secret Weapons of the Third Reich*, Schiffer Military and Aviation History, 1996.

[27]"The (NATO) Alliance's Strategic Concept Agreed by the Heads of State and Government Participating in the Meeting of the North Atlantic Council," Rome, 8 Nov. 1991, revised 1999.

[28]*World Atlas* 3rd, ed., Hammond World Atlas Corp., 2000.

[29]National Center for Geographical Information and Analysis, Univ. of California, CA, April 1995.

[30]Larson, Eric. V., and Kent, Glenn A., "A New Methodology for Assessing Multi-Layer Missile Defense Options," Rand Corp., Rept. ADA288328, 1994.

[31]Stockholm International Peace Research Inst., *The Problem of Chemical and Biological Warfare*, Humanities Press, 1971.

[32]Aldridge, Pete, "Remarks to Congress," Washington, DC, 17 March 2003.

[33]Miller, Maj. Walter F., "Density Altitude Maps of Iran and Iraq," U.S. Air Force, Rept. USAFETAC/PR-91/008, May 1991.

[34]Bate, Roger, Mueller, Donald, and White, Jerry, *Fundamentals of Astrodynamics*, Dover, New York.

[35]*Aviation Week and Space Technology*, 3 Feb. 2003, p. 34.

[36]Lloyd, Richard M., *Conventional Warhead Systems Physics and Engineering Design*, AIAA, Reston, VA, 1998.

[37]USAF Scientific Advisory Board, "Concepts and Technologies to Support Global Reach," Dec. 1992.

[38]Fleeman, Eugene, *Tactical Missile Design*, AIAA Education Series, AIAA, Reston, VA, 2001.

[39]"Budget of the US Government, Fiscal Year 2001," Government Printing Office, Washington, DC.

[40]*NATO Handbook*, 50th ed., NATO, 1998.

[41]International Inst. for Strategic Studies, *The Military Balance 1999/ 2000*, Oxford Univ. Press, 1999.

[42]Rumsfeld, Donald, "Allied Contributions to the Common Defense," Rept. to Congress, Washington, DC, March 2000.

[43]U.S. Navy, *Aircraft Cost Handbook*, 1977.

[44]Riddell, Frederick R., *Survey of Advanced Propulsion Systems for Surface Vehicles*, Inst. For Defense Analyses, Jan 1975.

[45]Armstrong, Neil, "Houston, Tranquility Base here. The Eagle has landed," Apollo II, Neil Armstrong's radio transmission to Mission Control, NASA Johnson Space Center, Houston, TX, 20 July 1969.

[46]Agrawal, Brij N., *Design of Geosynchronous Spacecraft*, Prentice–Hall, Upper Saddle River, NJ, 1986.

[47]*IEEE Spectrum*, Sept. 1997.

[48]Missile Defense Agency, Fact Sheets, March 2002.

[49]*Aviation Week and Space Technology 2003 Aerospace Source Book*, 13 Jan 2003, pp 141, 155–165.

[50]Hülsmeyer, Christian, "Hertzian-Wave Projecting and Receiving Apparatus to Indicate or Give Warning of the Presence of a Metallic Body, such as a Ship or a Train, in the Line of Projection of Such Waves," British patent No. 13, 170, 22 Sept. 1904.

[51]Skolnick, Merrill I., *Introduction to Radar Systems*, 2nd, ed., McGraw-Hill, New York, 1980.

[52]Inst. of Electrical and Electronics Engineers, Standard 521-1976, 30 Nov. 1976.

[53]USAF, Standard AFR-55-44, Oct. 1964.

[54]U.S. Navy, Standard OPNAVINST 3430.9B, Oct. 1964.

[55]"PAC-3, THAAD Costs Studied," *Defense News*, Vol. 14, No. 44, 8 Nov. 1999.

[56]Wright, T.P., "Factors Affecting the Cost of Airplanes," *Journal of the Aeronautical Sciences*, Vol. 3, Feb. 1936, pp. 122–128.

[57]"Tables of Actual and Projected Weapons Purchases," U.S. Dept of Defense, Procurement Document P-1, Washington, DC, 1997.

[58]*Aviation Week and Space Technology Source Books*, various years.

[59]*Desert Score*, Carroll Publishing, 1991.

[60]*Aviation Week and Space Technology*, 24 Feb. 2003.

[61]"DOD Appropriations," U.S. Dept. of Defense, Rept. to Congress, Washington, DC, 1983.

[62]Augustine, Norman, *Augustine's Laws*, AIAA, New York, 1983.

[63]"Airborne Laser: Makes Sense for Missile Defense," *Defense Week*, 17 Feb. 1998.

[64]"The Airborne Laser", *IEEE Spectrum*, Sept. 1997.

[65]"Tools from the US Arsenal," *The Seattle Times*, 19 March 2003.

[66]"The Defense Acquisition System," U.S. Dept. of Defense, DoD Directive 5000.1, Washington, DC, 23 Oct. 2000.

[67]"Operation of the Defense Acquisition System," U.S. Dept. of Defense, DoD Instruction 5000.2, Washington, DC, 5 April 2002.

[68]Packard, Deputy Secretary of Defense David, "Establishment of a Defense Systems Acquisition Review Council," Washington, DC, 30 May 1969.

[69]Carlucci, Deputy Secretary of Defense Frank, "Improving the Acquisition Process," Memo, Washington, DC, 30 April 1981.

[70]Carlucci, Deputy Secretary of Defense Frank, "Increasing Competition in the Acquisition Process" Memo, Washington, DC, 27 July 1981.

[71]Aldridge, Defense Secretary Pete, "Missile Defense Acquisition Policy," Testimony to Senate Armed Services Committee, Washington, DC, 13 March 2002.

[72]"DoD Financial Management Regulation," U.S. Dept. of Defense, DoD 7000.14-R, Vol. 2B, Chap. 5, Washington, DC, June 2002.

[73]"The Planning, Programming, and Budgeting System (PPBS)," U.S. Dept. of Defense, DoDD 7045.14, 22 May 1984.

[74]"An Analysis of Weapons System Acquisition Intervals, Past and Present," RAND, Rept. R-2605-DR&E/AF, Nov. 1980.

[75]"Report of the Acquisition Cycle Task Force," Defense Science Board, Summer Study, 1977, Final Rept., 15 March 1978.

[76]Mantle, Peter J., "Introducing New Vehicles," *Naval Engineers Journal*, Vol. 97, No. 2, 1985.

[77]"Handbook on the Phased Armaments programming System (PAPS), Vol. 1: PAPS Framework and Procedures and Vol. 2: PAPS Related Activities and Management Considerations" NATO, Rept. AAP-20, Feb. 1989.

[78]TRACE, U.S. Army Logistics Management Center, document, 1975.

[79]Air Force Systems Command Armament Div., document 1981.

[80]Bearden, David A., "When Is a Mission too Fast and too Cheap?, A Complexity-Based Risk Assessment of Low-Cost Planetary Missions," Aerospace Corp., 1999.

[81]Sarsfield, Liam P., "The Cosmos on a Shoestring," RAND Corp., Rept. MR/MR864, 1998.

[82]Navy Comptroller documents.

[83]U.S. Navy Comptroller Information System Historical File, 23 Feb. 1984.

[84]Miller, J. R., *Professional Decision Making*, Praeger, 1970.

[85]Raiffa, H., "Preferences for Multi-Attributed Alternatives," RAND Corp., RM-5868-DOT/RC, 1969.

[86]Saaty, T. L., *The Analytic Hierarchy Process*, McGraw-Hill, New York, 1980.

[87]Albert, K. J. *How to Solve Business Problems*, McGraw-Hill, New York, 1978.

[88]Ulvila, J. W., and Brown, R. V., "Decision Analysis Comes of Age," *Harvard Business Review*, Vol. 60, No. 5, 1982.

[89]Barzalai, J., "Understanding Hierarchical Processes," *Proceedings of the 1998 National Conference of the American Society for Engineering Management*, 1998.

[90]Kirkwood, Craig W., *Strategic Decision making*, Wadsworth, 1997.

[91]*Defense News*, 7 May 2001, p. 14.

[92]Bates, Roger R., Mueller, Donald D., and White, Jerry E., *Fundamentals of Astrodynamics*, Dover, New York, 1971.

[93]Hoerner, Sighard F., *Fluid-Dynamic Drag*, 1965.

[94]NACA RM A53D02, 1961.

[95]Diehl, Walter S., "Notes on the Standard Atmosphere," NACA TN 99, June 1922.

[96]*U.S. Standard Atmosphere, 1976*, U.S. Government Printing Office, Washington, DC, 1976.

[97]*Manual of the ICAO Standard Atmosphere*, International Civil Aviation Organization, Montreal, 1982.

[98]*Atmospheric Handbook 1984*, National Geophysical Data Center, Boulder, CO, 1984.

[99]List, Robert J., *Smithsonian Meteorological Tables*, Smithsonian Institution Press, Washington, DC, 1984.

[100]Ley, Willy, *Rockets, Missiles and Space Travel*, Chapman and Hall, Ltd., 1954.

[101]"Concepts and Technologies to Support Global Reach," USAF Scientific Board, Dec. 1992.

[102]Zarchan, Paul, *Tactical and Strategic Missile Guidance*, Progress in Astronautics and Aeronautics, Vol. 176, AIAA, Reston, VA, 1997.

INDEX